KU-687-453

Environmental aesthetics

Theory, research, and applications

Edited by

JACK L. NASAR

The Ohio State University

CAMBRIDGE
UNIVERSITY PRESS

Published by the Press Syndicate of the University of Cambridge
The Pitt Building, Trumpington Street, Cambridge CB2 1RP
40 West 20th Street, New York, NY 10011-4211, USA
10 Stamford Road, Oakleigh, Victoria 3166, Australia

© Cambridge University Press 1988

First published 1988
First paperback edition 1992

Printed in the United States of America

Library of Congress Cataloging-in-Publication Data
Environmental aesthetics: theory, research, and applications / edited
by Jack L. Nasar.
p.   cm.
Bibliography: p.
Includes index.
ISBN 0-521-34124-8
1. Environmental psychology. 2. City planning – Psychological
aspects. 3. Urban beautification – Psychological aspects.   I. Nasar, Jack L.
BF713.E7   1988   87–18774
720'.1 – dc19   CIP

British Library Cataloguing in Publication Data
Environmental aesthetics: theory, research,
and applications.
1. Human Ecology 2. Aesthetics
I. Nasar, Jack L.
304.2   HM206

ISBN 0-521-34124-8 hardback
ISBN 0-521-42916-1 paperback

LEEDS METROPOLITAN
UNIVERSITY LIBRARY

1700634731
EA
So 17028   30·3·93
16 . 4 . 93
304·2 ENV

To my parents
Frieda and Leon Nasar

Contents

Figures

xi

Tables

Contributors and participants

Participants at 1982 College Park, Maryland, sessions

Arnold Berleant   The American Society for Aesthetics, C. W. Post Center of Long Island University

D. Mark Fenton and **Joseph P. Reser**   Department of Psychology, Western Australian Institute of Technology

Barrie B. Greenbie   Department of Landscape Architecture and Regional Planning, University of Massachusetts

Tom F. Heath   School of the Built Environment, Queensland Institute of Technology

Thomas R. Herzog   Department of Psychology, Grand Valley State College

Rachel Kaplan   Department of Psychology and School of Natural Resources, University of Michigan

Stephen Kaplan   Department of Psychology and Department of Computer Sciences, University of Michigan

Jon Lang   Urban Design Program, University of Pennsylvania

Richard M. Locasso   Student Health Program, Southern Illinois University

Jack L. Nasar   Department of City and Regional Planning, Ohio State University

Brian Orland   Department of Landscape Architecture, University of Illinois

Wolfgang F. E. Preiser   Institute for Environmental Education, University of New Mexico, and **Kevin F. Rohane**   3-D International

Mollie Ridout   Department of Geography, Pennsylvania State University

Janet F. Talbot   School of Natural Resources, University of Michigan

Participants in 1983 Lincoln, Nebraska, sessions

Arnold Berleant   The American Society for Aesthetics, C. W. Post Center of Long Island University

Linda N. Groat   Department of Architecture, University of Wisconsin, Milwaukee

xvii

Fred A. Hurand   Department of Urban and Regional Planning, Eastern
Washington State University

Jon Lang   Urban Design Program, University of Pennsylvania

Jack L. Nasar   Department of City and Regional Planning, Ohio State
University

Jack L. Nasar, David Julian, Sarah Buchman, David Humphreys, and
Marianne Mrohaly   Department of City and Regional Planning,
Ohio State University

Werner Nohl   Landschafts-architektur, Technische Universität München

Fahriye Hazer Sancar   Department of Landscape Architecture, University
of Wisconsin, Madison

Contributors

Jay Appleton   Department of Geography, University of Hull

John E. Flynn   (deceased)   Department of Architectural Engineering,
Pennsylvania State University

Robert G. Hershberger and **Robert C. Cass**   Department of Architecture,
Arizona State University

Rachel Kaplan   Department of Psychology and School of Natural
Resources, University of Michigan, and **Eugene J. Herbert**
Siteplan, Queensland, Australia

Joyce Vielhauer Kasmar   Los Angeles

Eduardo E. Lozano   Lozano and White Associates

Anke Oostendorp   Surveys Analyst, Bell Canada, and **Daniel E. Berlyne**
(deceased)   University of Toronto

Kenneth T. Pearlman   Department of City and Regional Planning, Ohio
State University

James A. Russell   Department of Psychology, University of British
Columbia

Acknowledgments

This collection would not have become a reality without the efforts of a number of individuals. I owe a general debt of gratitude to two educators, Oscar Newman and Joachim F. Wohlwill. Oscar brought me in touch with the area of environment-behavior studies and environmental meaning, and Jack continued my education in the area of environmental aesthetics with his valuable insight and guidance during and after my dissertation years. Of more direct relevance to this collection, I am, of course, grateful to the over thirty individuals whose papers are presented in this collection for their contributions, patience, and responsiveness to my requests. The organizers of EDRA 13 – Polly Bart, Alexander Chen, Guido Francescato – and of EDRA 1983 – Doug Amedeo, James B. Griffin and James J. Potter – provided the opportunity for the Symposia and Workshops from which this book evolved. Louise DeMaseo typed several chapters of this book when needed (usually yesterday). I am grateful to the staff at Cambridge University Press: Susan Milmoe, who encouraged me to submit the book and who shepherded it through the early phases with Cambridge; J. Blake Bailey, her assistant; Jane Van Tassel, the production editor; and Irene Pavitt, the copy editor. Finally, my wife Judy and daughter Joanna were always there when needed. All of these people have earned my warmest thanks.

Preface

Background

Environmental aesthetics represents the merging of two areas of inquiry: empirical aesthetics and environmental psychology. Both areas use scientific methodologies to help explain the relationship between physical stimuli and human response. Empirical aesthetics is concerned with the arts (painting, music, literature, and dance), and environmental psychology is an applied field concerned with improving the quality of the human habitat. By combining a concern with aesthetic value, a problem focus on human habitat, and a methodological emphasis on applicability, environmental aesthetics becomes a unique endeavor. One notable factor is the use of a broader definition of aesthetics (see Wohlwill, 1976) to include environmental influences on the whole range of human affect. Thus central concerns in environmental aesthetics include understanding environmental influences on affect and translating that understanding into environmental design that is judged favorably by the public.

Although aesthetics is only one among a host of considerations in environmental design, it is an important one. The aesthetic quality of the surroundings may affect immediate experience – sense of well-being – in those surroundings; it may influence subsequent reactions to both the setting and its inhabitants; and it may influence spatial behavior in that individuals are attracted to an appealing environment and are likely to avoid an unpleasant one. With knowledge of the relationship between properties of the visual environment and human affect, design professionals can better plan, design, and manage settings to fit the preferences and activities of the users. This, in turn, may contribute to enhancing the quality of life.

The concern for understanding principles of aesthetics is not a new one. It has a long history in philosophy, design, and research. The aesthetic principle of "unity in variety," for example, was put forth by the early Romans (Bosanquet, 1892). In psychology, some of the earliest published experiments centered on aesthetics (Fechner, 1876). Mathematical approaches to aesthetics date back to ancient times and the golden section. In the twentieth century, architects such as Le Corbusier, with his "modular man," continued to consider mathematical solutions (Blake, 1960). Psychologists have also turned to mathematics for expression of aesthetic principles. Birkhoff (1933) gave mathematical expression to the role of order and variety in preference, and although he did not empirically test his theory, others (see Eysenck, 1941) did. In 1935, Koffka gave voice to the Gestalt school views on aesthetics. These views have subsequently been extended to the realm of environmental design (Arnheim, 1977; Hesselgren, 1975).

The past twenty-five years have seen a renewed interest in the empirical examination of aesthetics. Berlyne has published some of his seminal works on aesthetics (1960, 1971, 1974); Wohlwill (1968) has extended this work to large-scale environments; and design professionals such as Venturi (1966) and Rapoport (Rapoport and Hawkes, 1970; Rapoport and Kantor, 1967) have used principles from the empirical research to critique modern architecture and to advocate different kinds of solutions. Public policy has changed as well. Public-policy initiatives (Coastal Zone Management Act, 1972; National Environmental Policy Act, 1969) have recognized aesthetics as important and in need of quantification for environmental decision making.

Further evidence of the growth of interest in the area of environmental aesthetics can be seen in a review by Zube, Sell, and Taylor (1982). They refer to over 160 articles and 11 state-of-the-art reviews or bibliographies published in 20 journals between 1965 and 1980. Among the journals publishing work on environmental aesthetics are *Environment and Behavior, Journal of Environmental Psychology, Journal of the American Institute of Planners* (now *Journal of the American Planning Association), Landscape Planning, Landscape Research,* and *Environment and Planning A* and *B.* Professional magazines, such as *Planning, Landscape Architecture, Architecture,* and *Progressive Architecture,* have also highlighted empirical advances in environmental aesthetics. Additional state-of-the-art reviews in environmental aesthetics include those by Wohlwill (1976), Kaplan and Kaplan (1982), Porteous (1982), and Ulrich (1983).

That aesthetic quality of the environment is important to the public is evident. Homeowners become angry over a neighbor's aesthetic intrusion; citizens fight to protect an aesthetic resource or to remove an eyesore; cities try to control the visual quality of their frequently blighted commercial strips; and national policy attempts to reduce visual blight along highways. Empirical data substantiate such anecdotal evidence. A variety of studies examining subjective responses to environments (Canter, 1969; Harrison and Sarre, 1975; Hershberger and Cass, 1974; Lowenthal and Riel, 1972; Oostendorp and Berlyne, 1978a; Russell and Ward, 1981) indicate the importance of the evaluative or aesthetic dimension in response to the environment. Canter (1969), using factor analysis, found that for both architects and nonarchitects, the major factor in response to simulated environments was an aesthetic one – pleasantness; Lowenthal and Riel (1972), using the repertory grid technique, by which respondents generated their own descriptors, found aesthetic variables – beautiful–ugly and pleasant–unpleasant – accounting for most of the variance; Harrison and Sarre (1975), using personal constructs, found that for descriptions of both neighborhood and retail environments, descriptors relating to aesthetics – pleasant atmosphere and beauty – accounted for the largest portion of the variance; and Russell and Ward (1981), using a variety of techniques, found evaluation as one of two major dimensions in affective appraisals of the environment. In sum, individuals respond to their surroundings in highly evaluative ways. Visual quality is important.

Because of its importance, visual quality may well influence well-being and behavior. Empirical studies (Kasmar, Griffin, and Mauritzen, 1968; Maslow and Minz, 1956; Mintz, 1956; Moos, Harras, and Schonborn, 1969; Samuelson and Lindauer, 1976) demonstrate effects of variations in aesthetic conditions on well-being and behavior. Although some of these findings may be confounded by method artifact (Locasso, this collection), there is other evidence of aesthetic influences. Berlyne (1971), for example, documented physiological reactions to certain aesthetic variables, and Ulrich (1984) found that recovery time in a hospital was affected by the quality of the view from the room.

With regard to behavior, it is commonly held that the choice of a home depends, in part, on aesthetic factors. Thus, real-estate agents may recommend aesthetic improvements (painting, cleaning up, and planting) to help sell a house. Has empirical research documented any behavioral effects? Berlyne (1971) has demonstrated that the visual character of stimuli influences behaviors such as attention, looking time, or forced choice. More recently, researchers have found evidence of effects on spatial behavior. Newman (1972) cited aesthetics – image and milieu – in public housing as affecting sense of community and crime; Wener, Fraser, and Farbstein (1985) found that certain aesthetic variables influenced the success of correctional facilities; Downs (1970) found aesthetics to be one factor, admittedly a minor one, that influenced choice of a shopping center; and Ulrich (1973) found that shoppers drove out of their way to use a more visually pleasing route to shop.

At present, decisions about the visual quality of the environment are often made by design professionals. This is particularly true for large-scale facilities, such as offices, institutions, and commercial and recreational facilities. Because these facilities are experienced on a regular basis by large numbers of people, they may have a substantial influence on the evaluative image of a city. If professionals and the public share aesthetic values or if professionals could accurately gauge aesthetic needs of the public, then reliance on professional intuition might be acceptable. Unfortunately, the research indicates that professionals differ from the public in their environmental preferences (Canter, 1969;

Groat, 1982; Hershberger and Cass, 1974). Furthermore, such differences are
not trivial. They can result in widespread effects, as is demonstrated in the
controversy surrounding the Richard Serra sculpture and in the monumental
failure of Pruitt Igo. Serra won a National Endowment for the Arts competition
for a public sculpture to be placed in front of the Federal Building in New York.
The public regarded the sculpture as an ugly intrusion on their daily lives. Yet
many artists lined up in support of Serra and against the public. This was not a
museum piece that people could choose to see or to avoid. This was an object
placed in a major public plaza. What is more important in such instances, the
aesthetic values of the artist or those of the public who must live with the work on
a daily basis? It is noteworthy that this controversy took place over ten years after
Pruitt-Igoe. Recall that Newman (1972) argues that a difference in the aesthetic
values of the designers and the occupants was a factor contributing to this
monumental failure.

Although some designers disdain public values, many professionals want to
produce user-sensitive design. For the latter group, the research on environmental aesthetics may help inform their design decisions. Theoreticians and researchers can gain from understanding the practical constraints within which
decision makers operate. Inquiry into practical aspects of the application of
research findings to design and planning can help transform theory and research
into physical realities. It is through an understanding of theory, research, and
public policy that decision makers can be most effective in improving the quality
of the environment. Thus, for example, regulatory controls in sign ordinances
may rely on theory and empirical evidence about legibility and its influence on
safety. Through the accumulation of evidence, public policy will change. It is
hoped that this collection of papers can contribute to such change.

Evolution of the book

For the 1982 Environmental Design Research Association Conference, in College Park, Maryland, with the intent of sponsoring a symposium on environmental aesthetics, I put out a call for papers on new work in theory, research, or
applications in environmental aesthetics. I received over thirty abstracts and
papers. To accommodate this unexpected supply, the conference organizers –
Polly Bart, Alexander Chen, and Guido Francescato – allowed me to schedule
four sessions: one symposium on theory and three workshops – one on theory,
one on research, and one on design and planning applications. Many papers and
abstracts were distributed in advance to enable presenters to respond to one
another's papers during the sessions. At the conference, the large number of
presentations in each session required adherence to a strict schedule that prohibited discussion beyond the formal presentations. Clearly, a second meeting was
needed to explore the issues further.

Thus for the 1983 Environmental Design Research Association Conference, in Lincoln, Nebraska, I organized a second set of sessions on environmental aesthetics. With the cooperation of conference organizers – Doug Amedeo, James B. Griffin, and James J. Potter – two symposia and one workshop were scheduled. Independently, Linda Groat organized a session on contextual fit – an important concern in architectural aesthetics. Thus in 1982 and 1983, there were eight sessions defining the cutting edge in environmental aesthetics. Presenters included established leaders in the field and relative newcomers. Many papers were the first presentations of dissertation or thesis work, and one was selected by EDRA as a best student paper.

This collection evolved from those conference sessions. It attempts to organize the work to clarify issues, techniques, and directions for further inquiry in environmental aesthetics. The spirit of connecting theory, research, and application was maintained. Thus at the core of this book are some of the original ideas presented at the conferences. For this book, however, the contributors revised and improved their original conference presentations. The comments and discussions during and after the sessions influenced that process. Some papers in this collection represent revisions and updates of the session papers; others represent new work that grew from work presented at the sessions; and a third set (not from the conference sessions) are invited papers (some new and some previously published) that are important contributions to the work in the field.

Purpose and audience

The book is organized in three sections. The first section presents theoretical perspectives on environmental aesthetics. Included are papers on philosophy, psychology, and design. Each raises issues of relevance to basic and applied research, design education, and practice with regard to aesthetics.

The second section presents empirical studies. Examined are the nature of perception of physical environments, the nature of response to such environments, the linkage between environmental appraisal and features of the environment, and the effects of aesthetic surroundings on human inhabitants. The papers provide the researcher and designer with a basis for evaluating environmental quality or for conducting inquiries into the nature of environmental aesthetics. This section provides information on various kinds of environment, including interior, exterior, urban, and natural and rural. For each kind of environment, there are papers that examine the noticeable visual attributes and the relevant measures of emotional response. Readers can also glean directions for design, research, and theory through examining the relationships found between features of scenes and appraisals of those scenes.

The third section presents applied work. Readers are exposed to the mechanisms by which research findings and design guidelines can be converted into physical products. An understanding of these mechanisms can provide insight into new directions of applied inquiry.

The grouping of papers in any one category is, in fact, artificial, because much of the work contributes to theory, research, and application. Thus although the papers are classified as theory, research, or application, they may inform and enhance efforts in other categories. The papers share the goal of contributing to the enhancement of aesthetic quality of human surroundings. Thus they include theoretical work relevant to empirical testing and design, empirical work grounded in or affecting theory and design application, and application informed by and informing both theory and research. It is in the spirit of such inter-disciplinary cooperation that this book is offered.

The field of environmental aesthetics is a rich and varied one. By drawing from conference presentations, this book aims to present some of the latest advances. Nevertheless, it is inevitable that omissions due to limitations in space and time would be made. Consider, for example, the discussion of sociobiology by Edward O. Wilson and by Richard C. Lewontin; the phenomenological positions of David Seamons and Yi Fu Tuan; the ecological perspective of James J. Gibson; the methodological and empirical advances made by researchers such as David Canter, Kenneth Craik, Terry Daniel, Elwood Shafer, Roger Ulrich, Joachim F. Wohlwill, and Erv Zube and by Swedish researchers Carl-Axel Acking, Tommy Gärling, G. J. Sorte, and Richard Küller; or the ideas of professionals such as Amos Rapoport.

Integral to the interdisciplinary nature of the content of this collection is the view that readers will come from a variety of disciplines. The main audience is expected to include individuals who are interested in the applications of research to understanding and creating physical settings that appeal to the aesthetic values of the inhabitants. Specifically, the audience is expected to include both social and behavioral scientists and environmental practitioners (architects, landscape architects, and planners). The papers present information relevant to each field separately, and to the interdisciplinary fields of empirical aesthetics, environment-behavior studies, and planning. Thus, for example, some of the empirical studies have implications for the field of psychology; some of the theoretical papers have implications for the field of philosophy; and some of the papers on application have implications for design and planning.

The book is also aimed at the student with an interest in the conduct and application of environment-behavior research. Because such an interest might emerge at an undergraduate or a graduate level, the book is intended for use in specialized (environment-behavior) courses at either level. It is expected that both novices and advanced students in the area can gain from this book. Because such students and courses may appear in the design fields, the social and behavioral sciences, or philosophy, the book was designed to be a useful text for any of these areas. Selections from the book have already been used with success in undergraduate courses in architecture; in graduate classes attended by industrial designers, architects, landscape architects, and planners; and for doctoral students doing dissertation work on environmental aesthetics. These students found a theoretical, empirical, and applied grounding and have developed new ideas for research and application from that experience. It is to be hoped that other readers will glean similar benefits.

Section I

Theory

Editor's introduction

The subject of environmental aesthetics has at its core more than the monitoring of volatile tastes. Instead, researchers and designers seek universal principles that can explain commonalities and differences in response. The consideration of the theoretical underpinnings of environmental aesthetics can enrich the questions, solutions, and approaches considered by researchers, designers, educators, and others in the field of environmental-design research. The theoretical papers here thus present a framework for further inquiry.

Environmental influences on appraisals of aesthetic quality have two components: formal and symbolic or associational. Formal analysis of aesthetics focuses on the attributes of the object as they contribute to aesthetic response. Such an analysis may consider such properties as size, shape, color, complexity, and balance. Symbolic analysis of aesthetics focuses on factors that through experience produce connotative meanings such that the object implies something else. Thus despite similar formal attributes, a Mercedes and a Ford may produce different meanings, or an artificial flower (although it may look like a real flower) will likely call up different meanings when the observer realizes it is artificial. Symbolic analysis focuses on such things as style and context.

Heath presents a theoretical review of formal factors in aesthetic response. His review considers environmental features and behavioral and cognitive aspects of a situation. He describes the role of complexity and order in relation to two kinds of behavior: instrumental and diversive.

Lang examines symbolic meaning. He suggests five architectural variables that may carry symbolic meaning: building configuration, spatial configurations, materials, illumination, and pigmentation. He discusses models of cognitive consistency and the potential role of culture. Five models for the acquisition of symbolic meaning are presented.

In the first of two papers describing formal aesthetic preferences, Appleton argues that an evolutionary perspective may explain some preferences. In particular, he reasons that prospect (open views) and refuge (protection or opportunity for protection) have aesthetic value because such preferences would have en-

3

hanced human survival. Kaplan expands the evolutionary argument to suggest that aesthetic judgment is the product of two processes related to survival: one capturing the viewer's attention, and the other enhancing his or her comprehension. According to his framework, complexity, coherence, mystery, and legibility should enhance aesthetic value. In his second paper, Kaplan discusses the nature of aesthetic response. Is it an instantaneous affective response independent of cognition, or a function of cognitive choice? Seeing flaws in both views, Kaplan argues that aesthetic response is a complex interaction between cognition and affect in which the two components of affect – pleasure and interest – are influenced instantaneously by cognition.

The papers by Greenbie and Nohl focus in more detail on symbolic meaning. Greenbie applies an evolutionary analysis to environmental meaning. He argues that some aesthetic problems arise from the absence of expression of certain primitive symbolic forms. From an examination of animal territorial behavior, dominance, and mating habits, Greenbie finds certain forms that should be preferred because they have symbolic meanings grounded in evolution and survival. His analysis produces three hypotheses: (1) symbolic forms in the contemporary landscape have separated humans from nature, assertive symbols from sheltering ones; (2) built forms tend to suggest shelter; and (3) the most important aspect of built form is social neutrality.

According to Nohl, symbolic meaning results from three levels of response: perceptive cognition (which involves recognition, appraisal, and knowledge of a place), symptomatic cognition (in which objects disclose process behind them), and symbolic cognition (in which objects become symbols for something else). Using this framework, he argues that nature in urban open spaces is overdesigned and is lacking in variety, individuality, and historical context, such that these spaces fail as aesthetic objects.

Finally, Berleant challenges the traditional view of aesthetic response as involving "disinterested attention" and suggests that, at least for environmental aesthetics, a more appropriate model involves reciprocity and participatory engagement. According to this view, the environment is engaging and continuous with the observer. Such a transactional view has important implications for methods of research and design solutions.

Sancar's paper – although in Section III, Applications – explicitly addresses both theory and empirical research. She argues that "grounded theory," which integrates theory, research, and design, is needed.

In summary, these papers raise questions about the process of aesthetic experience and about the role of various formal and symbolic factors in that experience. Several of the empirical papers in this collection address the questions raised in these theoretical papers. With regard to design application, the theoretical papers offer the decision maker a framework of formal and symbolic concerns that are worth considering in attempts to control the visual character of urban areas. Where theory is supported by empirical data, planners, designers, and the courts have compelling evidence to guide public policy on aesthetics.

1 Behavioral and perceptual aspects of the aesthetics of urban environments

Tom F. Heath

This paper offers a theoretical framework that relates the physical and the psychological factors operating in the aesthetic experience of urban environments, and from which hypotheses about preferences can be derived. The environmental features considered are order and complexity. It is argued that the experience of these features is mediated by behavioral and cognitive aspects of the situation. From the behavioral point of view, the significant distinction is between instrumental and diversive behavior. Cognitive aspects are environmental comprehension, or cognitive mapping; the reinforcement or inhibition of formal or informal styles of behavior; and the support or contradiction of beliefs and values.

The need for such a theoretical framework in environmental psychology in general and in the environmental psychology of the city in particular has been well brought out by Proshansky (1978) in his paper "The City and Self-Identity." The aesthetic psychology of the urban environment is a narrower field, and it is therefore possible to propose a scheme rather simpler than Proshansky's concept of "place identity," although certain parallels will appear.

It is not necessary here to discuss in any detail the aesthetic features or qualities of urban settings or to review the contributions of Berlyne (1971), Arnheim (1977), Frances (1977), and others to our understanding of aesthetic psychology in general and of environmental aesthetics in particular. Wohlwill's (1976) pioneering review is still generally relevant, and we can summarize the current state of the art by saying that the importance traditionally assigned to order and complexity as the essential aesthetic features of the environment has been repeatedly confirmed, while very much greater precision has been given to these concepts. It is, however, necessary to say something about the precise position, in this context, of the collative variables novelty, complexity, conflict, and uncertainty.

Consideration of these variables raises an important theoretical issue. If we are to have a subject, aesthetics, then in general it must be concerned with the features or qualities of the material. It is easy to show that it cannot be concerned with preferences, since preferences arise for all sorts of reasons, many of which are quite clearly not aesthetic. Aesthetic psychology therefore must study the responses of people as part of the task of distinguishing and classifying the features of the material – in this case, the city – to which they are responding. The collative variables are not features of the setting; they are mediating or intervening variables that modify the aesthetic response properly so-called. We might call them features of an aesthetic *situation,* which includes the observer or subject.

This can be illustrated in relation to the "Preference Framework" proposed by Kaplan and Kaplan (1982). This divides the environmental features into those that are present, or immediate, and those that are future, or promised. The former are *coherence* and *complexity,* which are synonyms for what have here been called *order* and complexity. The latter, which involve the inferential extension of the present features into the future, are *legibility* and *mystery.* Legibility "is characteristic of an environment that looks as if one could explore extensively *without* getting lost" (Kaplan and Kaplan, 1982, p. 86). Mystery, on the contrary, "involves the inference that one could learn more through locomotion and exploration" (Kaplan and Kaplan, 1982, p. 85). The emphasis is on the extension of experience into the future. Past experiences and plans may equally affect aesthetic responses; the effect of novelty and familiarity on the aesthetic judgment of individual buildings is demonstrated by Herzog, Kaplan, and Kaplan (1976), and the effect of category formation, also an aspect of past experience, by Purcell (1984a, 1984b, 1984c). In general, we must see aesthetic experiences as being mediated by the more general experiences, plans, dispositions, and so on of the individual.

Unless we can clarify, order, and simplify the variety of these experiences, plans, and dispositions, the value of research for the practice of environmental design will be limited. Let us begin with plans or objectives – that is, with behavior. It may be that it is not the detailed content of the plan or objective, but the *kind* of plan or objective that is significant. We may, in fact, be able to divide objectives, from the aesthetic point of view, into just two main classes, based on Berlyne's (1960, 1971) original classification of exploratory behavior.

There are, first, the *instrumental* (or *specific*) objectives: getting from A to B, as in the journey to or from work; finding a suitable place to have lunch, set up a street stall, or get out of the rain. As long as people have such specific plans, the order or interest of the city is likely to be experienced only casually or momentarily. Instrumental values of convenience, comfort, and absence of distraction will dominate. Places will be valued for the features that contribute to the success of the activity being pursued: the sunny nook, protected from wind and crowds, for eating sandwiches, let us say.

These instrumental objectives do not occupy the whole of life, however. There is a different kind of behavior, the behavior of the tourist, the vacationer, the window-shopper, and the stroller. Having completed or set aside immediate tasks, they seek experience for its own sake. The various possibilities of the city for interest and excitement, for calm and order, or for some alternation of these are now positively sought out and appreciated rather than casually encountered. This is *diversive* behavior.

The hypothesis, then, is that instrumental behavior will inhibit aesthetic response, whereas diversive behavior will permit or even enhance it. There is not as yet much empirical support for this theory. A study by Peterson and Neumann (1968) distinguished two groups of beach users: one preferring "scenic natural beaches," attracted by trees and natural growth, and disliking crowding; the other preferring urban beaches and more interested in sand quality and "attractiveness" of the surrounding buildings. These data were interpreted in terms of environmental *personality,* but they may equally be understood in terms of differing objectives. A similar interpretation might be given to the factors found by Shafer and Thompson (1968) in responses to Adirondack campsites. There are no parallel studies of urban settings. It might prove fruitful to treat the specific objectives as a "set" that obstructs aesthetic interaction with the environment. The effects of set on environmental awareness and, specifically, on aesthetic awareness have been demonstrated by Leff, Gordon, and Ferguson (1974).

Way finding is a part of both instrumental and diversive activities. Aesthetic quality may be assumed to contribute to the character or identity of a place: its "placeness." Even this hypothesis has not been adequately tested, although it is supported by qualitative studies (Cullen, 1961; Lynch, 1960; Sitte, 1889/1965). Cognitive-mapping research suggests that our mental map of a city is constructed from landmark places (Devlin, 1976). This goes beyond our first getting to know a city; as Ittelson (1978) points out, exploration is a constant feature of urban existence.

We may then construct a second level of hypothesis: Places of high aesthetic quality will tend to become landmarks, over and above their specific roles in the activity system of the individual. (They cannot, of course, become landmarks unless they are encountered, so there is *some* dependence on the activity system.) A still further hypothesis is that where places of high aesthetic quality are also significant paths or nodes of the pedestrian or transportation network, their aesthetic affect will be reinforced (that is, more people will attend to their aesthetic qualities); but where they do not relate to way-finding activities, their aesthetic effect will be reduced or inhibited by the effects of set.

A city is more than a collection of paths, however well marked. It is a stage on which much of the drama of life is enacted. It provides, or fails to provide, the backgrounds and props that the various human activities it contains require to support them. As Proshansky (1978) says: "For each of the role-related specific identities of the person, there are physical dimensions and characteristics that help to define and are sustained by those identities.'' Here, again, the aesthetic qualities of the setting, it may be hypothesized, either reinforce or contradict the roles enacted, and conversely the aesthetic response will be either reinforced or inhibited.

The various place identities and roles of individuals will, once again, be very numerous and different. But here, too, it is possible to propose a simplifying hypothesis – that the most significant division of roles is that suggested by Goffman (1959) in his book *The Presentation of Self in Everyday Life: "front"* and "backstage.'' The dimension of behavior involved is formality–informality. "Front'' areas are those in which we know ourselves to be in a sense on show; in "backstage'' areas, we can unbutton and "be ourselves.'' As a matter of observation, cities have their "front'' and "backstage'' areas. The squares and boulevards of historic cities and the fashionable shopping areas and financial districts of modern ones are obvious "front'' areas. By contrast, service districts like Soho, factory districts, and dormitory suburbs are more or less "backstage,'' although there is some gradient of formality. This leads to the further hypothesis that a high degree of formal order is most appropriate in "front'' areas, where it serves to reinforce formal behavior.

In responding to or inhibiting behavior, the city expresses social values. Closely related but distinct is the expression of the social significance of places. As Ittelson (1978) puts it: "The city, like any environment, has the potential of enhancing value systems.'' It also has the potential of contradicting them. Social importance is generally linked to formality of behavior, which, it has been argued, is supported by order, or "good Gestalt,'' in the physical environment. However, observation of existing settings suggests that importance is also expressed by means of association and of interest, or formal complexity.

Importance is expressed by association through scale and quality of execution. The effects of scale are anecdotally obvious, but inadequately tested, although there is some evidence from way-finding and cognitive studies (Appleyard, 1969; Carr and Schissler, 1969). The size of a building or public space is historically strongly linked to its social significance, whether it is a medieval cathedral, the palace of Versailles, or "the tallest building in the world,'' on the one hand, or the Piazza San Marco in Venice, on the other. Conversely, the size of some modern office buildings may be perceived as an *undue* claim for significance and may account for part of their unpopularity.

The quality of material, execution, and detail – what Arnheim (1966b) calls "high definition'' (pp. 123–35) – are also commonly associated with importance. One thinks of the Erectheion or the Barcelona Pavilion by Mies van der Rohe; in the same way, the quality of paving, of street furniture (such as railings, lamps, and fountains), and of the surrounding buildings has often been used to mark the importance of public places. Again, we have only anecdotal evidence about people's response to these elements, although a study by Hall, Purcell, Thorne, and Metcalfe (1976) suggests a possible approach.

Interesting, or complex, forms make a call on our attention, and the elaboration of form that we often find in public buildings and places is therefore unsurprising: Complexity, too, marks importance. This is supported to some extent by the study by Herzog et al. (1976), in which judged complexity was shown to be significantly correlated with preference in relation to cultural buildings. This correlation was less strong in relation to buildings of other social content – commercial and entertainment – which might be interpreted as implying that the significance expressed by this sort of aesthetic quality was felt to be appropriate to cultural buildings but not to buildings of more utilitarian kinds.

In summary, the hypotheses proposed are that the experience of the aesthetic qualities of urban settings that are the objects of urban design, and the resulting expressions of preference, are inhibited by instrumental behavior but enhanced by diversive behavior. When major communication paths or nodes lack character, or aesthetic quality, aesthetic experience will, again, be inhibited. Correlation between order as an aesthetic quality and expectations of formal behavior will increase appreciation; contradiction will inhibit it. Finally, correlation between aesthetic qualities and the perceived social importance of a setting will enhance the aesthetic experience; contradiction will inhibit it. These are experimentally testable hypotheses and are important for our understanding of urban aesthetics.

2   Symbolic aesthetics in architecture: toward a research agenda

Jon Lang

Introduction

One of the fundamental goals of design has always been the aesthetic one–the creation of "delightful" rooms, buildings, townscapes, and landscapes. In attempting to understand the nature of the aesthetic experience, a number of people (e.g., Santayana, 1896) have found it useful to distinguish among sensory, formal, and symbolic interactions between people and their built environments. *Sensory aesthetics* is concerned with the pleasurableness of the sensations received from the environment. It involves the arousal of one's perceptual systems, is multidimensional, and results from the colors, odors, sounds, and textures of the environment. *Formal aesthetics* in architecture is concerned primarily with the appreciation of the shapes, rhythms, complexities, and sequences of the visual world, although the concepts can be extended to the sonic, olfactory, and haptic worlds. The appreciation of the associational meanings of the environment that give people pleasure is the subject matter of *symbolic aesthetics.*

The systematic, empirical, and often experimental study of aesthetics has been under way since the pioneering work of Fechner (1876). This research has focused mainly on the formal issues of pattern perception, sequential experiencing of vistas, definition of complexity and simplicity, and form empathy. There are many limitations to this research, but it has yielded a number of positive, if controversial, statements on formal aesthetics (e.g., Arnheim, 1977; Kepes, 1944). The research has not, however, yielded much in attaining an understanding of symbolic meanings, the importance of these meanings to people, or the enjoyment that people derive from them. The subject has, indeed, been largely neglected in environmental-design research. The study of symbolic aesthetics in architecture has been the purview of art historians (e.g., Gideon, 1941; Pevsner, 1936; Wöllflin, 1886). The focus in art criticism has been more on architects and their intentions than on the experiencing of the environment. Where the latter has been the focus, conclusions have been drawn from the introspective analysis of the critic. Thus we have many normative statements on symbolic aesthetics but

11

little in the way of a systematic *positive* theory. Scott (1935) collected a number of normative statements associated with the early years of the modern movement: Architecture must be "expressive of the national life"; "Architecture must be expressive of its purpose." More recently, there is Louis Kahn's statement on the design of a United States embassy: "I wanted to have a clear statement on a way of life." The positive basis required for understanding how such meanings are communicated is limited. We need better descriptions and explanations of the nature of architectural symbolism.

The recent growth of the desire to enhance designers' understanding of symbolic aesthetics coincides with three other developments:

1. The growth of the tertiary sector of the economy – the postindustrial society – and the search by architects for new architectural ideologies
2. The recognition that much of what has been built in the United States is rich in symbolism not recognized by the design ideology of the modern movement. The Levittowns and Las Vegases were seldom studied by architects until these places were brought to their attention through the studies of Venturi and his colleagues (Venturi, Brown, and Izenour, 1972)
3. The demand by clients that buildings be in styles with which they can identify.

Any design ideology with a humanistic basis thus needs a clear positive theory of symbolic aesthetics if it is to succeed in achieving its purposes.

The objective of this paper is to outline our present understanding of symbolic aesthetics and to suggest the lines of research necessary to enhance the ability of designers to create environments and/or environmental-design policies that better meet the symbolic needs of their clients – sponsors and users. The paper begins with a description of symbolic meaning and the purposes served by symbolic aesthetics. It then reviews our understanding of the architectural variables that carry symbolic meaning and their relationship to the affective dimension of experience. With this as a base, alternative theories of the acquisition of meaning can be explained. These reviews demonstrate the similarities and differences in the various theoretical suppositions regarding the nature and functioning of architectural symbolism. This, in turn, provides a foundation for a discussion of the research needed to build a positive theory of symbolic aesthetics in architecture.

Symbolic meaning

Different categorizations of meaning (e.g., Gibson, 1950; Hershberger, 1974) appear to have one similarity: They suggest that some meanings have to do with the potential instrumental use of an object or an environment, and other meanings have to do with the emotional qualities that an observer or a user reads into them. The latter meanings are of concern here. Of all the levels of meaning,

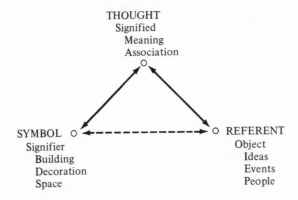

Figure 2.1. The basic semiological triangle.

symbolic meaning is the one least understood by modern architects, although almost all have referred to the symbolic content of their work. A similar confusion exists in the behavioral sciences.

The terms *image, symbol,* and *sign* are often used interchangeably. An *image,* it will be assumed here, is an imitation or a reproduction or a similitude of something. "The image of St. Peters is an image of St. Peters and nothing more; if it suggests Rome or the Holy Catholic Church, the image becomes a symbol" (Gibson, 1966). A *symbol* is something that stands for something else. It may do this as the result of an association, a convention, or even an accident (Burchard and Bush-Brown, 1966). A symbol is the result of a cognitive process whereby an object acquires a connotation beyond its instrumental use. An "object" may be an environment or a person as well as a material artifact. Its meanings are derived from what an observer imputes to them (Kepes, 1966). A *sign* in contrast, is a conventional figure or device that stands for something else in a literal rather than an abstract sense.

Semiology

Semiology is a young field very much concerned with the nature of symbols. It has already influenced the thinking of a number of architects (e.g., Jencks and Baird, 1969). The concerns of the field are summarized in the basic semiological triangle (Figure 2.1), which specifies a relationship among symbol, thought, and referent.

A symbol in the built environment consists of a structure of surfaces of various materials, pigmentation, and illumination levels. This structure is the signifier. The thoughts, or meanings – the signified – associated with it may vary from individual to individual or from group to group because the referent is different. If one is dealing with language, one finds considerable agreement on what is signified. Architecture is only partially analogous.

The built environment conveys symbolic meaning in subtle ways. The correspondence between a building pattern or set of patterns and what is signified has to be learned. Sometimes this is done consciously, but often it is unconsciously. Architects, among others, often attempt to establish new symbol systems. To get them accepted, they have to educate others about the set of associations between the new patterns – the symbol – and the signified. This may involve advertising, polemic writing, or direct teaching. Within any field, elite groups are likely to control some of this process (Barthes, 1967), but other meanings are largely unconsciously developed.

If one accepts that symbolic meanings are primarily socioculturally determined, then people who do not understand the "language" being used to convey meaning cannot appreciate the environment in the manner that an architect might intend. They assign meanings to the environment given their own symbol system. This is the potential trap that architectural historians and critics face in their interpretive analyses of buildings, particularly those of cultures other than their own. It is also clear that meanings of specific building patterns depend on their context.

Morris (1938) suggested that there are three levels of semiological meaning: syntactic, semantic, and pragmatic. *Syntactic meaning* results from the location of a building or decorative element in its surroundings. *Semantic meaning* refers to the "norms, idea or attitude that an element represents or designates" (Rainwater, 1966). Thus much public housing in the United States represents an idea of a small group of people (see also Wolfe, 1981) on how the working class should live. *Pragmatic meaning* relates the symbol to those who use it (see also Norberg-Schulz, 1965).

Two other psychological processes explain some of the confusion over the interpretation of the symbolic meaning of specific patterns of built form. These are the processes of *stimulus* and *response generalization*. It is known that the same architectural variables (e.g., façades of buildings) may have different meanings for different people. Different forms may also communicate the same meanings. The associations that people have with specific patterns may also change over time in a manner that is difficult to predict. Chain-link fences, block walls, and asphalt landscaping may be associated with institutions now, but this may not endure.

The ability to predict how people will interpret the symbolism of an environment is limited by the designer's lack of a positive theory of architectural symbolism (Hershberger, 1974). This is particularly true in dealing with groups whose values are different from those of the designer. In such situations, the designer cannot rely on his intuitive knowledge because that is drawn from his own experience and not that of others. To understand the importance of symbolism in people's lives, one has to understand the purposes served by symbols.

The purpose of architectural symbols

"Human beings are symbol mongers" (Langer, 1953). One way in which people communicate with one another is via symbols. Architectural symbolism is one of a set of nonverbal mechanisms that people use to communicate messages about themselves, their backgrounds, social statuses, and world views to others. Other material artifacts of everyday life that carry such symbolic meanings include automobiles, clothing, furniture and furnishings, and even household pets.

Some people find it very important to communicate meanings about themselves to others through the material artifacts they own or use. Others do not. A tentative set of explanations for this can be derived from an examination of Maslow's (1954) model of human motivations. Because the model is widely known, a brief overview of it will suffice here.

Maslow suggested that human needs can be arranged in a hierarchical fashion, with strongest-level needs taking precedence. His hierarchy, in descending order, is as follows: *physiological needs,* such as hunger and thirst; *safety needs,* such as security and protection from physical and psychological harm; *belonging and love needs,* which concern the relationship of responsive, or affectionate, and authoritative needs; *esteem needs,* or those needs of an individual to be held in high esteem in his own eyes as well as those of others; *self-actualization needs,* which represent the desire to fulfill one's total capacities; and *cognitive and aesthetic needs,* such as the thirst for knowledge and the desire for beauty for their own sakes (see also Lang, Burnette, Moleski, and Vachon, 1974, p. 84). In using this classification, it must be recognized that an individual's or a group's perception of its needs cannot be simply correlated with socioeconomic status. Even for people of the lowest socioeconomic levels, the need for belonging or for esteem is often very important. Some hypotheses can, nevertheless, be suggested about the relative importance of architectural symbols for people with different basic needs.

When people are struggling for survival, the symbolic aesthetics of the environment will not be the focus of attention. The physical character of the environment will still communicate messages about the status of the people concerned; they are likely to be well aware of this, but they will have little energy and thus inclination to act to purposively change the symbolism. For people whose prime concern is with safety, architectural variables – particularly those associated with symbolic barriers representing territorial demarcations – become more important, but it is in fulfilling belonging and esteem needs that architectural symbols are particularly important.

The symbols that people choose to have around them may reflect their perceptions of who they are or may reflect their perceptions of who they aspire to be or may simply reflect a rejection of the past. If one aspires to be a member of a group, then the symbols associated with that group become particularly important. It should be noted, however, that the perception of important symbols associated with a group might well differ between those outside the group and those who are members. If we have full membership in a group − be it socioeconomic, cultural, or ethnic − the symbols of membership become less important, and the environments chosen are more likely to reflect personality or other idiosyncracies. This is also true for those whose needs are primarily self-actualization and cognitive and aesthetic ones.

At each level in Maslow's (1954) hierarchy, needs are largely fulfilled through social and cultural mechanisms not related to elements of the architectural environment. The attributes of the built world are, however, important, so it is necessary to recognize the variables that can carry symbolic meaning.

Architectural variables that carry symbolic meaning

A content analysis of the writings of architects, artists, and art historians and of the few studies that have dealt empirically with the symbolism of the architectural environment enables the compilation of a tentative list of variables that have symbolic meaning.

Building configuration

The shapes and patterns that an architectural style comprises carry meaning. In certain cultures, specific shapes, such as a circle, or particular patterns, such as symmetry, have associational meanings themselves; but these meanings have been largely lost in the Western world, except in certain religious institutions where the linkage between pattern and meaning has become a social convention. In architecture, it is principally the style of a building that carries symbolic meaning.

In this category are included classical motifs to represent democratic ideals, simple clear shapes with few historical allusions to represent the machine age and modernism, and the emerging complexity of shapes to represent postindustrial society. In terms of housing, the meanings of the rectangular, single-family structure are mentioned frequently in both the architectural and the social-science literature. The association, it is suggested, is with a life style and with individualism of action. The roots of these meanings are said to lie deep in the

Anglo-Saxon psyche (Handlin, 1972). The meanings are, however, also recognized by almost all the people of the United States (Cooper, 1974; Hinshaw and Allott, 1972), including those who choose other environments for themselves (Michelson, 1966).

Spatial configurations

The volume, degree of enclosure, and proportions of enclosed space also carry meaning. The consumption of space, per se, is an important symbol. Higher-status people in most organizations, formal or communal, inhabit physical settings that are larger than those inhabited by those of lower rank. People have little difficulty in recognizing this. The meanings of other spatial relationships have hardly been studied rigorously. In one study (Beck, 1970), five simple, dichotomous spatial variables were identified as potential carriers of architectural meaning. They are diffuse versus dense space, delineated versus open space, verticality versus horizontality, right and left in the horizontal plane, and up and down in the vertical plane. A summary of one of these, delineated versus open space, will give a feeling for Beck's goals: "Delineated space refers to bounded, constricted, contained, contracted or centripetal space; open space suggests inward and outward movement, spatial penetration, liberty and freedom." His research shows that different people have different associations with architectural spaces located at different points on those scales, but it is not developed enough to derive a series of clear positive statements on the link between spatial patterns and symbolic meanings for different people.

The same is true of Norberg-Schulz's (1971) speculations on the nature of existential space. He refers to architectural space as the "concretization" of existential space. Existential space "is a psychological concept denoting the schemata man develops interacting with the environment in order to get along satisfactorily." Norberg-Schulz is thus concerned with linking concepts in the mind with spatial configurations. Although the concept of a schema has been part of psychological theory for a long time, its nature and how it actually guides the perception and interpretation of meanings in the environment is still highly conjectural (see Neisser, 1976).

Materials

A plain wooden interior may be chosen for a ski shop, marble for the Kennedy Center in Washington, D.C., or metal for a museum of technology. These materials may be chosen partly for their technical attributes, but also for the associations they afford. Not only the visual character, but also the sonic, haptic, and sometimes olfactory nature of a material have associations. Certain materials have, through use, become associated with building types. The result is that there is often a clash between technical sensibleness and symbolic aesthetic requirements in design. Practicing architects have an intuitive understanding of many of these associations within their own culture.

The artificiality or naturalness of materials have associations for some people. Izumi (1969) believes that the use of plastics to simulate wood and other materials sets up doubts in a viewer because our perceptual systems are fooled and/or because there is an incongruity between what a material looks like and what it feels or sounds like.

The nature of illumination

The effects of the directionality, source, color, and level of illumination of a behavior setting have long been regarded as fundamental variables of architectural composition. There is a considerable, if scattered, body of introspective and interpretive analyses of the psychological impact of the use of light as a compositional element. For instance, Gropius (1962) wrote: "Imagine the surprise and animation experienced when a sunbeam shining through the stained glass window in a cathedral wanders slowly through the twilight of a nave and suddenly hits the altar piece. What a stimulus for the spectators."

On a more mundane level, there are strong correlations among the level of illumination, types of light fittings, the nature of illumination, and certain behavior settings. For instance, a student study of restaurants in Philadelphia revealed a correlation between illumination type and restaurant type. The less expensive restaurants, which cater to a large number of patrons daily, have high levels of illumination from overhead fluorescent lighting, whereas the more exclusive restaurants tend to have much lower levels of illumination from incandescent light fittings on tables or walls.

Although such anecdotal statements exist in abundance and there is much data on the human-engineering aspects of light in architecture, there needs to be more research on the use of light to meet the symbolic needs of people. How do the directionality, color, level of illumination, and contrasts between light and shade convey meaning in different situations in different cultural contexts?

Pigmentation

The colors of buildings, surfaces, and smaller artifacts carry symbolic meaning – often by explicit social conventions. These conventions may be understood by broad segments of a population, even though the antecedents of the convention may be unknown. Color conventions differ from society to society. In traditional Beijing, bright colors were reserved for palaces, temples, and other buildings housing rituals; ordinary buildings were artificially made as colorless as possible. Color was a symbol of status.

Colors seem to be associated with building types and population groups in a complex set of ways, but the evidence for these links is highly contradictory (Hayward, 1974; Porter and Mikellides, 1976). Often it seems that it is not specific colors that carry messages, but deviations from customs. The effect of deviations from norms has to be understood as much as the affective meanings of colors themselves.

Non-physical variables that carry architectural symbolism

To make matters more complex, the symbolic meanings of specific environments are not dependent simply on their architectural qualities. They are dependent on such things as the names of places, as the designers and developers of new residential areas well know (Michelson, 1976), the perceptions of those involved in the decision-making process, and the people in and activities taking place within the setting. Buildings and urban designs reflect the way decisions are made and the perceptions of those who made them. They act as symbols of the process. Goodman (1971) is among those who now perceive the urban-renewal projects of the 1950s and 1960s in the United States as symbols of the effort by architects and politicians to impose their values on others and the perpetuation of the oppression of the poor in cities. Other people may see the same artifacts as symbols of concern for the poor.

Some places are peculiarly associated with certain people or events. A particular setting may have symbolic meaning not because of its physical attributes, but because of the events that took place there. The building becomes a symbol of the events. These events may have been recurrent ones, or there may have been a single event. Independence Hall in Philadelphia and the Anne Frank House in Amsterdam act in this way. The form of the buildings is largely irrelevant in terms of their associational meaning (Rapoport, 1977). This type of symbolism is thus beyond the control of the designer; it is acquired over time or through an idiosyncratic occurrence.

The aesthetic dimension

Affect is a general term for "emotion" or "feeling." An understanding of symbolic aesthetics involves an understanding of the positive and negative attitudes that people have about the symbolic meanings available in the built environment. An *attitude* results from combining a *belief* about something with a *value* premise about it.

There are a number of definitions of the word *belief*. Most social psychologists consider a belief to be an assertion about an *associative* rather than a defining characteristic of a thing. Thus buildings with pointed arches may be a defining characteristic of Gothic architecture; "such windows go well in ecclesiastical architecture" is an associative characteristic. Many such beliefs are verbalized in architectural writings, but many can be inferred only by observing what architects create.

Values are related to motivations, for they define the attractive and repulsive elements of the world. In architecture, anything a person desires or compliments has a positive value for that individual; anything that is disdained has a negative value. Values represent a link among a person's emotions, motivations, and behavior.

Attitudes toward specific built environments arise from the attribution of a value to a belief. As in other aspects of life, attitudes vary in direction, saliency, strength, cognitive differentiation, action orientation, and verifiability.

Direction refers to the evaluation of an object; saliency, to the extent of preoccupation with the object; strength, to the degree of resistance to change of an attitude; cognitive differentiation, to the degree of clarity of an attitude; action orientation, to the degree with which components of action are part of the schema of the attitude – because one likes a particular set of symbols does not necessarily mean that one will use them – and verification, to the degree to which a belief can be tested. Many differences in attitude toward the symbolic content of buildings rest on fundamentally different beliefs about the nature of aesthetics. Such beliefs, like religious beliefs, are difficult to verify. They have to be taken on faith.

The nature of cognitive consistency – having mutually supportive attitudes – has been the subject of much recent research. A number of models of cognitive consistency help architects understand the vagaries of aesthetic analysis. These models are important to architectural theory because they help us understand how likes and dislikes are developed and maintained.

The simplest model is that of *balance theory* (Heider, 1946). If one person has a positive attitude (likes, promotes, seeks) toward another person (referent), who, in turn, has a set of positive attitudes toward an inanimate object (symbol), such as a building or a set of ideas, then the first person's attitudes would have to be positive toward the object or ideas for the system to be in balance. The referent need not be another person, but may be a system of ideas. Figure 2.2(a) graphically portrays this relationship. Figures 2.2(b) and 2.2(c) show other consistent relationships; Figure 2.2(d) illustrates an inconsistent one.

The Osgood and Tannenbaum (1955) *congruity model* and the Rosenberg–Abelson (1960) *psychologic model* build on Heider's (1946) formulations. Both deal with mechanisms for the maintenance of the congruity of attitudes. The congruity model recognizes that two people may have very different reasons for

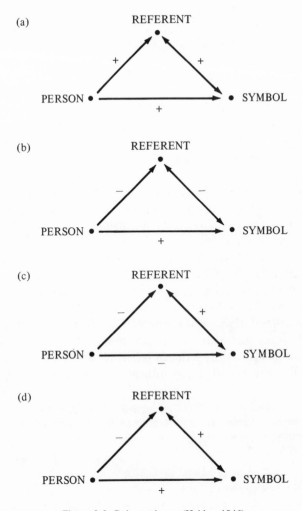

Figure 2.2. Balance theory (Heider, 1946).

holding the same attitude. In addition, some attitudes are more strongly held than others and thus are less likely to change based on the influence of other people than are weakly held ones. The psychologic model recognizes that people isolate attitudes that are not consistent with others and subtly refuse to acknowledge the discrepancy. Thus one has to recognize the greater complexity of attitudes than that suggested by Heider's simple balance-theory model. The Heider model, however, still presents a point of departure in discussing symbolic aesthetics.

Whether one regards the symbolism of Levittown or Las Vegas positively or

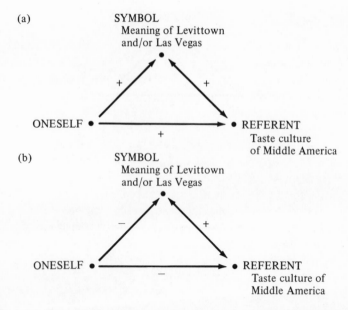

Figure 2.3. Two consonant positions on the symbolism of Levittown and Las Vegas.

negatively can be explained by Heider's (1946) balance theory or any of the elaborations of it. Venturi et al. (1972) note:

The content of the symbols, commercial hucksterism and middle-class social aspirations, is distasteful for many architects. . . . They recognize the symbolism but they do not accept it. To them the symbolic decoration of the split level suburban shed represents the debased, materialistic values of a consumer economy where people are brainwashed by mass marketing and have no choice but to move to tickey-tackey, with its vulgar violations of the nature of materials and its visual pollution of architectural sensibilities and surely, therefore the ecology.

People strive for a consonance of their attitudes (Festinger, 1957). The symbolic values of a Levittown or a Las Vegas or a commercial strip on the outskirts of most American cities are found to be quite acceptable by many people. Figure 2.3 places the symbol, taste culture, and referent at the corners of a triangle of relationships. If a population likes the symbolism of Levittown's housing, an individual (referent) can maintain consonance by having positive values on all three relationships, as in Figure 2.3(a), or a negative value toward the population and its taste culture, as in Figure 2.3(b). In making such observations, it becomes clear that culture plays a major part in aesthetic analysis.

Culture and affect

Much of human behavior is governed by *culture* – the system of shared attitudes and symbols that characterizes a group of people. The culture of a people is a shared schema that designates regularities in a group's thinking and behavior. Individuals are socialized within a culture, but their behavior also shapes the culture so that it is not something static, but something that evolves over time. Each culture is unique because it has its own history. This does not mean that certain values are not held by many cultures, but that each culture is a result of the past efforts of a people to deal with its physical and social environment.

People can deal with their own cultures in an unconscious manner. As a result of being socialized into a culture, an individual has the ability to know the appropriate behaviors. This holds for professional as well as societal cultures.

Architects are members of at least two cultures embedded in each other. Each has its own socialization processes. One culture is the professional culture, and the other is that of the broader society within which the professional culture exists. The professional society has its own norms of what is acceptable, including attitudes that define appropriate symbols for the built environment. These norms can be extremely coercive (Montgomery, 1966).

Architects have long attempted to influence cultures through the buildings they design by creating symbols for society. Their ability to do so depends on the strengths of attitudes toward existing systems and the architects' ability to convince others of the symbolic meaning of new architectural forms. Minoru Yamasaki has used essentially the same façade patterns on a synagogue, on the World Trade Center in New York, and for the Dhahran Airport in Saudi Arabia. "Yamasaki must be a very good explainer," wrote critic Thomas Hine (1978).

If architects adhere to existing symbolic norms, the symbols become incrementally more coercive: Conformity adds to the stability of the system of symbols for others. Consistency over time is heavily conditioned by social reinforcement, according to behaviorist theories of psychology. The reinforcement comes from the response of others both inside and outside the profession. There are, however, conflicting interpretations of how these processes occur.

The acquisition of symbolic meanings

A number of coexisting theories explain how the environment possesses or communicates symbolic meaning. Possibly, these theories explain processes that act simultaneously. Each explains one aspect of symbolism. If so, there needs to be an integrative model. If not, there needs to be a clarification of the explanatory power of these theories.

The theory of physiognomic properties

One theory of symbolism is that buildings have physiognomic, or expressive, properties that are directly understood by the observer. This is the "symbolism through perceptual dynamics approach" of Arnheim (1977). Arnheim calls this "spontaneous symbolism," which he contrasts with "conventional symbolism" (for example, the thirty-six columns of the Lincoln Memorial, which represent the thirty-six states at the time of Lincoln's death, or the fish-shaped Presbyterian church designed by Harrison and Abramowitz in Stamford, Connecticut).

Spontaneous symbolic meanings are said to arise from a directly perceivable analogy between the visual structure of an object, such as a building, and a corresponding generic characteristic of shapes, such as "height or depth, openness or closure, outgoingness or withdrawal." Many of these empathies are said to develop from observation of the natural world, including the human body (Wölfflin, 1886), or the building process (Rasmussen, 1959).

Some psychologists, such as Gregory (1966) and Gibson (1966, 1979), have severe doubts about Arnheim's Gestalt psychology approach to visual expression. This doubt is also expressed by Colquhoun (1967), who is cited by Venturi et al. (1972) to justify their theory of symbolism: ". . . the arrangement of forms such as found in a painting by Kandinsky is, in fact, very low in content, unless we attribute to them some conventional meanings not inherent in the forms themselves."

The Jungian approach to symbolism

The Jungian approach is best expressed in Cooper's (1974) paper "The House as Symbol of Self." Jung postulated that people have a collective unconscious linking them to past eons. In this unconscious are "bundles of psychic energy," termed archetypes – one of the most fundamental of which is the self. Difficult to grasp, people express archetypes in architectural, among other physical, symbols. People attempt to express their personalities and aspirations through the environments they select for themselves. While this may well be true, it does not specify how the associations are developed. This is what the behaviorist theory attempts to do.

The behaviorist model

The behaviorist approach to the understanding of symbols suggests that because elements of the built world and meanings have become associated in situations that people find rewarding, they continue to seek these associations. What may

REFERENT $\xrightarrow[\text{convention}]{\text{Social}}$ SYMBOL $\xrightarrow[\text{association}]{\text{Psychological}}$ THOUGHT

Figure 2.4. A model of symbolic meaning (Gibson, 1966).

have been a deterministic relationship between "function" and form, for instance, is carried on as a symbolic association because the form has come to provide people with a sense of identity – a sense of social and physical place in the world.

Anne Tyng's model of form empathy

Tyng's (1975) work represents an attempt to link concepts of universal symbols to the unconscious. She is concerned with the symbolism of specific forms themselves. These may be aspects of a whole building form or of its elements. She extends the Jungian concept of *individuation* (von Franz, 1968) by attributing an empathy to a category of forms for each stage in the process.

In Jungian psychology, the growth of personality is regarded as a movement in which the individual is able to integrate greater and greater amounts of the unconscious into his or her conscious life. Thus an individual (or a group of people) increases his or her ability to deal with the unconscious until a "rebirth occurs." Rebirth involves a restructuring of what is conscious. The process is accompanied by an empathy for increasingly complex forms until the rebirth occurs, when the empathy returns to simple forms.

Ecological models of perception

The correspondence between a particular pattern of architecture (referent) and a particular thought has to be learned. Sometimes, the process follows a model proposed by Gibson (1966), as shown in Figure 2.4. When it is, the social conventions usually occur naturally through the processes of socialization. The psychological association is also learned. Gibson's model stresses the cultural basis of architectural symbolism.

Conclusion: toward a research agenda

This paper has presented an overview of the nature of symbolic aesthetics in architecture as presently understood. A number of concepts developed by social scientists are helping to give clarity to architectural arguments about the nature and purpose of symbolic aesthetics. At the same time, much remains unclear. Although many architects are willing to specify normative positions on symbolic aesthetics, there is little concerted effort to build a body of positive principles on the subject. This needs to be changed if the design professions are to better understand their roles in the service of humans and to better debate normative positions on the future.

A set of research questions on issues of concern to architects come to mind:

1. What are the variables of design that carry symbolic meaning?
2. Are they the variables described here?
3. How important is it that the associational meanings of the environment are congruent with the schemata of a particular population?
4. If they are not, so what? Will the lack of congruence increase stress levels or lead to conflicts in self-identity?
5. What are the associational meanings of specific patterns?
6. How have these meanings changed over time?
7. How stable are they now?
8. What is the method of identifying symbolic meanings of concern to specific cultures?

The research of Venturi et al. (1972) is important not only because it brought attention to issues of symbolic aesthetics long scorned by the design professions, but also because it dealt with issues at a scale appropriate to architectural theory. While it is easy to criticize the study because of its lack of scientific rigor, it is the type of research with which architects feel comfortable even if they do not care for the normative issues of design that it raises. If the environmental-design-research community can be encouraged to engage in theory building at the ecological level at which design decisions are made, our knowledge of the symbolic dimension of aesthetic experience will be rapidly advanced.

3   Prospects and refuges revisited

Jay Appleton

Figure 3.1. Barradine–Narrabri Road, New South Wales, Australia. The vista directs attention to a point on the horizon.

It is questionable whether authors ever ought to go into print in defense of their own work. Objective criticism from others is much more likely to hit the mark. I am persuaded to put pen to paper now partly because the editors have specifically asked me to review my theories after a decade;[1] partly because I have no intention of making it a "defense" in the sense of a refutation of all criticism, but, I hope, a realistic appraisal; and partly because developments since that time, both in my own thinking and in the publications of others, have altered the situation in various ways.

Ten years ago, I saw the discipline of geography as emerging from an ordeal of self-examination on which it had embarked in the belief that some drastic reshaping of its image was necessary for its academic survival. This was not the

This paper was first published in *Landscape Journal* 3 (1984): 91–103.

first occasion on which its prophets had proclaimed the need for a kind of scientific purity (Hartshorne, 1939), but whereas individual voices had once been raised in the cause, in the 1960s the main stream of the subject was diverted into a new channel. Under the philosophical influence of logical positivism, the quantitative revolution established itself. As new techniques were discovered, much effort was directed to their refinement and perfection, and they were assiduously applied to those areas of investigation in which they seemed to have the best chance of success. These tended to lie in the field of locational analysis, the study of spatial relationships mathematically expressed.

Visual aspects of the environment, such as the study of what places look like, seemed at first to lend themselves much less promisingly to the new methodology, although from the late 1960s onwards, this impression began to be corrected – some might say overcorrected – by a profusion of statistically based exercises in "landscape evaluation" that proceeded from a restricted methodological base. Although the techniques employed varied in detail, almost all of them consisted of the setting up of scales of values derived from panels of observers, scales that were then used to measure the merits of particular landscapes (Penning-Rowsell, 1973).

By this process, the visual aspect of the environment was certainly reintroduced into geographical inquiry, albeit in a very limited way. The search for other approaches was being pursued by the humanistic geographers, the phenomenologists, and others, and it was against this background that there seemed to be a need for a holistic theory that would help us to see, if only dimly, how landscape preference or landscape taste might be interpreted in the light of all those different kinds of evidence that the heritage of recorded human experience presented to those who had the patience to fit them together.

"Prospect-refuge theory" was the rather pretentious name that I eventually chose for what is in effect an extremely simple idea, although, like all simple ideas, it lends itself to almost any degree of elaboration. I was looking for a simple model that could relate the idea of preference to a typology of landscapes through the medium of the biological and, more particularly, the behavioral sciences.

I was well aware of the risks involved in attempting this. All simplification is necessarily oversimplification. Put in another way, theory isolates the common characteristics of a set of circumstances and ignores the others, and this inevitably distorts the total image of each individual circumstance. One of the most common kinds of criticism of prospect-refuge theory clearly arose from a failure to understand its role as an agent of simplification for explanatory purposes. Many other theories have suffered from similar misunderstandings, misrepresentations, or misuses.

Consider, for instance, Davis's (1899) "geographic cycle,"[2] which became the basis of much geomorphological explanation for well over half a century. Davis's theory was, in essence, that because the shapes of landforms resulted from the processes that formed them, the nature of these processes could be inferred from these resultant shapes. Their evolution was to be interpreted in terms of "structure, process and stage": the structure of the underlying rocks, the processes of denudation operating on those rocks, and the stage that had been reached in those processes.

As a general model, this "Davisian cycle" offered a clear, simple explanation of the development of river valleys, river systems, and so on – in short, the physical basis of most terrestrial landscapes. Problems arose, however, as soon as one attempted to use it to achieve more than this, and it is unlikely that a modern "process geomorphologist" would find it of much value in devising field experiments. Davis's "stages" consisted of "youth, maturity and old age," and while these categories could be broadly recognized by their visual attributes, the demarcation of precise boundary lines between them was almost impossible. It was the attribution of particular cases to their appropriate categories that proved an insurmountable difficulty, and without this the validity of the model could not be thoroughly tested. That it survived for so long as an explanatory model was due to the fact that it seemed, in the light of experience and common sense, to offer a credible interpretation of a very large number of environmental observations. If it has now been superseded in the teaching of geomorphology, this is not because it has been proved wrong, but because it has appeared inadequate, and to that extent irrelevant, in pressing forward the frontiers of research. Most geomorphologists, if they are honest, will admit that it played a crucial role in identifying in broad terms the processes that they are now measuring with meticulous care and great gusto, while human geographers, concerned with a less sophisticated and less exacting study of geomorphology per se, but requiring a simple descriptive terminology for explaining the interaction of people with their natural environment, still find it useful for this purpose.

It is understandable that authors who put forward theoretical ideas wish to think of them as "solving" problems, establishing once-and-for-all explanations of the truth. Yet they know very well, if they pause to think about it, that this never happens. Theory is constantly modified by verification as the answers to each question provide the premises for the next. "Knowledge" is advanced in steps, and at the beginning of the twentieth century, the need among geomorphologists was for a simple formula that would present a framework of explanation for morphological phenomena in terms of their probable evolution. Until the problem of explaining the configuration of landforms and land surfaces was conceived in this way, it was premature to move on to the sort of work that currently composes most geomorphological research.

Some three-quarters of a century after Davis (1899) put forward his model, it seemed to me that the study of landscape had reached a kind of milestone, not unlike that which geomorphologv had reached at the turn of the century, in that it needed some coordinating system that would enable its participants to see the significance of any one component of the landscape in terms of all the other components. This is what lay behind the idea of prospect-refuge theory. So how did it set about achieving this?

I think, in retrospect, that it aimed at four objectives as means to the end just outlined. First, it attempted to set the aesthetics of landscape within the context of a biological interpretation. The argument was that preference for particular environments is a part of a pattern of behavior, the meaning of which is to be sought in environmental adaptation. In this sense, the concept of aesthetic value, which traditionally has been more closely associated with the arts, is approached on the assumption that it is also susceptible to examination by the methodology of the sciences (Appleton, 1980). We must understand at the outset that this assumption may prove to be invalid, but if this is so, its invalidity must emerge from the argument and from the evidence and must not be assumed as a starting point for the inquiry. This was not, of course, an original philosophical position, and due recognition was paid to its earlier protagonists, notably Dewey (1929, 1934), who, however, had not followed the implications of his hedonistic philosophy very far into the field of *landscape* perception.

Second, and following directly from this, prospect-refuge theory laid great emphasis on the need to elicit evidential support from as wide a range of human experience as possible. It turned for such support to the arts and to the sciences and even to the circumstantial evidence afforded by the everyday experience of the man and woman in the street. Inevitably, an investigation as broad as this had to be superficial. It had to be argued at a level comprehensible to the non-specialist (it had, indeed, to be argued at a level comprehensible to the author!). Furthermore, it was very clear that even if investigation could be pursued in parallel within the arts and the sciences – even, that is to say, if two contrasting types of methodology could be made to coexist – it was by no means a foregone conclusion that ways could be found to integrate the findings of one with those of the other. It was equally clear, however, that this must be an important part of the objective.

Third, the theory attempted to reduce those concepts that emerged from this complexity to a relatively simple form, even perhaps "formula," and simplification, as I have already pointed out, is, by its very nature, a risky thing to attempt; yet without such a reduction, there can be no progress in explanation. The important thing is to be aware of the particular dangers that attach to any particular theory and the limitations that must be observed in its application.

Fourth, having achieved this reduction, it was to be available as a potential explanatory model for the use of anyone interested in applying it in any particular area of investigation or in any particular method or technique.

Figure 3.2. Brownsea Island, Dorset, England. A deflected vista arouses curiosity and stimulates anticipation.

These, then, were the objectives of prospect-refuge theory; but what did it actually say? It began by identifying environmental perception as the key to all adaptive behavior. Without environmental information, so ran the argument, there can be no adaptation. It is reasonable, therefore, to suppose that individuals within each species of animal, including *Homo sapiens,* are motivated to perceive their surroundings in such a way that environmental information is acquired and stored in a form in which it can be efficiently and quickly retrieved when needed to ensure survival. Men and women, in short, perceive their environment in much the same way as other animals perceive their habitat, and to this idea I gave the name "habitat theory."

Every aspect of the habitat as a theater for survival is important, but some kinds of information are likely to be needed more *immediately* than others, particularly those that relate to self-preservation from sudden, unexpected danger. Because the sort of knowledge that permits rapid strategic adjustments is so immediately important, we have to pay particular attention to it; nature ensures that we do so by providing us with a powerful craving to satisfy our curiosity first and foremost about our environment as a theater for survival.

There are certain ways in which we can improve our chances of survival by paying attention to certain kinds of environmental opportunity, and two of these emerge as of particular importance. The first is the opportunity to keep open the channels by which we receive environmental information. (All the senses are

Figure 3.3. Queens' College, Cambridge, England. The refuge symbolism of the building is strengthened ("reduplicated") by that of the trees that partially screen it. The River Cam lends itself to exploratory environmental adaptations of many kinds.

Figure 3.4. Wind erosion at Watamolla, New South Wales, Australia. Discovering the limits of safety (and therefore danger) is an important part of adaptive behavior and leads to fascination with hazard symbols. This Australian student got it right, but he was lucky that there is not a fault in the Hawkesbury Sandstone.

involved; but in considering "landscape," we are naturally most concerned with the sense of sight and can therefore be justified in using the word *seeing* to describe this process.) The second is the opportunity to achieve concealment. These two opportunities give us the twin basis of our simple classification of "prospect" and "refuge."

It is apparent, with hindsight, that certain aspects of prospect-refuge theory were not explained as clearly as they might have been. Misconceptions began to appear in the first reviews of *The Experience of Landscape*. Some of these arose from the fact that some reviewers had read only part of the book and that others had difficulty in bringing themselves to accept ideas that they found distasteful, but I must certainly accept responsibility for some of the discrepancies that arose between my own ideas and the impressions of those ideas that were formed in the minds of readers. For instance, one reviewer referred to my having "dismissed Darwin along with Freud," whereas the whole argument was rooted in Darwinism. Another referred to my "brusque dismissal of the eighteenth-century ideas of the Sublime, the Beautiful and the Picturesque." What I (Appleton, 1975a) actually wrote was "It is, I think, fair to say that the prolonged controversy about the Beautiful, the Sublime and the Picturesque is widely regarded today as somewhat tedious and not very directly relevant to modern aesthetics . . . " (p. 38).

I went on: "This is, I believe, an extremely ungenerous view which arises from too direct an assessment of the achievements of the participants in terms of their own avowed objectives . . . " (pp. 38–39).

A little later: "But if the eighteenth-century aestheticians failed to defend convincingly the positions they had taken up, we have failed for another reason to appreciate the importance of what they achieved" (p. 39).

And a little later still:

When we look at the "picturesque" controversy and its preamble from Shaftesbury to Uvedale Price and Repton, we shall find a number of ideas to which, I believe, we must turn back if we are to make the kind of rapprochement between landscape and aesthetics at which we are aiming. We shall then discover not only that the thinkers of the eighteenth-century were aiming their enquiries more directly towards our target than were most of their nineteenth-century successors in Britain (though not, perhaps, in America), but also that many of the ideas which they introduced or developed contained the germ of what we may well regard as a modern view. (p. 39)

This must surely be one of the strangest "brusque dismissals" of all time.

Perhaps the most common misconception about prospect-refuge theory arose from a failure to understand the point I have already made – that it was a reductionist tool, attempting to identify primitive patterns of behavior and particularly of perception as the origins from which the highly complex preference patterns of *Homo sapiens* had grown, certainly not as subsuming those prefer-

Figure 3.5. New Porter's Pass, South Island, New Zealand. Sunshine, snow-white skylines, atmospheric clarity, and the absence of prominent refuge symbols accentuate the high prospect values of this winter scene.

Figure 3.6. Niagara Falls, Ontario, Canada, and New York State. Falling water is one of the most powerful hazard symbols.

ence patterns in their maturity or their entirety. "There is more to primitive man than primitive man in a three-piece suit," wrote one reviewer.

There is no use in an author's throwing up his hands in despair when confronted with interpretations of his work that seem to point to the exact opposite of what he is trying to say. The message is clear enough. He ought to make himself more clear, and certainly there are aspects of prospect-refuge theory, as described in the book, that I can see very well, with hindsight, were obscure or evasive. Obscurity and evasion arose not from any intention to wriggle out of a tight corner, but partly because of a failure to anticipate precisely those areas that needed fuller explanation and partly because of lack of room for development. (Some 20,000 words had to be cut out, many of which amplified ideas that were subsequently misconstrued.)

Thus one reviewer, a philosopher, wrote: "He alludes to the eighteenth-century treatment of the sublime . . . but he does not himself offer an explanation, in terms of habitat theory generally or prospect-refuge theory in particular, for the aesthetic pleasure we can find in such landscapes and landscape-paintings" (Crawford, 1976, p. 369). There is no misconception here: The reviewer was absolutely right. The explanation has chiefly to do with the symbolism of the hazard, which is a constantly recurring theme throughout the book, but the amplification of the argument about *how* this works appeared in one of the passages I decided to eliminate through pressure of space. Only when I read the reviews did I realize how unwise and unhelpful the excision had been. *Mea maxima culpa!*

Another line of criticism that might have been avoided by a fuller and more careful exposition of the argument was that which was based on too literal an interpretation of the concept. Thus Brotherton (1979) postulated that there could be only two possible mechanisms by which the "prospect-refuge 'hang-up' " had evolved:

Either man evolved in environments with strong p-r [prospect-refuge]; or individuals that used such prospects and refuges as existed in their environment to help them to "see without being seen" survived and reproduced better. One or other of these propositions must be sound if the theory of prospect-refuge is to stand. (p. 13)

Not unnaturally, Brotherton had little difficulty in raising serious doubts about both mechanisms, but I suspect he had not fully grasped the idea that the terms *prospect* and *refuge* refer to *concepts* rather than objects, and that they find expression only in the very *basic* processes of environmental perception. They have to do with noticing environment in terms of opportunities for behavioral response, and there is no reason to suppose that such evolutionary changes as have taken place either in landscape or in the perceiving creature, "man," have altered them fundamentally. The landscape architect's "voids and masses" embrace a vast range of landscape phenomena in which, however, the essential concept remains constant; in a similar way, sophisticated civilized humans, in their responses to sudden warnings of danger, still show every sign of responding to symbolic indications of opportunities to take evasive or protective action.

Another problem with which the theory has had to contend is that of dissemination. In fact, I think there are really three problems here. The first arises from the multidisciplinary nature of the argument. This cuts both ways. The book was reviewed in periodicals covering many areas of human experience, but within each area there was, at best, a minority of readers interested in broad concepts of this sort. Almost all the reviewers were experts in much more restricted fields.

This was the second problem. Even highly reputable scholars tend to read into innovative ideas what they expect to find there and what fits into their own established habits of thought. Perhaps the more glaring misrepresentations are attributable to this cause. One reviewer took me to task for not having touched on the paintings of Courbet, who was his favorite landscapist. Almost the whole book could have been illustrated from the works of Courbet or of many other painters who received no mention at all; but the point of mentioning any painter was to illustrate some theoretical point, not to make a critique of any one artist's work in terms of prospect and refuge. I wondered whether one might fault a book on the circulation of the blood that failed to mention Virgil, George Washington, or the Duke of Wellington, all of whom, presumably, had blood that circulated!

The third problem, which was invariably tied up with the first two, was that the very possibility of explaining aesthetic values in biological terms is still anathema to the romantic mind. I found myself likened to Robert Ardrey and Desmond Morris, and one cannot be damned more severely than that – unless, of course, those authors should turn out to be right!

The consequence of this was that prospect-refuge theory had a very wide range of receptions, from outright rejection ("a prime example of an insidious idea which has fed an obsession to a point where a great all-embracing structure is built upon it without adequate justification"), through varying shades of indifference, to qualified enthusiasm ("perhaps the most stimulating writing on the theory of landscape since Lord Clark's *Landscape into Art*").

How, then, has the theory fared in the first decade or so of its existence? Toward the end of *The Experience of Landscape,* I posed two questions: "Can it be substantiated, and is it of any use?" (p. 260). I went on to explain that I had been attempting "to bring the argument to a level of plausibility at which scholars competent to pursue . . . further enquiries may conclude that there is enough *prima facie* evidence to warrant their attention" (p. 262). Several authors have risen to this fly, and with respect to both of the questions, we are beginning to get a little feedback.

Figure 3.7. Jasper National Park, Alberta, Canada. Trees on the lower ground create a powerful impression of refuge, but the distant peaks seen from a high vantage point make this a prospect-dominant view.

Perhaps the most systematic and direct testing of the validity of the hypothesis was that carried out by Clamp and Powell (1982). They set out specifically to test the theory by subjecting four questions to the assessment of four judges. In brief, these questions tested (1) uniformity and variability of response; (2) validity of prospect, refuge, and hazard, and the balance among them, as predictions of judged quality; (3) applicability to landscape, not merely as perceived from a particular viewpoint; and (4) the *relative* significance of prospect, refuge, and hazard.

Apart from establishing a majority preference for prospect-dominant landscapes, Clamp and Powell's test failed "either to support conclusively or to negate the central claim of prospect-refuge theory" (p. 8). They offer a number of possible explanations of this failure by reference to methodological problems: the smallness of the panel of judges (only four), the possibility of mismatching actual photographs with conceptual categories, and the like. They kindly conclude that the theory may have "a more subtle value which is not to be revealed by the kind of empirical public examination employed here" (p. 8).

Without wishing to prejudice any future studies of this kind, I am bound to say that I would go along with this last opinion. The first major problem encountered by Clamp and Powell (1982) is closely analogous to that which geomorphologists have encountered in *testing* the "Davisian cycle." It is a problem of attributing cases to categories. Where do we draw the line between youth and maturity, and on what basis do we decide how much of the symbolism of prospect or refuge a landscape expresses? Statistical analysis absolutely requires that we be able to make such attributions positively, precisely, and self-confidently, and the authors themselves admit that they were not able to do this. The test *must*, therefore, be inconclusive, which does not mean that it was an unprofitable exercise to carry out. If they achieved nothing else, Clamp and Powell alerted the author to certain areas of misconception that have clearly arisen from inadequacies in the explanation of the argument as originally set out. I shall refer to this later.

Another study in which an attempt was made to introduce prospect-refuge theory into landscape evaluation was undertaken by Woodcock (1982) at The University of Michigan, under the influence of S. and R. Kaplan, who have made a number of references to the theory in their own work (e.g., Kaplan and Kaplan, 1982). Although there were several hypothetical correlations between preferred landscapes and the symbolism of prospect, refuge, and hazard for which Woodcock was not able to find empirical support, there were others in which very clear patterns began to emerge. This obviously leads one to ask why he was able to find the theory more promising than did Clamp and Powell.

I think one important reason is that whereas Clamp and Powell (1982) simply attributed symbolic values subjectively to particular landscapes and then sought to compare them with judged preferences of those landscapes, Woodcock (1982) set out to achieve a different objective. His concern was to examine the evidence for supposing that environmental preference is *derived* from environmental-perception habits as essential components of hominid survival behavior, that it is a product, in other words, of evolution. To test this, he interposed between his preference-ranked landscapes and his attributions of symbolic (prospect, refuge, hazard) values an intermediate concept, that of the biome, a visually recognizable type of environment distinguished chiefly by its natural vegetation, (e.g., savanna, forest, desert). There is a certain loose way in which biomes can be seen to reflect values of prospect, refuge, and hazard, which may be used to explain (partially, at least) correlations between expressed preferences and biomes. The correlations themselves, however, are unaffected by the attributions of these symbolic values. It is in the *interpretation* of these correlations that prospect-refuge theory may prove helpful.

Perhaps another way of expressing the difference is to say that whereas Clamp and Powell (1982) explicitly set out to *test* prospect-refuge theory, Woodcock (1982) set out to *use* it. The two questions that I posed – "Can it be substantiated, and is it of any use?" – ought ideally to be resolved in this order. But paradoxically, many of the discoveries of mankind have been made in the reverse order. We find things useful before we are able to validate the reasons why. Many drugs, for instance, had been used in the treatment of diseases long before

Figure 3.8. Near Fulda, West Germany. Houses in a wooded valley symbolize refuge, but the sweeping view with multiple horizons seen from a high vantage point creates an overall impression of a prospect-dominant landscape.

their biochemical functions were demonstrated or understood. It is therefore not surprising to find that the most favorable reception of prospect-refuge theory seems to have come from people who found that, like Davis's (1899) "geographic cycle," it presents familiar phenomena in an unfamiliar *explanatory* framework. In this way, it suggests relationships that are at first apprehended rather than comprehended, a distinction that may perhaps be more easily accepted in the arts than in the sciences. My impression is that, while I would be the last person to maintain that these two fields of human experience are fundamentally different (Appleton, 1980), the theory has so far proved more promising in the methodology of the former than of the latter.

This conclusion receives some support from an examination of other studies that have come to my notice. A kind of halfway house between an "arts" and a "science" treatment of the subject may be found in the work of Heyligers, particularly in his paper on dune landscapes (1981). Heyligers is a botanist who, as a member of the research team that produced the voluminous documentation of the landscape of south Australia (Laut et al., 1977), had been particularly intrigued with the aesthetic qualities of the environment he had been scientifically studying, especially on the Coorong coast to the east of the mouth of the Murray River.

By selecting dune landscapes, Heyligers was able to eliminate the grosser variables that make difficult a comparison between mountainous and flat land. The dunes themselves afforded a basal surface consisting of a combination of flat and gently undulating parts, the latter being further divisible into convex and concave landforms. These last obviously suggested prospect-dominant and refuge-dominant values, respectively. This basal surface was covered in places by vegetation and elsewhere left bare, and the taller components of the shrub vegetation were tall enough to evoke strong feelings of refuge.

Heyligers (1981) concludes that "the examples discussed above show that prospect/refuge symbolism is applicable at various levels of abstraction" (p. 11), but continues: ". . . the aesthetic appreciation, the integration of an environmental stimulus into one's own perceptual framework, will necessarily remain a personal experience and, after all, beauty remains where it always has been: in the eye of the beholder!"

It is, then, in the literature of the arts that I suggest we look for the more promising signs of the usefulness of prospect-refuge theory. As an example of the theory's application to landscape description, let me refer to Solar's (1979) work on the panorama. Solar was particularly concerned with the panoramic prospects obtainable in the Swiss Alps and, even more particularly, with the interest that the early-nineteenth-century diplomat Hans Conrad Eschen von der Linth displayed in these panoramas; but his approach was much broader and touched on the whole question of a search for some theoretical framework of explanation. This he seemed to find in *The Experience of Landscape*. I suspect that the terms *turning point (Wendepunkte)* and *breakthrough (Bruch)*, which he employed, were undeservedly generous, but the point is that through the application of prospect-refuge theory, Solar was able to find a simple way of fitting Eschen von der Linth's concept of the alpine panorama not only into the historical development of taste and fashion, but also into a functional interpretation of the role of long-distance views in environmental perception.

Prospect-refuge theory, as an exploratory tool, has begun to be pressed into service in both literary and art criticism. One example of each must suffice. In literature, I have particularly noted a book by Brownlow (1983) on the poetry of John Clare. Clare (1793–1864) was a very controversial English poet whose stock has been steadily rising and who has inspired a number of authors to examine his work in the past few years. Probably the most stimulating, and certainly the best known to students of landscape, is that by Barrell (1972). Brownlow approaches his subject differently from Barrell but includes a chapter that draws heavily on prospect-refuge theory when writing about hitherto unexplained facets of this highly perceptive but sadly disturbed personality. The fen-margin landscapes of Helpston, with their sweeping, open, flat panoramas and contrasting patches of close woodland, afford as interesting an environment as Heyligers's sand dunes and one that is equally capable of simplification into units to which prospect and refuge values can be meaningfully attributed without too much difficulty. But it is not only in the actual landscape (which was, perhaps, as important to him as ever a "place" was to any poet) that Clare lends himself to interpretation in prospect-refuge terms, and I know of no poem that better exemplifies the *concepts* of prospect, refuge, and hazard, particularly the last two, than his charming little verses about the ladybug, or, in the Northamptonshire vernacular, the Clock-a-Clay. Here is a tiny creature in a Lilliputian landscape communicating through the poet feelings of what we might prosaically call environmental interaction, which acquire a new meaning simply because they have been translated from the human to the minuscule scale.

Let me turn to the use of prospect-refuge theory by an art historian. By far the most persuasive example I have yet encountered is to be found in the work of Ronald Paulson. Perhaps it is significant that Paulson is a professor of English whose interests expanded from eighteenth-century writers to encompass engravers and painters and led him to participate actively in the work of the Mellon Center for Studies in British Art, at Yale University. From Swift and Fielding, he moved on to Hogarth and Rowlandson, always on the lookout for symbolic meaning, whether in the written word or on the canvas or engraver's plate. This was specifically the subject of a book (Paulson, 1975) published in the same year as *The Experience of Landscape*.

In a review essay, Paulson (1976–77) summarizes prospect-refuge theory and then employs it in explanation of his own new interpretations of the paintings of Constable. I am not suggesting that without the theory, he could not have expressed these new ideas, but at least he seemed sufficiently satisfied with this approach to repeat and expand it in a later book (Paulson, 1982).

Although the theory has not figured widely in the literature of the field, my impression is that landscape architecture is the field in which prospect-refuge theory has made the greatest impact. Certainly, I have received more invitations to speak on it at schools of landscape architecture than in any other disciplines, including my own, geography. This is, perhaps, not surprising. Landscape architects themselves tell me that whereas they are reasonably confident that they have the mastery of the scientific and economic bases of their profession (plant materials, soil science, microclimatology, quantity surveying), as well as of the graphic techniques to represent their intentions to clients, contractors, and others, all in the furtherance of the realization of their artistic ideas, the question of where those ideas come from is another matter. There seems to be among landscape architects a keen awareness that some strengthening of the theoretical underpinnings of their art would be welcome, and I know of a number who have found the theory to be of some practical use, at least to the extent of clarifying their own thoughts about their own feelings. Certainly, I have found students of landscape architecture in schools I have visited in Australia, New Zealand, Canada, the United States, and Britain to be more receptive, more perceptive, and more easily aroused to discussion than any other category of student.

42   JAY APPLETON

Figure 3.9. Versailles, France. The symbolism of prospect and refuge is strait-jacketed in the geometric regularity of André Le Nôtre's formal design.

Figure 3.10. Sheffield Park, Sussex, England. An early-twentieth-century designed landscape in which reflecting water surfaces help open up a prospect within a cozy nest of refuge symbolism.

Obviously, the past ten years have not passed without my revising or amplifying many of the ideas that went into the *The Experience of Landscape*. I have given much further thought to the wider evolutionary context of habitat and prospect-refuge theory. The charge, referred to and admitted earlier, that I had

LEEDS METROPOLITAN
UNIVERSITY
LIBRARY

not offered an adequate explanation of the aesthetic pleasure to be found in landscape prompted me to read a paper in Edmonton, Alberta, in 1978, subsequently published in Sadler and Carlson (1982). This turned the spotlight on the pleasure principle itself as the motivating drive lying behind all survival behavior and, therefore, behind exploratory behavior, which is no more or less than purposeful, investigative environmental perception. I have also dabbled in expressing the ideas of prospect-refuge theory through other media – painting (not, for me, a vehicle for publication!) and poetry (Appleton, 1978). I have, needless to say, been much influenced by what I have read since 1975 and by discussion with colleagues and, more particularly, with researchers with whom I have worked (especially Colin Laverick, Ian Smith, Gary Armitage, and Jane Gear), so that I cannot always find it easy to say which are my own innovative thoughts and which are theirs. But I have not had occasion to weaken in my conviction that the biological basis of landscape aesthetics as suggested in *The Experience of Landscape* will eventually prove to be aiming in the right direction.

What I *do* regret are the ambiguities and imperfections of expression in *The Experience of Landscape* that have led to misinterpretation. Let me give one example. One of the shortcomings of the empirical study by Clamp and Powell (1982), outlined earlier, is largely attributable to their assumption that "balance between prospect and refuge, when neither dominates the scene, make[s] it more attractive, as indeed Appleton postulates" (p. 7). They go on to cite the authority for this: " . . . a landscape which affords both a good opportunity to see and a good opportunity to hide is aesthetically more satisfying than one which affords neither . . . " (p. 8). True, this is what l wrote, but I failed to make clear that this balance does not *necessarily* have to be achieved in a single momentary view from a particular viewpoint, any more than a balanced diet demands a perfect mix in every mouthful! Perhaps I had not fully anticipated the heavy reliance that researchers into landscape evaluation were to make on still photographs as sources for empirical research. Very often the balance that can be achieved from serial vision, involving the successive experiences of exposure to strongly contrasting landscape types – strong prospect and then strong refuge – is more potent than that which comes from trying to achieve a balance all at once. I think I would have written a whole chapter on this if I had known then what I know now!

What, then, of the future? I suppose that prospect-refuge theory may become obsolete when either it is proved invalid or it has outlived its usefulness. The first fate I think is unlikely to befall it. The second, which is the fate that overtook Davis's (1899) "geographic cycle," is perhaps more likely, but by that time it may have served, like Davis's model, to further our general understanding of landscape during a phase of inquiry in which a more general model of this sort may be appropriate. We must not shrink from putting forward models, theories, or explanations merely for fear that their success may bring about their own obsolescence. Perhaps a more likely fate still will be one of modification and refinement as a result of the experience of trial and error.

Of one thing I am certain. While in general it is essential for scientific investigations into landscape preference to adhere strictly to the rules of scientific method, *some* academics must be prepared (if I may employ a mixture of metaphors) to fly kites, to cross boundaries, to stick out necks, to publish theories without waiting for them to be proved (provided, of course, that they do not represent them as though they *were* proved), and to see their own personal reputations, if necessary, demolished with their own theories in the greater interest of the pursuit of truth, to which they all protest allegiance. Perhaps it is only we elderly academics, put out to grass, who can really afford to do these things.

Notes

1. "Prospect-refuge theory" was first described in *The Experience of Landscape* (Appleton, 1975a), the first draft of which was completed in the autumn of 1973.
2. There is some controversy over the origin of Davis's theory of the "geographic cycle," alternatively known as the "geomorphic cycle." It was developed during the early years of the twentieth century, but its main essentials were described in Davis's paper of 1899.

4   Perception and landscape: conceptions and misconceptions

Stephen Kaplan

It would seem that the psychology of perception should have something useful to contribute to landscape aesthetics. Certainly, students of landscape aesthetics make assumptions about the nature of perception. Although certain of these favorite assumptions are probably false, there was a time when it was not obvious that the psychology of perception had anything better to offer. Fortunately, a number of recent developments shed considerable light on the significance and functioning of the aesthetic reaction to landscapes.

From the perspective of landscape aesthetics, perhaps the single most salient theme in recent work in perception is what Gibson (1977) has labeled "affordances." An *affordance* refers to what a perceived object or scene has to offer as far as the individual perceiver is concerned. Perception is viewed as not merely dealing with information about the environment, but also yielding information about what the possibilities are as far as human purposes are concerned. In addition to Gibson's contribution to this topic, this emphasis on the function that an object or environment might serve for the perceiver has appeared in the work of Gregory (1969) and S. Kaplan (1975).

One can go a step farther than the perception of affordances per se. As Charlesworth (1976) has pointed out, a species not only has to be able to recognize the sorts of environments in which it functions well, but also has to prefer them. Animals have to like the sort of settings in which they thrive. Ideally, they would not have to learn such an inclination. It could be costly for an animal to spend years barely subsisting in unsatisfactory environments in order to learn that such environments were in fact unsatisfactory. And to the extent that erroneously choosing certain environments could be a fatal error, such a bias would ideally be innate and immediate. Hence, one can view preference as an outcome of a complex process that includes perceiving things and spaces and

This paper was first published in *Proceedings of Our National Landscape Conference* (USDA Forest Service, General Technical Report PSW-35) (Berkeley, Calif.: Pacific Southwest Forest and Range Experiment Station, 1979), pp. 241–8.

reacting to them in terms of their potential usefulness and supportiveness. In this perspective, aesthetics must, at least to some degree, reflect the functional appropriateness of spaces and things.

It should be noted, however, that this view of preference as an expression of bias toward adaptively suitable environments has no necessary connection to what is *currently* functional. What was functional during the evolution of a species is presumably what would be preferred, quite independent of what might be functional today.

Some comments on preference

In the context of an evolutionary perspective, it is hardly surprising that human preference would have some relationship to those environments in which survival would be more likely. This context, however, does not appear to be characteristic of discussions of landscape preference. Indeed, a significant number of students of landscape aesthetics view preference with alarm, or at the very least, distaste. There are many who feel that preference judgments are bound to be arbitrary, idiosyncratic at best, and perhaps even random. In part, the distrust of preference judgments probably stems from the fear that aesthetics will be debased by stooping to a popular consensus. Hidden in this fear is a profound irony. It implies that there is no basic consistency, no underlying pattern characteristic of preference judgments. Without such an underlying basis, however, aesthetics becomes trivialized. If aesthetics is not an expression of some basic and underlying aspect of the human mind, then it is hard to see why it is of more than passing significance. It is reduced to mere decoration, as opposed to being something with pervasive importance. In the concern to preserve aesthetics untarnished, it makes it at the same time inconsequential.

If indeed aesthetics has no deep underlying significance for our species, then one would expect preference judgments to vary randomly from one person to the next. But this is not a matter of opinion; it is an empirically testable hypothesis. This conjecture, as well as the others implicit in the rejection of preference, are not only testable; they have been tested many times. These fears receive no support from many studies. (For a discussion of a number of such studies, see S. Kaplan, 1979b.) Preference judgments are neither random nor highly idiosyncratic. Neither are they debasing; many of the rules that preference follows turn out to have correlates in the classic aesthetic and landscape-architecture literature. At the same time, their divergences are thought-provoking and instructive.

To summarize the ground covered so far, recent work on perception views the perceptual process as inextricably connected with human purposes, and perhaps with human preferences as well. Preference, in turn, is a far better behaved measure than is often feared. The remainder of the paper is based on the premise that preference judgments are not antithetical to aesthetics. Rather, they are seen as providing a powerful tool for understanding the patterns underlying what we consider to be aesthetic. Such judgments may also point to the underlying significance of the aesthetic in the larger human scheme of things.

Making sense and involvement

If people's reactions to things and spaces depend on people's purposes, then understanding preference requires that we first understand what these purposes are. Because different people pursue different purposes and because any one individual will pursue different purposes at different times, an analysis in terms of purpose might appear to be at best unpromising. Fortunately for both science and practice, however, human purposes are by no means totally scattered and idiosyncratic. In fact, our research on preference has over the years repeatedly pointed to two underlying purposes that people are concerned with throughout their waking hours. These two purposes probably had an important impact on the long-term survival potential of the individual. Their pervasive influence seems appropriate because they are necessarily vital to any specific purpose that an individual may choose to pursue.

We have come to call these persisting purposes "making sense" and "involvement" (R. Kaplan, 1977a). *Making sense* refers to the concern to comprehend, to keep one's bearings, to understand what is going on in the immediate here and now and often in some larger world as well. *Involvement* refers to the concern to figure out, to learn, to be stimulated. At first glance, these two purposes may seem to be contradictory, or at least at opposite ends of a continuum. But on closer examination, this turns out to be a misconception. Certainly, there are environments that one can comprehend and at the same time be stimulated by. Likewise, there are environments that offer neither possibility. In fact, all combinations are possible; knowing that an environment makes sense tells one nothing about whether it will be involving.

Although our realization of the centrality of these two purposes arose in the context of research on preference, there are sound theoretical grounds for believing that they would be necessary to the survival of an information-based organism. There are also parallels in the psychological literature – order, security, closure, and the like, on the one hand, and curiosity, challenge, stimulation, and so on, on the other. (Limitations of space permit only a brief mention of these themes here; for a more extensive discussion, see Kaplan and Kaplan, 1978.)

If making sense and involvement are indeed pervasive purposes for humans, then environments that support these purposes should be preferred. The term *support* here refers to whatever an environment might afford that makes a particular purpose more likely to be pursued to a successful conclusion. For making sense, this refers to the perceived structure of the environment. It takes in anything that would make the environment easier to map, easier to characterize, to summarize to oneself. It involves those affordances that increase one's sense of comprehension.

For involvement, on the contrary, the supportive environment is one rich in possibility. In a sense, the affordances for involvement entail the raw materials for thinking about and coming to understand. The issue here is to be challenged, to have to call on one's capacities in order to process the information successfully. Thus a poem or a landscape that is "simple-minded" or "obvious" fails to offer affordance for involvement. (It should be noted that the "raw material" that challenges one's capacities need not actually be present; it constitutes an effective challenge even if it is only implied or suggested.)

The visual array

In reacting to the visual environment, people seem to relate to the information they pick up in two quite different ways. They react both to the visual array, the two-dimensional pattern – as though the environment in front of them were a flat picture – as well as to the three-dimensional pattern of space that unfolds before them. The idea of the visual array is easiest to think of in terms of a photograph of any given landscape. The pattern of light and dark on the photograph, the organization of this "picture plane," constitutes the basis of this level of analysis.

Just as the surface of a photograph can have much or little to look at, scenes can vary in involvement at this level of analysis. Comparably, the pattern of information on the surface of a photograph can be easier or harder to organize, constituting the making sense aspect of the visual array. Let us examine each of these components in somewhat greater detail.

Complexity is the involvement component at this surface level of analysis. Perhaps more appropriately referred to as "diversity" or "richness," this component was at one time thought to be the sole, or at least the primary, determinant of aesthetic reactions in general. Loosely speaking, it reflects how much is "going on" in a particular scene, how much there is to look at. If there is very little going on – for example, a scene consisting of an undifferentiated open field with horizon in the background – then preference is likely to be low.

Coherence is the making-sense component at this surface level of analysis. It includes those factors that make the picture plane easier to organize, to comprehend, to structure. Coherence is strengthened by anything that makes it easier to organize the patterns of light and dark into a manageable number of major objects and/or areas. These include repeated elements and smooth textures that identify a "region," or an area, of the picture plane. Readily identifiable components aid in giving a sense of coherence. It is also important that a change in texture or brightness in the visual array is associated with something important going on in the scene. In other words, something that draws one's attention within the scene should turn out to be an important object or a boundary between regions or some other significant property. If what draws one's attention and what is worth looking at turn out to be different properties, then the scene lacks coherence.

People can hold only a certain amount of information in what is called their "working memory" at one time. Research on this phenomenon suggests that this limit in capacity is best understood in terms of a certain number of major units of information, or "chunks." Thus rather than being able to remember a certain number of individual details or facts, people seem to be able to hold on to a few distinct larger groupings of information. The current evidence suggests that most people are able to hold approximately five such chunks, or units, in their working memory at once (Mandler, 1967).

It follows from this that anything in a scene that helps divide it into approximately five major units will aid the comprehension process. All the various factors that contribute to coherence tend to do this. The greater the complexity of a scene, the more structure is required to organize it in this way, or, in other words, for it to be coherent as well.

Three-dimensional space

The analysis at the level of the visual array, of the picture plane itself, is important to the viewer but at the same time is limited. Landscapes are three-dimensional configurations, and it was in that third dimension that our ancestors functioned or failed to function, survived or perished. It is hardly surprising that people automatically interpret photographs of the environment in terms of the third dimension as well.

As we might expect, given the evolutionary importance of space, humans are highly effective at perceiving depth. Perhaps the most central issue in analyzing a scene involves the three-dimensional space and its implications. As Appleton (1975a) points out, there are implications both in terms of informational opportunities and in terms of informational dangers. The informational opportunities he calls "prospect." This idea of being able to gather new information has a kinship with the involvement side of our framework. In particular, the opportunity to gather new information in the context of an inferred space is what we have come to call "mystery."

Mystery. One of the most striking aspects of people's reaction to landscapes that suggests a three-dimensional interpretation is their preference for scenes where it appears as though one could see more if one were to "walk into" the scene. Strong as this involvement component of the spatial interpretation is, it has been frustratingly difficult to name. We have decided on *mystery*, a term long ago used in the context of landscape architecture to refer to an essentially similar idea (Hubbard and Kimball, 1917).

Some investigators in this area have assumed that we were referring to surprise or novelty. But *novelty* implies that one is perceiving something new, and a scene high in mystery may have nothing new present (and, conversely, a novel object present in the scene in no way ensures mystery). Likewise, *surprise* implies the presence of something unexpected. Mystery involves not the *presence* of new information, but its promise. Mystery embodies the attraction of the bend in the road, the view partially obscured by foliage, the temptation to follow the path "just a little farther." While the "promise of more information" captures the essential flavor of this concept, there is actually more to it than that.

Scenes high in mystery are characterized by continuity; there is a connection between what is seen and what is anticipated. While there is indeed the suggestion of new information, the character of that new information is implied by the available information. Not only is the degree of novelty limited in this way; but there is also a sense of control, a sense that the rate of exposure to novelty is at the discretion of the viewer. A scene high in mystery is one in which one could learn more if one were to proceed farther into the scene. Thus one's rate and direction of travel would serve to limit the rate at which new information must be dealt with. For a creature readily bored with the familiar and yet fearful of the strange, such an arrangement must be close to ideal.

Another area of potential confusion should perhaps be mentioned. The word *Mystery* to some people connotes the ambiguous, even the incoherent or impossible to understand. Although it is in some way true that anything that makes no sense is mysterious, the term is intended in a far more limited sense. Admittedly, it implies uncertainty. But here the uncertainty is thoroughly constrained and bounded. It is of a limited degree, and its rate of introduction is under control. It is by no means beyond comprehension; rather, it is possible to anticipate to a reasonable degree. Mystery arouses curiosity. What it evokes is not a blank state of mind but a mind focused on a variety of possibilities, of hypotheses of what might be coming next. It may be the very opportunity to anticipate several possible alternatives that makes mystery so fascinating and mind filling. The human capacity to respond to suggestion is profound.

Legibility. The other aspect of landscape stressed by Appleton (1975a) concerns safety in the context of space. While he terms this component "refuge," emphasizing being able to see without being seen, from an informational perspective, safety encompasses considerably more than this. This broad conception of safety

Table 4.1. *Preference matrix*

Level of interpretation	Making sense	Involvement
The visual array	Coherence	Complexity
Three-dimensional space	Legibility	Mystery

closely parallels the making-sense side of our framework; we have chosen the term *legibility* to refer to the possibility of making sense within a three-dimensional space.

Like mystery, legibility entails a promise, a prediction, but in this case not of the opportunity to learn, but to function, It is concerned with interpreting the space, with finding one's way, and, not trivially, with finding one's way back. Hence it deals with the structuring of space, with its differentiation, with its readability. It is like coherence, but instead of dealing with the organization of the picture plane, it deals with the organization of the ground plane – the space that extends from the foreground to the horizon.

A highly legible scene is one that is easy to oversee and to form a cognitive map of. Hence legibility is greater when there is considerable apparent depth and a well-defined space. Smooth textures aid in this. So, too, do distinctive elements well distributed throughout the space that can serve as landmarks. Another aspect of legibility involves the ease with which one can perceive the space as divided into subareas, or subregions. There is a strong parallel here to what makes a scene coherent, but coherence differs in referring to the organization of the visual array rather than to that of the three-dimensional space. Coherence concerns the conditions for perceiving, whereas legibility concerns the conditions for moving within the space.

It must be emphasized that the interpretation of a scene in three dimensions is, like the analysis of the visual array, or picture plane, an automatic and generally nonconscious process. People tend not to know that they are doing this. It characteristically happens very rapidly and effortlessly. Although the basis for hypothesizing such a process comes from data on preference, this is precisely the sort of processing of affordances, of what the environment offers, that one would expect of a far-ranging, spatially oriented species.

Overview of the preference matrix

These two large domains of human preference, considered in terms of the visual array and in terms of three-dimensional space, can be summarized by a 2 × 2 matrix (Table 4.1).

The entries in the table should be thought of as broad concepts, each subsuming a variety of components. Some of these specific components, in turn, influence more than one cell in the matrix. Smooth textures, for example, tend to enhance both coherence and legibility. Comparably, the factors that tend to make a scene appear to have greater depth enhance both legibility and mystery.

Although both the surface and the three-dimensional levels of analysis are represented in the matrix, these two levels may not have comparable weight. It is necessary for a scene to have at least a modicum of coherence and a modicum of complexity to be preferred, but high values of these components do not necessarily lead to high preference. By contrast, legibility and especially mystery seem to influence preference throughout their entire range.

It should be reiterated in the context of Table 4.1 that making sense and involvement are independent aspects of a scene. Although a scene of high complexity can lack coherence, it can also, like the Taj Mahal, possess a great deal of coherence. Likewise, the presence of high legibility, as in a scene with a well-structured space, does not prevent the partial obscuring and the opportunity for exploration that is characteristic of a scene rated high in mystery. In more general terms, one can think of these issues from the perspective of the kinds of information that are required for making sense and the kinds of information that enhance involvement.

On the making-sense side of the ledger, structure is required in a scene both to comprehend what is where in the visual array and to interpret the larger spatial configuration. At the same time, there may be few different things in the scene (low complexity) or many. The more complex scene increases the possibilities for what one could look at, and hence in a sense increases the uncertainty. But the structure is not thereby decreased.

Likewise, there may be much suggested by the scene as being available but beyond one's present view. Here, too, there is an increase in uncertainty that in no way contradicts or undermines whatever structure the scene possesses.

Readers familiar with an earlier version of this matrix (S. Kaplan, 1975) might find the current configuration a bit disorienting at first glance. Note, however, that three of the four constructs (complexity, coherence, and mystery) have remained the same and have the same relationship to one another. The major additions, the making sense–involvement distinction and the two levels of interpretation, have a number of advantages over their predecessors, both intuitively and theoretically.

Some common misconceptions

Idiosyncrasy in perception

There is a great deal of concern and confusion over the problem of idiosyncrasy in perception; it becomes particularly acute when the issue of preference is included. After all, everyone knows how much taste differs from one person to the next. And if taste is purely personal – or if not personal, at the whim of culture – then decisions about preference will be at best arbitrary.

This concern, while understandable enough, is misguided. Perception, and preference for that matter, are no more variable than any other aspect of human experience and human behavior. As with everything else, there is regularity and there is variability. As with anything else, identifying and understanding that regularity is crucial to appropriate policy- and decision making. It has been the purpose of this paper to identify some of these regularities and to suggest some ways in which they might be understood.

While idiosyncrasy per se turns out to be not that much of a problem either theoretically or empirically, some of the ways proposed to deal with it have created problems that are rather more serious. One reaction involves relegating all the variability to some postperceptual process. Thus it is asserted that perception is the same for all people, while interpretations vary. The difficulty with this solution is, as we have seen, that perception and interpretation are inseparable. The perceptual process is itself influenced by all those cultural, experiential, and individual factors that are supposed to underlie interpretation.

A similar approach has been to create variables for people to judge that are "objective," in order to get around the subjectivity of preference. Sometimes the "objectivity" is achieved by having people rate scenes in terms of landscape features rather than in terms of preference. Agreement using such a procedure has not, however, been particularly impressive (R. Kaplan, 1975). This is perhaps not surprising. People make judgments of preference quickly and easily. By contrast, judgment of the presence or absence of certain landscape features may seem to many people to be unnatural or forced. Unlike preference, it is not a judgment they make frequently and intuitively.

There is thus the temptation to dispense with ordinary people altogether, to rely on selected judges to make the ratings. And often what they are asked to do is make "assessments of aesthetics" rather than rate the scenes for preference, From a psychological perspective, the distinction between these two kinds of judgments is difficult to justify.

Although perceptions are not all the same, there are some remarkable communalities, perhaps in part because of our common evolutionary heritage. And while there are indeed certain cultural differences, these may involve differential emphasis on the components of preference discussed earlier. Some of the most reliable differences among groups as far as preference is concerned turn out, interestingly enough, to be between experts and everyone else (R. Kaplan 1973; Anderson, 1978). This makes good sense psychologically. There is now substantial evidence that experts perceive differently from other people (S. Kaplan, 1977; Posner, 1973). This is restricted to their area of expertise and, in fact, constitutes an important facet *of* their expertise. But the fact remains that experts, often without realizing it, do not see their part of the world the way anyone else sees it. Although experts are invaluable resources when used appropriately, they are a dubious source of "objective" judgment about what people care about in the landscape.

What is really *important in landscape*

There have been numerous efforts to identify the crucial aspects of the human reaction to landscape, to get to the heart of the matter, as it were. One such approach is both impressive in its directness and disturbing in its implications. This approach is based on ferreting out the unique. It is argued that the more unusual the scenery, the more valuable it is. This is essentially an economic argument, relying exclusively on scarcity. While it is undoubtedly true in some limited sense, it is equally false in a larger sense and certainly a caricature of the multiplicity of factors influencing preference.

People's reaction to nature is an example of a noneconomic need (see Hendee and Stankey, 1973; Kaplan and Kaplan, 1978). It is not something to be exchanged for something else, but an intrinsic reaction. People value even rather common instances of nature (R. Kaplan, 1978b; Kaplan, Kaplan, and Wendt, 1972). At the same time, certain rare, nonnatural elements (a unique statue, for example) are not valued at all.

There *is* a sense in which uniqueness is valued. That is when a place has a distinctiveness, a "sense of place," that makes it possible to know where one is whenever one visits that place. This is a rather special meaning of uniqueness that has little relation to the visual-scarcity interpretation.

Another derivative of the uniqueness idea that might be worth studying is the uniqueness of a place in terms of access. Thus the only park that one can get to for lunch within walking distance of downtown is unique in an important way, although not in the traditional sense of "unique content."

Another influential scheme for landscape analysis focuses on the four factors of form, line, color, and texture. This approach conflicts with the realities of human perception in a number of respects.

First, these properties tend to emphasize the two-dimensional picture plane. Clearly, this is not irrelevant to landscapes; they, too, can be analyzed as a visual array. But much more important, a landscape is a three-dimensional space. Since a substantial portion of the human response to landscape turns out to depend on the sort of space involved and the way the individual envisions moving in that space, any approach that emphasizes the picture plane is bound to miss much of what matters most to people in landscapes.

Second, these four factors do not provide even an adequate sampling of the key properties of the picture plane. They are heavily weighted toward visual contours and toward discreteness. Recent work on visual functioning suggests that a far more global kind of processing may be at least as important, especially as far as spaces to walk in are concerned. An ancient component of the human visual system that has been referred to as the "location processing system" (S. Kaplan, 1970) is so closely tied to the capacity to wander through three-dimensional space that it has also been identified as the system that makes possible "ambient vision" (Trevarthen, 1978). For this system, the size and rough spatial arrangement of elements on the picture plane interact with texture to provide a global overview of the situation. To the extent that this system plays a role in preference, it will be necessary to take a fresh look at the factors that are most salient at the level of the picture plane.

This paper has dealt with a variety of factors that play a role in human preference for landscapes. It has also attempted to deal with some widely held misconceptions about preference and about perception. But the purpose of the paper is not to propose a new set of factors to take the place of a traditional set. The purpose is, rather, to describe a different way of thinking about people, a new way of conceptualizing what goes on in people's heads when they react to a landscape or another environment. What I would like to propose is a functional approach, a view of what people are trying to do. When people view a landscape, they are making a judgment, however intuitive and unconscious this process may be. This judgment concerns the sorts of experiences they would have, the ease of locomoting, of moving, of exploring – in a word, of functioning – in the environment they are viewing.

5 Where cognition and affect meet: a theoretical analysis of preference

Stephen Kaplan

Introduction

It is becoming increasingly common to use scenic-quality or -preference ratings as a way to incorporate a psychological component into design, management, and planning decisions. This component has proved to be an effective tool in such settings. It thus may be rather surprising to discover that the role or theoretical status of preference within psychology is by no means a settled question.

One view of preference is as an indicator of aesthetic judgment. Research carried out in this framework has tended to focus on stimulus properties, with the primary emphasis being on stimulus complexity (Berlyne, 1960). In this context, the motivational impact of the stimulus is the central control; cognition is not assumed to play a significant role.

An alternative view of preference involves decision making and choice. Perhaps a preference judgment reflects the complex calculations assumed to be involved in any process of choosing among alternatives. According to the influential rationality model, one multiplies perceived value and subjective probability to determine the desirability of a given choice. The choice with the highest value would then be the most preferred. Although desirability of the alternative would seem to play a central role, attempts have generally been made to hold this portion of the equation constant by assigning dollar values. Thus the emphasis in research carried out in this framework has been on the probability component, with particular interest in how people deal with the concept of risk (Slovic, Fischhoff, and Lichtenstein, 1976; Tversky and Kahneman, 1981).

The two contrasting views suggest that understanding preference involves an analysis of the relationship between cognition and affect. The "aesthetic" point of view seems to imply a purely affective role for preference, while from a decision-theory point of view, considerable analysis and calculation, and hence cognition, precede the affective outcome. There are, however, independent

This paper was first published in P. Bart, A. Chen, and G. Francescato (Eds.), *Knowledge for Design* (Washington, D.C.: Environmental Design Research Association, 1982), pp. 183–8.

grounds for suspecting that preference is the outcome of a far more complex interaction between cognition and affect than either of these positions imply. The purpose of this paper is to propose yet another interpretation of what preference judgments reflect. This view is based on an analysis of the role of preference responses not only in humans, but in other mammalian species as well. Both the fact of preference and an examination of the settings that are preferred point to the potential evolutionary significance of such reactions. Such an interpretation would speak both to the importance of preference and to the likelihood of there being substantial communality in preference among individuals despite differing backgrounds. It would also provide a means for analyzing various categories of the interaction between cognition and affect; by looking at the adaptive advantage of such patterns, it may be possible to bring disparate observations together into a more unified framework.

Preference and cognition

A major step in the theoretical analysis of preference was taken by Zajonc (1980) in his thought-provoking paper "Feeling and Thinking: Preferences Need No Inferences." The central theme of this analysis is that preferences are not the product of rational calculation. Indeed, preference judgments are often made so rapidly that they precede rather than follow conscious thought. This analysis provides an important correction to a widespread misunderstanding about the relation of thought and affect. At the same time, the conclusion that preference is therefore independent of cognition is itself misleading. There is surely more to cognition than conscious thought. Categorization, assumption, and inference often occur without awareness; consciousness is not a necessary condition for information processing.

An example of the intimate relationship between preference and cognition involves our research on the role of mystery in landscape preference. Mystery is not a surface property of a photograph that represents a scene (S. Kaplan, 1979a). It requires, first, an interpretation of the three-dimensional properties of the scene. Then it is necessary to determine if it would be possible to enter the scene, to move with reasonable ease across the ground plane. Finally, there is the critical issue of whether one would be likely to learn more as one progressed into the scene. All this inference, of course, takes place not only unconsciously, but also very rapidly. But unobtrusive as this cognitive process may be, it appears to play a major role in preference for outdoor environments. Mystery has been a major predictor variable in studies covering a wide range of environments (e.g., Hammitt, 1978; R. Kaplan, 1975).

Although the role of mystery in preference provides a useful example of the intimate relationship between cognition and affect, it by no means exhausts the possible kinds of relationships. In order to obtain a fuller picture of the possibilities, it is necessary to make some further distinctions.

The affective domain: pleasure and interest

The first of these distinctions involves two sorts of affective reactions that, although they often occur together, can occur independently of each other. First there is the pleasure–pain, or like–dislike, component. This has been a common component of discussions of affect in vastly different contexts. Olds and Milner (1954) made an important contribution to an understanding of the pleasure–pain aspect of affect with their discovery of both pleasure and pain cells in the mammalian brain. Schlosberg (1954), discussing ways of understanding the facial expressions of emotion, proposed pleasure–pain as one of the "three dimensions of emotion." Osgood, Suci, and Tannenbaum (1957), in an analysis of semantic space, also arrived at a three-dimensional solution, one of which was labeled "evaluation."

Several of these analyses of affect propose a component in addition to pleasure. This has been referred to variously as "interest" or "potency." It is closely related to fascination (S. Kaplan, 1978) or involuntary attention (James, 1892/1962). Although it is undeniable that something can be both pleasant and interesting, some things that arrest the attention – that are difficult not to look at – are by no means pleasant.

The cognitive domain: content and process

Another distinction that must be made in order to conceptualize the relationship of affect and cognition is in the cognitive domain. Sometimes, the tie to affect is through the contents of cognition; at other times, it is cognitive processes themselves that carry affective implications. Certain cognitive contents, through either learning or genetics, are associated with affective reactions. One example of such content coding is a cognitive map of a city; places to be avoided can evoke negative feelings merely by thinking of them. Another example, in a domain where the sociobiologists have provided vivid imagery, is our reaction to those who are related to us in some way.

The impact of the cognitive process on affect, while not as widely recognized as the role of content, may be equally pervasive. Basic processes such as recognizing and predicting are, under appropriate circumstances, highly affect laden. A failure to recognize, for example, can be confusing to the point of being painful, and highly interesting at the same time. Conversely, being able to recognize despite difficulty and uncertainty can be a source of considerable pleasure. Bird watchers constitute a class of individuals who specialize in the

Table 5.1. *The cognitive and the affective in people's interaction with the environment*

	Facets of cognition	
Facets of affect	Constant	Process
Pleasure–pain	Good and bad things	Managing uncertainty,
Interest	Interesting things	recognizing, predict-
		ing, evaluating, acting

pleasures of recognition. Comparably, gambling capitalizes on the fascination of predicting under uncertainty and on the (occasional) pleasure of achieving a successful prediction. It should be emphasized that here, again, the typical case probably combines elements of content and process.

Cognition as a source of affect

As we have seen, one can distinguish at least two components of affect. Cognition is a potential source of each of these. By looking at cognition as content and cognition as process, a space is created that suggests the richness of the relationship between cognition and affect (see Table 5.1).

Content and pleasure. Certain cognitive contents give rise to pleasure. Some would be tempted to be highly cautious in positing innate components here. Certainly, sweet things seem to be preferred without much help from education or socialization. Midgley (1978) suggests that there are numerous examples of this kind. Orians (1980) has extended this to landscape preference, arguing that the foliage patterns preserved in the Japanese garden are expressions of preference for patterns originally characteristic of the African savanna, where humans are believed to have evolved. His analysis of the extensive parallels in patterns observed in these two landscapes tends to support this hypothesis. Woodcock (1982) has pointed out that innate preferences of this kind would not be surprising, since many mammalian species demonstrate a strong inherited bias to select the sort of environments in which they would be most likely to survive. This bias, usually referred to as "habitat selection," has been extensively studied; the evidence suggests that this inclination tends to entail both an inherited and a learned component, where the role of the individual's experience varies from species to species.

Content and interest. The impact of cognitive content on interest involves those things that are involuntarily interesting. In a memorable passage in his discussion of attention, James (1892) proposed some of the things that fall in this category. His list includes "strange things, moving things, wild animals, bright things . . . " (p. 23). As with habitat selection, some of the pertinent patterns have their effect through learning, whereas others may be innate. Novelty and potential danger are two large categories of patterns that hold one's attention without effort. An interesting example of the role of such patterns in preference is Appleton's (1975a) "hazard" category: Potential dangers – at a suitable distance – are sufficiently attractive that they have become a recurring feature in landscape painting.

Process as a source of affect. The cognitive processes that have affective implications do not lend themselves as well as cognitive contents to being categorized in terms of pleasure–pain or interest. They tend to combine these affective components, with the emphasis depending on the particular situation. Probably central is the degree of uncertainty involved. Recognizing, for example, becomes increasingly interesting as the uncertainty increases. When something is exceedingly difficult to recognize, though, the high interest may no longer be accompanied by pleasure. The "What the hell is that?" reaction expresses such a state. Comparably, predicting and evaluating and acting are more interesting when the uncertainty increases. These cognitive processes, too, are more frustrating than pleasurable when the uncertainty level is exceedingly high.

The evidence for the affective payoff of these basic cognitive processes is all around us, although only limited experimental work has been done on this problem. It is widely recognized that people work jigsaw puzzles, struggle with crossword puzzles, and read detective stories. It is misleading to attribute the attraction of these activities to novelty or curiosity; something much more specific is going on. People are carrying out basic cognitive processes; they are actively engaged in recognizing, predicting, and the like. The level of uncertainty in such circumstances is controlled; it tends to remain at a challenging but manageable level.

Some functions of affect

The central assumption of an evolutionary perspective on preference is that preference plays an adaptive role; that is, it is an aid to the survival of the individual. For this to be the case, then for every aspect of preference we can identify, there should be a discoverable benefit or payoff. Thus preference might lead to an individual's learning what is appropriate to learn. In addition to its role in the acquisition of information, preference might influence performance. Thus it might "pull" the individual in preferred directions, affecting what he or she actually does in a given situation. At a higher level of abstraction, preference for the processes of exploring and figuring things out, and for doing things that one is good at doing, could aid in the acquisition of useful skills and knowledge. Thus preference may be involved in a number of ways in enhancing the individual's survival potential. Indeed, the very complexity of the preference phenomenon may be due in part to the multiple adaptive roles that preference plays.

To understand preference in such an adaptive perspective, it seems appropriate to look at the payoffs for learning, for performance, and for skill and knowledge. Once again, it is useful to distinguish between the two affective components that ultimately determine preference – pleasure and interest.

Pleasure and interest as factors in learning

Let us look at the issue of learning. The impact of pleasure and pain on learning has been studied extensively by experimental psychologists (Bugelski, 1956; Hilgard and Bower, 1966). The issues are complex, but it is clear that pleasurable events tend to reinforce or strengthen learning. Although pleasure and pain may seem to be opposites, pain does not weaken learning; indeed, the memory for painful events may be just as strong as that for pleasurable ones. Both pleasure and pain can become associated with certain previously neutral events, so that the events are remembered as being either positive or negative. Such motivational coding is essential in guiding an organism's future behavior.

The relation of interest to learning involves focus. In any given situation, there are many different possible patterns of information that could be taken in and learned. Interest determines which of these patterns is in fact processed by the system. Those aspects of the environment that are inherently interesting are likely to become landmarks in people's cognitive maps. If they happen to fall near choice points, the map will be functional. If not, navigational errors will be more likely. Appleyard's (1969) article "Why Buildings Are Known" suggests the role of dimensions of interest in people's learning about the built environment.

Pleasure and interest as factors in performance

The affective components that guide learning also have a central role in the ongoing functioning or performance of an organism. The guidance of the organism's moment-to-moment behavior is substantially influenced by affect. The direction of locomotion, the choice of foods, the selection of a mate – are all influenced by the "attractiveness" of the stimulus patterns. Attractiveness, in turn, can be analyzed in terms of the pleasure and interest elicited by the particular pattern.

This analysis has brought us full circle, for a preference rating must be an indication of perceived attractiveness. Thus guidance of performance, and of locomotion in particular, is a recurring theme in preference research. Preference for outdoor scenes has been shown to be influenced by such factors as apparent ease of locomotion, depth, safety, and the possibility of controlled acquisition of new information (S. Kaplan, 1978, 1979b; Kaplan and Kaplan, 1982). In other words, these are appropriate factors in the decision about where to go next. The guidance of locomotion also constitutes the closest parallel to the habitat-selection research that has looked at the preference behavior of nonhuman mammalian species.

It is not the case, however, that performance is influenced by only pleasure and interest working together. Certain patterns in the environment are interesting without being coded for pleasure or pain. Strange or unusual objects, or even potential dangers at a distance, may elicit close attention without any indication of what the individual's ultimate reaction will be. Attending to such patterns allows the individual to track the ongoing course of potentially important events.[1]

Pleasure and interest as factors in skill and knowledge

Not only are learning and performance aided by affect, but the development of skill and knowledge is itself greatly influenced by affective factors. There is pleasure in acquiring information. There is also pleasure in being able to recognize what is hard to recognize and, likewise, in being able to predict, evaluate, and act when it is challenging to do so. In this way, both extending one's cognitive map and practicing the capacities that the map makes possible are experienced as satisfying. Because threats to survival often must be dealt with rapidly and decisively, prior acquisition of knowledge and prior practice in using it would make survival more likely.

The role of interest in the development of skill and knowledge is complementary to that of pleasure. Interest is heightened by discrepancies, by exceptions, by surprises. In other words, the instances in which one's cognitive map fails to match reality are matters likely to receive careful scrutiny. Attention is held by whatever goes counter to our expectations. In this way, the cognitive map is updated and corrected. Not being able to make sense of some trivial matter can disturb one's daily activities in a frustrating way, leading one to wonder how useful such a mechanism really is. But the calm acceptance of small discrepancies in one's map can, over a period of time, lead to a widening gulf between the map and reality. The motivational pressure for keeping one's cognitive structure updated must in the larger picture be highly adaptive.

Preference in evolutionary perspective

Preference is sometimes viewed as a highly idiosyncratic and probably unimportant property of members of our species. The cliché "There is no accounting for tastes" implies that it is probably not very important to do so anyway. Viewed in the context of evolution, however, preference appears in a quite different light. From this perspective, preference guides behavior and learning. Preference fosters the building and use of cognitive maps. Environmental preference is the outcome of what must be an incredibly rapid set of cognitive processes that integrate such considerations as safety, access, and the opportunity to learn into a single affective judgment.

Thus an evolutionary analysis achieves a number of objectives:

1. It points to the importance of preference in the human psychological make-up.

2. It provides a basis for expecting an underlying commonality in preference across individuals: a commonality that has, in fact, been found in research on environmental preference.

3. It suggests that preference research has substantial theoretical as well as applied interest; it provides a powerful tool for enhancing our understanding of the relationship between cognition and affect.

4. It identifies a class of variables that is likely to be effective in the prediction of preference. There is much to be gained from becoming sensitive to variables that may be indicators of a favorable human habitat. Utilizing this rich source of hypotheses in subsequent research could contribute not only to more accurate prediction, but also to an understanding of how preference functions in the human psychological make-up.

Note

1. This might seem to be a simple case of curiosity or exploratory behavior, but that would be a somewhat misleading interpretation. Some high-interest situations do in fact encourage active investigation, as the research using incongruous stimuli has demonstrated (Berlyne, 1960). But this requires that the threat level be low; many times, taking a "close look" can be highly risky. The behavior in such situations is more striking for the rapt attention that results, for the cessation of all other behavior, than it is for its exploratory aspects.

6   The landscape of social symbols

Barrie B. Greenbie

Figure 6.1. Vancouver, British Columbia.

Most discussion of landscape aesthetics deals with the relationship of humans to nonhuman nature. In this paper, I consider the subject from the point of view of what the landscape means to us in our relations with one another.

    When we speak of landscape, we often assume that it refers to something "out there," separate from and independent of our personal existence. To some extent, this must be so, or there would be no commonly recognized features of the landscape. But what we perceive as the environment is, in fact, a synthesis in

This paper was first published in *Landscape Research* 7 (1982): 2–6.

which our current perceptions of what is actually out there are combined with a complex tapestry of associations based on our experience both of the physical world and of other people. This applies to even the most detached and rational inventory of landscape features. When it comes to analyzing subjective responses to an existing or a proposed landscape design, the problem of logical analysis becomes almost insurmountable. It has eluded philosophers for millennia. And yet such analysis is precisely what landscape architects and other environmental designers are required by their professions to achieve, and they have no choice but to continue in the attempt.

Sociobiological approaches to landscape analysis

One of the most promising approaches currently being taken by a few – still too few – students of landscape aesthetics follows the discipline known as sociobiology (Dawkins, 1976; Wilson, 1975). This is an essentially Darwinian science that seeks to explain at least some aspects of the behavior of animals, including humans, in terms of evolutionary adaptation. Perhaps the most concrete attempt to apply this view to landscape aesthetics has been made by Appleton (1975a). Appleton considers the human aesthetic response to have its origin in the atavistic impulse of animals to respond to two characteristics of the habitat, which he calls "prospect" and "refuge." He suggests that the kind of landscapes that are most satisfactory to us are those that once would have afforded us, as individuals involved in the struggle for survival, the opportunity to see without being seen, to eat without being eaten, to produce offspring that survive. He does not suggest, of course, that the human aesthetic response to the landscape involves conscious perception of danger, rather that those elements that once provided prospect and refuge now underscore our sense of pleasure in the landscape. He documents his view very convincingly. His theory is offered not as a substitute for, but as an augmentation of, other theories of landscape aesthetics; and he finds remarkably little conflict with them.

Another, somewhat similar approach to landscape aesthetics can be found in the preference studies by the environmental psychologists S. and R. Kaplan. For many years, they have sought to identify and classify the elements that lead to a human preference for one kind of landscape rather than another (Kaplan and Kaplan, 1978). The Kaplans view the human being as an information-seeking animal, reasoning that survival has depended to an unusual degree on obtaining information from and about the environment. Natural selection, they argue, would favor individuals who develop a love of information for its own sake. The most preferred landscapes, according to the Kaplans, are those that, among other things, have elements of "mystery" and "involvement" (Kaplan and Kaplan, 1978, 1982). In these landscapes, more information is promised than is actually revealed. This corresponds with at least some aspects of Cullen's (1961) view of "townscape." The Kaplans' theory puts more stress on prospect than on refuge, but it is wholly compatible with Appleton's.

These two approaches are basically ecological. They concern the relationship of individuals to the environment in which they live. But that environment is also part social (the intraspecific relationships), and my particular concern has been with the question: What does a given landscape mean to us in terms of our relationship with other human beings? My own hypotheses (Greenbie, 1974, 1976, 1981) are somewhat more complex than Appleton's (1975a) prospect-refuge theory and the Kaplans' (1978, 1982) work on preference, but are compatible with them. Indeed, all concern our control over our immediate environment, a vital consideration because a landscape in which we feel out of control is unlikely to be perceived as pleasant or beautiful. This paper argues the importance of social symbols in landscape aesthetics.

The significance of social symbols

Most higher animals have evolved complex patterns of social organization. It is the origin of these patterns that sociobiology seeks to explain through individual and, sometimes, group selection. In essence, those individuals whose body forms and behavioral traits enable them to produce more surviving offspring are, to the extent that those forms and traits are heritable, favored by selection; the characteristics displayed by these individuals increase in the population and, indeed, define the population. These characteristics are often markedly different in males and females, reflecting differing roles in reproduction and other activities. In many higher animals, including humans, intraspecific competition among males for social dominance or some other enhancement of reproductive success seems to link male morphology and behavior with aggressive assertion, whereas female morphology is linked with nurturing (which may, nonetheless, have its own form of aggression).

This intraspecific aggression among animals is often ritualized in a way that obviates fatal conflict. Higher animals communicate with one another through various signals, displays, and behaviors: They interpret their environment symbolically. The symbols that animals use are generally direct products of their bodies, including gestures, sounds, scents, and body markings of great variety and complexity. Humans have learned to hide or disguise many of these symbols by clothing themselves and in other ways. In turn, clothing and other artifacts become social symbols with communicative value, although they may become exceedingly variable with culture. But most important in terms of this discussion is our ability to extend all sorts of symbolic representations away from our bodies and into the landscape itself. Such symbols take on generalized abstract meaning, often representative of collective social orders rather than individuals. These landscape symbols may be modifications of the natural environment or constructions placed on it by humans. A predominance of the former will be called rural and of the latter, urban. But it is important to recognize that both are landscapes, and both project a continuous stream of symbols, for better or worse.

Symbolism in the landscape

I want to argue the particular significance of sexual symbolism in the landscape. There can be little doubt of the profound influence of reproductive success in determining body form and behavior pattern. But I must underscore at the outset that my references to gender pertain to abstract forms, not to the behavior of actual men and women. Even among nonhuman animals, it can hardly be said that females are always gentle and males always aggressive. We are talking of the neurologically primitive bases of human perception, not of actual human responses in all their rich complexity and ambiguity. Furthermore, in suggesting sexual symbolism in landscape forms, there is no inference that for most of us most of the time this symbolism is consciously erotic. As with the predator–prey relationships suggested by Appleton (1975a), we are dealing with very primitive predispositions to respond, not to conscious attitudes or feelings.

The association of the female body with the earth is very old, and may indeed be universal in human imagery. To my knowledge, most cultures refer to "mother earth." The land itself is known by the term *motherland* (whereas the word *fatherland* is used to imply the social dominance structure of a nation). And, indeed, the rounded forms of most earthscapes do suggest the female body rather than the male, and there is a special sense of protection in the convoluted forms of clefts, caves, and coves. Life emerges out of earth, as it emerges out of the female body. Such exceptions as promontories and jagged mountain peaks are associated, respectively, with defense and the more terrifying power of the elements, which in most mythology invokes the male gods.

In a classic work that should be required reading in all schools of landscape architecture, Scully (1962) has traced the evolution of the Greek temples. They range from the earth-goddess cults of the Stone and Bronze ages to the more confident, male-dominant view of the world of later periods. He shows how in both cases, the peculiar magic of each temple lies in the association of the built structure with the landscape around it – not only from the architecture, but also from the intense spirit of place. Scully stresses the symbolic association of such earth forms as clefts and hills with female body features. He deals relatively little with male forms, but phallic symbols of all sorts, usually man-made – such as obelisks, towers, and weapons – have been associated from time immemorial with both aggression, on the one hand, and protection, on the other.

Figure 6.2. Temple of Poseidon, Greece.

Figure 6.3. Mount Jouctas, home of the earth goddess, aligned with the palace at Knossos, Crete.

It seems to me that as an atavistic substratum of the human aesthetic response to the landscape, all this can be extremely significant. It leads to the conclusion that we are likely to find in the response to natural forms in the landscape, especially the controlled natural forms of the pastoral landscape, a symbolic association with the female body, with nurturance, with tranquillity, and with absence of aggression. By contrast, we are most likely to find in the built environment symbols not only of man's attempt at dominance over nature, but also of man's dominance over man. The most assertive, phallic forms tend to be the work of humans: They are constructions on the landscape, rather than products of it.

Most preindustrial architecture, however, blends the more assertive external aspects of the phallic and angular male forms with the internal and rounded female forms. In the Gothic cathedral are the upward-thrusting spires (symbolizing God the Father) modulated by arches, rose windows, and serene internal spaces (symbolizing Mary the Mother). In the Moslem mosque are even more clearly expressed the enfolding curves of arch and dome – internal spaces – with the assertive minarets standing guard outside. In contrast, industrial landscapes show increasing emphasis on harsh, unmodulated shafts and towers – chimneys, cranes, skyscrapers – with overwhelming dominance of scale. In Appleton's (1975a) terms, we might say that the industrial landscape is largely a landscape of prospects without refuges, often exciting, dramatic, and dynamic, but physically overpowering. My own viewpoint stresses the competitive and aggressive symbols that stream from these male-dominated landscapes and that lack the moderation of more tranquil forms.

Three hypotheses

I draw three hypotheses from these and related arguments. The first is that the contemporary landscape has too severely isolated the assertive, aggressive, and dynamic symbols both from the sheltering domestic symbols made by humans and from nonhuman features. It is the segregation of humans from the rest of nature, not any inherent virtue of one or the other, that causes the main aesthetic dissonance. My second hypothesis is that although built forms can symbolize both aggression and shelter from aggression, for industralized peoples, nonhuman nature more generally suggests the latter. My third hypothesis is that for increasingly crowded and overstressed urbanites, the most significant aspect of nonhuman nature is its social neutrality.

Why do we value nature?

The idealization of nature and, much worse, the sentimentalization of nature, which are so prominent in the urban cultures of our time, are relatively recent.

Figure 6.4. An Indian
totem pole, Vancouver,
British Columbia.

The idealization of "wild" nature as superior to "human" nature goes back to
Rousseau but not much farther. Primitive societies both fear and revere but
cannot afford to idealize nature, although agricultural societies may show more
confidence. But until the end of the Renaissance, most cultures looked toward
human settlements for security and well-being and viewed untamed wilderness as
threatening. This was true in the East as well as in the West (Tuan, 1974).

The idealization of anything depends on its distance from everyday reality.
Industrialization has made modern people increasingly dependent on large numbers of their fellows. Few of us now meet our survival needs by acting directly on
our own environments. The cooperative interlocking systems of modern technology require that other people constantly mediate between us and the rest of
nature, which demands little of us personally. Our survival depends not on
influencing nature directly, but on influencing the behavior of a long chain of

Figure 6.5. A church in Bergen, Norway.

Figure 6.6. Manhattan, New York City.

other people who are expected to "do something" about whatever it is that is needed. Thus the more threatening and violent aspects of nature are hidden from us, except in catastrophes like floods or earthquakes. Indeed, nature appears motherly, nurturant, supportive, and benign, whereas danger, indifference, and capriciousness appear to come mostly from other human beings.

A common ideal of overstressed urbanites toward nature is that of solitude, of "getting away from it all" – "it" being competition with other people, not inappropriately called the "rat race." Studies of rat behavior (Calhoun, 1971) show that rats, too, need space to get away from one another. Current political rhetoric, especially among academic youth, constantly calls for more involvement or interaction. But the truth is that individuals vary widely in their social needs: Many of us have too much involvement with others. This is especially true of the imaginative people who start the cultural innovations required for adaptive behavior in an environment of dense, demanding humanity. For those who can go a long way with a little social activity, natural environments are a means to come to terms not only with the cosmos, but also with other human beings, by putting social relationships in a larger perspective.

It seems to me no coincidence that American Earth Day, which heralded the popular acceptance of what scientists had been saying for a century, took place in the spring of 1970. This was shortly after astronauts had landed on the moon. We had our first glimpse of our earth as a small orb in infinite space, thus completing a revolution in human consciousness that began with Copernicus and Galileo and has since been eroding the Judeo-Christian hubris, which gave humans their delusion of dominance over all living things. The view of our small planet alone in space does not encourage us to feel dominion over anything, except perhaps our own behavior. And even that must now be understood in natural terms. It is ironic, but appropriate, that it took the ultimate technological feat of our era to achieve such humility, far from universal though it is.

The idea of nature as wilderness is itself merely a symbol, however. Very, very few of us experience an actual wilderness, even in the United States, which is blessed with more of it than most countries. Even fewer of us would idealize it if we had to contend with it unaided by the instruments of civilization. While there are many grounds for preserving wilderness not limited to aesthetic ones, its aesthetic significance is largely symbolic. My argument is that symbolism is the most important aesthetic aspect of any landscape. But there is danger in thinking of nature only in terms of that which is unmodified by humans. Rousseau extolled wildness, but the romantic landscape movement that he helped to found praised the pastoral, and not only the pastoral but also the highly contrived at great social cost. The later movement to contrive ways of introducing nature into cities is more democratic and much more to the point. We desperately need to reintroduce natural elements into our built landscapes on a far greater scale, and this requires understanding both humanity and nature without idealizing either. I believe, in particular, that the social neutrality of nature has great value in societies where people are engaged in increasing competition with one another for space and resources.

This is not to suggest that such neutrality, or serenity, can be found only in spaces that are full of natural forms. The serenity of a cathedral suggests otherwise. But generally speaking, the restfulness of urban environments is enhanced by the presence of living things that are not human. And although a special kind of tranquillity suffuses certain urban spaces that are full of human beings with whom we can feel at one without involvement, that is another line of thought I have explored at length elsewhere (Greenbie, 1981).

Theory and practice

Obviously, such theories as those discussed here are merely aids to understanding for the landscape architect, and they offer no formulas. Even if the fundamental symbolism of various landscape forms could be scientifically established, the responses of particular individuals and populations will often differ. They will depend not only on native culture and social class, but also on individual life experience, life style, and life situation at a given moment. To the corporate executive on the fortieth floor of an office tower, the urban landscape may not be aggressively threatening but aggressively exciting, a symbol of human purpose and power. He may enjoy the presence of trees on the street far below, but that enjoyment may have very low priority with him, especially if he has a lush home in a green suburb. To the clerk leaving an anonymous desk on a lower story, a walk through a park on the way home may be restorative; but he may prefer to elbow his way home by the shortest route to an apartment with a television set and a few potted plants on the window sill, or stop off in the purely social landscape of a local bar. The problem for designers is to determine the particularities in all cases, and to do that they must have a disciplined design imagination, supported by an understanding of not only natural ecology, but human nature as well. As there is no substitute in art for intuition, so there is no substitute for empathy in designing for others.

7   Open space in cities: in search of a new aesthetic

Werner Nohl

Figure 7.1. Munich, 1982. (Photo: K. Neumann)

Open spaces, including playgrounds, schoolyards, playing fields, open areas around housing, and even public parks – all usually designed by professional landscape architects – have received increasing public criticism in recent years. One common complaint is that nature is being destroyed by design. The well-known French cultural magazine *Traverses* published a double edition in 1978 with the theme "jardins contre nature," or "gardens against nature."

At the same time, a new field of environmental research has developed in the past twenty years to examine the beauty of landscapes. Today, when we use

This paper was first published in *Landscape* 28 (1980): 35–40.

numbers to evaluate success in many aspects of life, some researchers are trying to measure the beauty of open spaces – more precisely, they are trying to objectify and quantify it. The reasons are obvious. As Carlson (1977) explained:

If the aesthetic quality of environment is to be seen as a resource and the demand on it is to be weighed against that on other resources, then we must have an objective basis for comparison. And since objectivity has been achieved in regard to other resources by means of quantification, this seems the natural approach to take in regard to aesthetic quality. (p. 135)

This desire for objectivity and quantification apparently has led many investigators to throw out all dimensions that are difficult to translate into numbers. The analysts have limited their studies to perceivable form. The aesthetic quality of the environment has been construed only in terms of formal characteristics, such as shapes, lines, colors, and textures, or combinations of these properties, such as complexity and novelty, or order and unity.

Studying form is not in itself wrong. The problem arises when formal dimensions are considered exclusively. Aesthetics comprises both form and content. If we remove the content, we destroy the communication between the object and the viewers. The viewers lose their points of reference and cannot decode the message. Then they must rely on sheer emotional reaction, which may leave them susceptible to manipulation. This is how aesthetics is often used today in advertising, as well as in politics and administration.

The symbolic–aesthetic meaning of nature

If we ask people why they like open spaces in cities, sooner or later they answer that a park, for example, is a counterworld of the almost completely built urban environment; as such, it is a symbol of nature. And if we then ask them what nature means to them, we get a whole range of answers. People associate a multitude of qualities with nature, including health, peace, loneliness, freedom, and originality. In our society, we have many meanings for nature, and therefore each of us has different needs for experiencing nature.

But one interpretation of nature is especially relevant to the creation of a nature aesthetic. As culture develops, people become increasingly divorced from nature. The imposed division of labor brings repression, exploitation, and alienation. As our relationship with nature disintegrates, we are confronted with nature as an object that we have renounced, and we form a sentimental attachment to it. Schiller (1967) remarks in his essay *Naive and Sentimental Poetry:*

When man enters the cultural state and when art [in the sense of civilization] has laid its hand on him, then he loses all sensuous harmony and is then only able to express himself as a moral unity, that is to say as something striving for unity. (p. 697)

Schiller asserts that before humans started on the course toward civilization, they lived in a peaceful community of free and equal individuals, in harmony with nature. The costs of progress were the loss of these very characteristics of life, and civilization was achieved only at the price of alienation and exploitation. If we include the concept of natural beauty in a general aesthetic theory, then Schiller's words explain why we still desire contact with nature. He points out that we have managed to hang onto our memory of this blissful way of life (one that may never have existed) in the form of our need for natural beauty in a repressive society.

Seen in this way, the idea of natural beauty can symbolize a consummate

Figures 7.2 (*left*) and 7.3 (*above*). Playgrounds for children can be highly designed or unkempt and unstructured. The more natural environment may offer a greater opportunity for participation and enjoyment.

society in which subject and object, man and nature, individual and society are reconciled with each other. Natural beauty indicates the "dimension of the contrasting," in the phrase of Marcuse (1964); in other words, it allows us to anticipate in a historico-philosophical sense, as discussed by Bloch (1979). Of course, the ideologies that guide our behavior toward nature and our fellows in a particular period of history are also expressed in the forms of natural beauty, because nature is always altered nature. Yet natural beauty can at the same time presage, or anticipate, an ideal end to human history by displaying real possibilities. In this "prospective horizon," natural beauty acquires a political dimension. Because nature offers us the opportunity to transcend historic realities in a symbolic–aesthetic manner, we have a strong need to experience nature. Because it offers a concrete utopian prediction, natural beauty does not awaken simply inconsequential illusions, but "anticipations which point towards their possible fulfillment," in the words of Bloch (1979). A nature aesthetic for urban open spaces must take this symbolic–aesthetic dimension as its starting point.

These considerations are not merely abstract speculations. Their reality is illustrated in part by studies of people who live permanently in motor homes or pursue other countryside leisure activities. The participants reveal not only their wish for open-air recreation and change of scenery, but also their desire for

e, freedom, and self-realization. These feelings are expressed not
cated people, who have had the opportunity to develop such attitudes
rary examples, but also by a greater selection of people. Advertising
homes has long made use of these desires.

Levels of cognition

To clarify the role that the symbolic–aesthetic meaning of nature plays in our
aesthetic perception of open spaces, we must study aesthetics in terms of cognition. Most of us would agree that aesthetics involves discovery as well as both
sending and receiving messages. Thus aesthetic perception is a common means
for us to judge our environment, seeing it as a collection of aesthetic objects
according to their significance in our daily life. Yet the meaning of an aesthetic
object, such as a gnarled tree, is not always apparent. We frequently find the
meaning in a camouflaged form that requires decoding. The French sociologist
of art Bourdieu (1974) has pointed out that every aesthetic object offers meanings
at several levels. He distinguishes between the primary and the secondary level.

At the primary level, the most elemental characteristics of objects are experienced. We recognize colors, shapes, and structures that can be perceived with
our senses. Transposed into psychological qualities, these provide the pleasure of
simple perception. At the secondary level, however, lie the symbolic meanings
of objects. To interpret them correctly, we need a certain degree of acquired
taste, as Bourdieu (1974) makes clear. In studying open spaces as aesthetic
objects, two levels are not enough to give order to the repertoire of possible
meanings. By using an idea put forward by Langer (1963), who differentiates
between symptoms and symbols, I want to introduce three levels that deliver to
the viewer information about the aesthetic object: the perceptive, the symptomatic, and the symbolic levels.

At the level of *perceptive cognition,* the sensual perception of the aesthetic
object or arrangement takes place. We recognize that a meadow is green, a lime
tree is round, a fountain consists of three shells, and a flower smells. But at this
level, perception is not limited to objective properties; subjective, emotional
interpretations are involved, too. Thus we may experience the meadow as
soothing, the lime tree as attractive, the fountain as exciting, and the flower as
alluring. At this level, we develop knowledge about the characteristics of things
and the ways in which they affect us. Studies rooted in the field of psychology
focus on this level of cognition. This level is especially basic because without it
no meaning could be conveyed. Yet if we do not investigate other levels of
cognition, our research is extremely limited.

At the level of *symptomatic cognition,* objects not merely are pictured in our
consciousness, but also point to further content. The form becomes an indica-

tion, a symptom, for the uses of open spaces. Thus a park bench may prompt us to think of resting people; a beaten path, of an important direction; and a playground, of active children. Our aesthetic enjoyment increases immensely not only if we perceive colors, shapes, and smells – the perceptive level – but also if the objects disclose the processes behind them, whether past, present, or future. In order for us to think at the symptomatic level, we must have gathered information at the perceptive level, but the reverse is not true.

At the level of *symbolic cognition,* we move beyond the level of symptoms, which operate on the basis of habit or conditioning. Here we bring our reflective ability and our power of anticipation into play. The perceivable form of an object becomes a symbol for something else. Because we always include hope in the process of aesthetic appreciation, symbols usually point to a better human life and a perfect society. The most effective symbol is the one that suggests that the goal indicated by the symbol has a high probability of being realized. For medieval people, the Gothic cathedral stood for a better life in the Kingdom of God; for the freemen of the Greek polis, the agora represented a highly developed democratic society in an otherwise barbarous world. We experience a park as being beautiful if information from the perception of form, transformed into our emotional response, can be linked to our knowledge about its use and, finally, to symbols of real possibilities for a harmonious, or in the words of Aristotle, a eudemonic, human society.

This system of cognition levels works for all aesthetic objects, but there are very different classes of objects. Differentiating among artistic beauty, natural beauty, and technical beauty is essential. In the case of open spaces, which belong in the category of natural beauty, we discover symbolic–aesthetic meaning of nature at the level of symbolic cognition.

The totality of nature

Until now, we have not posed the question of what qualities nature must possess in order to fulfill the symbolic–aesthetic role. The heavy criticism of modern urban open spaces indicates that people often feel cheated out of what they call "nature." Indeed, construction methods and costs in our society have reduced the layout of green spaces to a few, excessively large components. These open spaces lack variety, local individuality, and historical context. They cannot stimulate a symbolic response in us; they cannot fulfill their role as symbols of a better life.

According to Schiller (1967), natural beauty is expressed only where the totality of nature is apparent. This may explain why so many city dwellers are drawn out into the countryside, where the landscape, if it has not been converted to an industrial tract, is neither purely nature nor strictly man-made. Nature and

Figures 7.4 (*top*) and 7.5. Nature can be allowed to assert itself, or it can be tightly controlled, as in the swimming-pool complex. Less design may encourage more satisfying activities.

culture appear reconciled with each other, and this ideal of reconciliation under-lies our concept of natural beauty. (The phrase "totality of nature" refers to the visible relationship between people and nature, whereas when we use "natural beauty," we mean that the symbolic–aesthetic reference is invoked by the total-ity of nature.) This rural idyll is deceptive because it ignores the real human condition in the countryside. Adorno (1983) emphasizes that such cultural land-scapes always bear the marks of "sorrow" and "restriction." But I do not intend to assert that a better society exists in the countryside, rather that the landscape in the countryside serves as a symbol of ideal civilization.

The role of the totality of nature in natural beauty and the importance of landscape as a symbol of a consummate society have recently been stressed by Ritter (1974):

In this way, in competition with the objective world of the natural sciences, which has distanced itself from metaphysical ideas, the aesthetics of nature have assumed the role of presenting man with subjectively based "graphic" images of the totality of nature and the "harmonious unity of the cosmos" and making them relevant for him. . . . (p. 153)

Natural beauty

Although dramatic rural landscapes may be important to the aesthetic experience of people visiting open spaces outside the city, they offer few ideas to help us discover successful designs for green spaces within the city. Yet, the basic pattern that allows us to experience the totality of nature is transferable. City dwellers find green spaces beautiful when they represent the reconciliation of culture and nature. It is only in combination with human design that nature can appear as a contrasting element, and this combination is necessary to convey the characteristic of totality. In these green spaces, the aggressive dominance of people over nature, which is apparent everywhere else in the city, is relinquished or reduced to a level that ensures nature a life of its own. The totality of nature thus created can then induce in us aesthetic images that transcend their outward appearance and point to a concrete and indestructible utopia.

In the countryside, our uses of nature, such as farming and forestry, are productive. But in urban open spaces, our activities are recreational. If users of open spaces are to develop aesthetic images, their activities should not be pre-scribed or predetermined. Instead, activities should be independent, creative, educational, and cooperative and should involve the environment. This type of activity can be described by the term *appropriation*. Appropriation does not lead to the destruction of nature because it is oriented to the functional value of the space and because it depends on agreement with other users.

Activities involving appropriation range from altering the environment to en-joying the beauty of nature. They require an incomplete landscape – that is, an open space that has not been designed in every detail and that is not perfectly maintained. Only under these conditions is independent involvement, creativity, and environmental learning possible. Like other uses, these activities leave visible traces, but they can be erased by nature. Indeed, they must be continually wiped away by nature because open space is meaningful only when it can be used every day.

To cultivate the right landscape for appropriation, we should allow nature to develop unhindered wherever possible. We can let weeds invade the cracks in the sidewalk, edges of paths become overgrown, patches of grass mature without mowing, plants spread freely, dead leaves collect where they fall, and pools of water silt up. We should use sparingly hard materials that constrict nature. Concrete and asphalt might then be superfluous: Paths could once again be surfaced with gravel; riverbanks could slope naturally; and slabs edging lawns could be eliminated. In this way, we could develop a natural environment that would be satisfying to us and, conversely, a pattern of human use that would restore the natural environment. This is the outline for an environment that would provide a glimpse of the totality of nature as well as of its own distinctive interplay between human use and the forces of nature.

Today, most open spaces are designed in detail. As a result, people have little opportunity for imaginative activities, and nature has little opportunity for unhindered growth. We are frustrated by the facilities that dictate our activities, and the spread of vegetation is thwarted by constant maintenance. The appropriation of open spaces by nature or by users is also made difficult because the design produced by the landscape architect is often regarded as a work of art and because implementing and maintaining the design is expensive. As a result of high maintenance costs, authorities often set up constraints to prevent the public from determining how it will use open spaces and to restrict the growth of plants.

Open spaces that are presented as valuable and unalterable works of art will always remain somehow alien. Only when we can experience landscapes through our own actions will they become everyday living spaces with personal associations. And environments in which we do not allow nature to intervene continuously may express the *genius architecti*, but the *genius loci* will certainly be absent. It is the interplay between users and natural processes that gives a place its special character. Together, they successfully produce an impression of the totality of nature in urban open spaces, which, in turn, offers the basis for satisfying our symbolic–aesthetic needs.

Toward a new environment

I have tried to present a workable aesthetic concept of nature for designing open spaces in postindustrial capitalist cities. We urgently need an environment that people may appropriate whenever and however they wish. Because we must *use*

Figure 7.6. In a no man's land, such as this one in Vienna, children and adults may leave their marks without guilt. Nature will erase them. (Photo: Werner Geim, 1982)

open spaces in order to enjoy them aesthetically, we must have free and autonomous use of open spaces in the city. As residents of cities contribute to parks and playgrounds, they will become both recipients and producers of aesthetic experiences, and it will become easier to fulfill our objective of providing satisfying open spaces.

Until now, only landscape architects have officiated as designers, and we have become dependent on them and subject to their concepts of order. Likewise nature, which has been an object of the design proposals, has too often lost its special character within a so-called work of art. In contrast, the nature aesthetic, which I have implied more than described, restores the rights of both people and nature, because the reconciliation found in the totality of nature permits neither unnecessary dominance of one person over another nor excessive control of nature.

This harmonious totality of nature, which allows us the anticipation of a perfect society, must exist in a variety of forms. We need some open spaces in which activities take preference over nature because we want to play football on the grass as well as enjoy rose beds. However, this is no reason to abandon a new approach to a nature aesthetic, because there are many other places where nature can dominate. Although more untouched nature in cities is needed, we must also provide for access to and use of overgrown areas. City dwellers would gain little if the restrictive aura of art was simply replaced by that of nature.

8   Aesthetic perception in environmental design

Arnold Berleant

I

For well over two centuries, the doctrine of the disinterestedness of aesthetic perception has stood as a dogma of Western philosophy. Developing in the eighteenth century in the works of Shaftesbury and succeeding British writers, emerging as the theoretical centerpiece in Kant, elaborated by Schopenhauer, and refined by Bullough, Bergson, and Croce, aesthetic disinterestedness stands as a "major watershed in the history of aesthetics" (Stolnitz, 1961, p. 139). Indeed, up to the present time, the tendency to affirm this attitude has dominated most theoretical discussion in the English-speaking world.

Let me recall the claims of this idea by citing Kant's (1790) classic formulation:

Taste in the beautiful is alone a disinterested and *free* satisfaction; for no interest, either of sense or of reason, here forces our assent. . . . *Taste* is the faculty of judging of an object or a method of representing it by an *entirely disinterested* satisfaction or dissatisfaction. The object of such satisfaction is called *beautiful*. (First Moment, 35).

Although there is much in the Kantian aesthetic that continues to reflect our experience and claim our assent, such as his contention that the judgment of taste concerns pleasure that is unintellectual and immediate, there are facets of his view that reflect a tradition in aesthetics that is open to serious question. This is because the doctrine of disinterestedness and the correlative ideas that it fosters do not derive from a close examination of the conditions and features of our aesthetic experience of art and nature. They come instead from an uncritical allegiance to an intellectual and cultural tradition that embodies preconceptions about the meaning, order, and value of the various domains of experience, including the aesthetic (Berleant, 1970).

This raises issues that are both basic and broad, and it would be helpful to begin with a particular feature of perceptual experience that exemplifies the ways in which this attitude is employed. It is a feature that also appears in some form or other in nearly every art and is central to environmental design in general: the

84

factor of space. In fact, disinterestedness is often presented in a spatial metaphor through the notion of distance. Yet as I hope to show, a phenomenological study of spatial perception in aesthetic experience offers convincing evidence for replacing the concept of disinterestedness with a quite different one. After exploring the perception of space, I shall turn to the more general subject of aesthetic experience and show how a clearer grasp of such experience suggests alternatives for shaping the human environment.

The spatial mode of classical physics takes on certain characteristic forms when applied to aesthetic perception. Space, as an abstract, universal, and impersonal medium in which discrete objects are placed, assumes the form of aesthetic distance, embodied both in the attitude of a perceiver and in the domain of the art object itself. This distance develops not only as a space between perceiver and object, but also as a division between them. It may be expressed by describing the space of the perceiver as real space and that of the object as illusory, or by using a notion such as psychical distance to represent a psychological separation between viewer and object. The perceiver becomes primarily a visual awareness, adopting a contemplative attitude toward an isolated object. From the physical arrangements for aesthetic perception provided by museums, theaters, and concert halls, to the arrangement of space within a painting as a self-contained entity and the disinterested attitude that the perceiver is expected to adopt, this is the dominant and all-embracing model.

A phenomenological approach to the perception of space proceeds along a sharply different path. It takes its cue from the descriptive observation that objects are not independent of the perceiver but rather that "consciousness always means and bears in itself its object" (Husserl, 1931/1960, p. 33): To be conscious is to be conscious of something. This is what is meant by the intentionality of consciousness. Rather than being a strange and unaccustomed condition, it is held to emerge from an unassumptive examination of how consciousness actually works. The claim is that such perception is original and primary.

Applying this approach to space gives us not a neutral and objective medium, but one continuous with the act of perception. Space is both shaped by and inseparable from the perceiver. This conception bears a striking resemblance to the space of modern relativity physics, in which matter and space are "fused into one single dynamical reality" (Capek, 1961, p. 176). This reality takes the form of a dynamization of space. Space is incorporated into "a type of becoming which, together with its dynamical temporal unfolding, still possesses a certain, so to speak, *transversal* extent or *width*" (Capek, 1961, p. 161). Thus the coalescence in modern physics of matter, space, and time and their continuity with the observer combine with the assimilation by consciousness of its object in phenomenology to produce a realm that harmonizes the dimensions of the human

Figure 8.1. A panoramic
landscape.

Figure 8.2 (*below*). A participatory landscape.

world into a single, total domain. Another expression of the same idea is to speak of the human world as an inclusive perceptual field in which perceiver and object are conjoined in a spatiotemporal continuum.

There is, then, the classical account of visual space – space as it is known in reflection, and "the vision that really takes place" (Merleau-Ponty, 1964, p. 176). Similarly, the space in which I live is not the "network of relations among objects" witnessed by an outside observer or formulated by a geometrician. "It is, rather, a space reckoned starting from me as the zero point or degree zero of spatiality. I do not see it according to its exterior envelope; I live in it from the inside; I am immersed in it" (Merleau-Ponty, 1964, p. 178). Depth is no longer a third dimension, but

the experience of the reversibility of dimensions, of a global "locality" – everything in the same place at the same time, a locality from which height, width, and depth are abstracted, of a voluminosity we express in a word when we say that a thing is *there*. (Merleau-Ponty, 1964, p. 180)

These two conceptions of the perception of space carry consequences of major significance, both for aesthetic experience and for aesthetic theory, since they lead to vastly dissimilar ways of ordering the world. Their directions can be traced in many forms, but perhaps none is more apparent than in environmental perception. Applying these different modes of apprehending space to landscape painting and environmental design, two alternative schemata emerge. One is a predominately visual idea of landscape, which we can call the panoramic landscape; the other, the more intimate, participatory landscape.

The panoramic landscape emphasizes physical distance and breadth of scope. It is a primarily visual experience that carries a sense of separation between viewer and landscape. This landscape involves different spatial modalities, that of the viewer and that of the viewed, and they are kept quite distinct, incommensurable, and unrelated. By separating sharply the space of the viewer from the space of the landscape, the panoramic landscape possesses little or no visual continuity; instead, an abrupt break occurs between the spatial location of the viewer and that of the scene. Nothing leads the eye smoothly into the picture space or into the actual space of the landscape. The viewer is a remote observer, disengaged and gazing contemplatively on a landscape from which he or she is utterly removed. Moreover, the experience is primarily visual: The landscape is something to be seen, and its appeal lies in its great breadth of scope, a grand spatial array into which the viewer is not tempted to set foot. When the visual object is man-made, such as a building or a garden, the emphasis is often on monumentality, symmetry, a geometric balance and harmony that is instantly apparent.

While there are numerous examples in painting of the panoramic landscape, they are relatively infrequent, especially among landscape painters of the first rank. Albert Bierstadt's *Scene in the Tyrol* (1864) is typical. In the right fore-ground lies the edge of a hilltop meadow with grazing cattle, bounded by a ruin. Beyond and below in the middle distance lie in succession a wooded area, a plain, and a lake; in the far distance are snow-capped mountains. There is little continuity between the meadow and the remainder of the painting, which looks like a painted backdrop for the set of an early film. What is most pertinent is that the scene in the foreground is quite removed from the viewer of the painting, both in the direction of one's gaze and in the plane of the ground. The painting offers a sentimental scene wholly discontinuous from the viewer, a fitting object for romantic contemplation. Such depiction of grandiose landscapes typifies the work of the Hudson River School.

Unlike examples in landscape painting, instances abound of the panoramic landscape in environmental design. Here the prevalent ideas in recent Western cultural history are not usually corrected by the experiential eye of the artist. The scenic outlooks along highways that pass through hilly areas are places designed for travelers to pause in their journey in order to enjoy the view of a distant landscape. Generally bounded by a barrier wall, they offer at their best an impressive view of the countryside from a commanding height. Yet the land-scape is an inaccessible region open to visual regard alone, with no continuity with the observer or means of access. Again, government buildings often assume a stolid monumentality, and an escarpment is often selected as a site because it places a building in a position of lofty dominance. Such structures frequently oppose the viewer with a grandiose symmetry, as does the Parliament Building in Victoria, British Columbia. On a far more modest scale stands that monument to petit-bourgeois aspiration, the development house, its brick façade limited to the front that is visible from the street, separated from its close neighbors by a fence or hedge, and gazing back with a picture window that reciprocates its visual isolation.

The participatory landscape encourages a quite different mode of experience. Unlike the panoramic landscape, which is more a product of intellectual history than of environmental experience, the participatory landscape develops a spatial continuity with a person. One is no longer a disinterested spectator, and the appeal of the landscape is not exclusively visual. What is most striking is the difference between the two landscapes in the use of space. Here, space is not opposed to an observer but reaches out to encompass the viewer as a participant. One does not contemplate such a landscape; one enters it. This recalls Kan-dinsky's (1913/1964) artistic goal: "I have for years searched for the possibility of letting the viewer 'stroll' in the picture, forcing him to forget himself and dissolve into the picture" (p. 31). The participatory landscape does not exert an appeal that is exclusively visual; it draws on kinesthetic responses, the apprehen-sion by the body of mass, of texture, and of the various sense qualities that constitute the perceptual experience of the spatial environment. Movement and time are essential components of such experience, and the homogeneity of such experience renders them inseparable from space. There is, then, a continuity between the conscious human body and its perceptual world. Consider James's (1904) account:

> The nucleus of every man's experience, the sense of his own body, is, it is true, an absolutely continuous perception; and equally continuous is his perception (though it may be very inattentive) of a material environment of that body, changing by gradual transition when the body moves. (p. 543)

The design of environment can recognize, extend, and develop the possibilities of such experience.

The participatory landscape is known to environmental planners, too, although not universally or consistently. In the design of environments lies a unique opportunity to integrate the "viewer" into the landscape through paths that beckon one onward through field, marsh, and woodland; highways that caress the contours of the land rather than push blindly through them; city plazas that, like the unsurpassable Piazza San Marco, require the participating pedestrian to render it complete; gardens that lead the stroller through a succession of discoveries; buildings that welcome and cooperate with their users.

These kinds of landscape, the panoramic and the participatory, do not determine alternative spatial modalities by the choice of subject matter or the use of particular devices or techniques alone. What makes them truly distinct is that they require and evoke different modes of perception. It is with the contemplative stance of disinterested awareness that we view the panoramic landscape, where measure, control, and distinct boundaries are enhanced by the objectivity of distance, both physical and psychical. In contrast, a clearly formulated aesthetic has only recently begun to emerge, an aesthetic that defines the perceptual mode appropriate for the participatory landscape, although its characteristic experience has been recognized in the West since classical times. Yet this touch of the Dionysian has been suspect in the Apollonian tradition, which has dominated Western thought, and the history of aesthetics has labored under a rationalistic burden. Perhaps we can describe the participatory landscape as requiring an aesthetic of engagement, in which the experience of the body as it moves in space and time creates a functional order of perception that fuses participant and environment.

This analysis holds practical as well as theoretical consequences, and the implications of an aesthetics of engagement for environmental design are manifold and various. They cover the entire range of human environments, from museum design to highway engineering, from domestic architecture to city planning, from park to wilderness management. The employment of an aesthetics of engagement requires imagination by those expert in each sphere of design, informed by detailed knowledge and experience. Let me indicate some possible consequences of this aesthetic by suggesting the direction its application can take.

Museum interiors and the installation of exhibits are characteristically treated as problems in interior decoration, in which art objects are arranged as elements in a total design scheme to form a pleasing décor. Instead of using paintings to ornament rooms by arranging them on walls in the manner of formal compositions, an aesthetics of engagement implies that rooms be built around paintings in a fashion that encourages the best position of the viewer for participating in the space of the paintings. A Cézanne landscape, for example, typically requires placement at a point of great distance, perhaps on a far wall opposite the entrance, just as Cézanne's own position when painting landscapes was usually far removed from the subject. Monet often demands middle distance, although sometimes, as with *The Manneport, Etretat II* (1886) and his late works, the optimum distance is so great that most galleries cannot accommodate it properly. It is more usual, however, for landscapes to require surprising intimacy, and exhibition rooms could be divided by panels so that viewers would be obliged to stand at distances from the painting no greater than about 2 feet. Paintings would then be grouped and placed according to their optimum distance for viewing and at an eye level that would be conducive to making the space of the viewer continuous with that of the painting, even though this may result in irregular and uneven patterns on the walls of exhibition rooms.

Other applications are equally innovative. Highways should respond to the geological and topographical features of the countryside, conveying its contours to the driver and encouraging responses of respect for and participation in the rhythms of the landscape. The experience of motion and its fusion of visual and topographical traits must be the primary goals of the highway engineer, instead of considerations of speed, technical simplicity, and geometrical directness. This would help make highway travel an experience of landscape and not an exercise in endurance, suffered as a means for gaining some distant destination. In yet another case, an aesthetics of engagement would encourage us to develop housing as a bond to a place, reflecting its physical, historical, and regional characteristics, and not as the impersonal imposition of stock patterns on an available tract of land. Local materials, regional designs, sensitivity to the features of a particular site, adaptability to individual uses, and distinctive modifications are some of the considerations that would help in achieving engagement.

Cities, too, would be seen as opportunities for the unique cultural experience of different social groups and traditions and not principally as economic arrangements shaped by political forces. More than a field for economic and social processes, however, urban experience can extend beyond the functional to offer opportunities for grasping a variety of connections. We can approach the metaphysical grounds of human existence through encounters with the sacred and the artistic. We can discover an opportunity for playful imagination and fantasy in the color, surprise, and unexpected richness of urban activities. We can affirm our cosmological bonds with the universe through the effects on the city of the changing positions and light of the sun and moon, the transition of the seasons, and the distinctive geographical traits of the urban site (Berleant, 1973, 1978; Maruyama, 1979). Parks, playgrounds, and wilderness areas need not be places on which predetermined human uses are imposed. Instead, they could be designed to offer opportunities for responsive contact with the natural landscape in settings that vary from personal to social.

Some interesting empirical questions can be formulated from this contrast between imposed and responsive landscapes, although one must be sensitive to the dangers in attempting to quantify intuitive perceptions and qualitative responses. There are fruitful opportunities for research in first identifying instances of these two types of landscape. What collection of traits characterizes each? Do these types occur in sufficient numbers to lend themselves to investigation? If they do (and I suspect that they are likely to appear in the contrast between slowly evolving vernacular landscapes and those engineered according to supposedly rational design criteria), are there any other unanticipated features that also occur? More important, however, are the qualities of human experience each encourages: creative or routine, impulsive or regimented, filled with wonder or boring, generating surprises or predictable. Most crucial are the conditions of people's lives that emerge: conditions of belonging and an identity with place or of alienation from it; of ease or anxiety; of delight or despair; of free and friendly human exchange or of isolation. To turn these questions into a research agenda itself requires imagination and sensitivity to the values involved and to the dangers of trivializing them through blithe unawareness of the reality and importance of the elusive qualities of human perception and consciousness.

The fact is that design choices are available in all these cases, and an aesthetics of engagement suggests definite directions in which to make them. However, these choices can never be made mechanistically. On the contrary, they call for the continuing sensitivity of the designer to the changing needs of social groups, the constant transformation of human cultures, and the natural resources that are the ultimate source for all choices and our final limit.

II

This discussion of the influence of different modes of aesthetic attitude on spatial perception and particularly on the landscape leads to the more general issue of aesthetic experience, and we can get at the notion of aesthetic engagement from this direction as well. Accounts of such experience have appeared in many forms, and environmental perception offers a particularly rich and challenging field for developing and testing the concept. Moreover, a clear idea of the aesthetic dimension in environmental perception will better guide its design. In environmental aesthetics, we can identify three theoretical models of aesthetic experience in common use.

The contemplative model is so securely established as to assume the status of an official doctrine. Resting on a philosophical tradition that extends back to classical times but first formulated in the eighteenth century, the doctrine emerged that identifies the art object as separate and distinct from that which surrounds it, isolated from the rest of life. Such an object requires a special attitude for its appreciation, and we have already seen how the attitude of disinterestedness arose, an attitude that regards the object of art in the light of its own intrinsic qualities with no concern for ulterior purposes.

When this doctrine of separation and distance is applied to environmental experience, we recognize a conception of space modeled on the space of the physicist – more specifically, the eighteenth-century physicist. The first part of this essay contrasted the space of classical physics with the phenomenological conception of space, pointing out how the classical view considers space to be an abstraction, a medium that is universal, objective, and impersonal. This provides the grounding for the contemplative model of experience. Here objective space leads to the objectification of things that are situated in and move independently through it, and these things are then regarded from the stance of an impersonal observer. We noted earlier how landscapes that assume this notion of objective space depict a scene observed from some vantage point in which the observer is removed from the scene and is contemplating it from a distance, what we called the panoramic landscape. The usual definition of a landscape painting as "a picture representing a section of natural, inland scenery" reflects the conception of a landscape as "an expanse of natural scenery seen by the eye in one view" (*Webster's New World Dictionary*, 1959).

There have been attempts since the eighteenth century to develop alternatives to the contemplative model of aesthetic experience. They offer to overcome the passivity and separation of the standard theory by depicting the aesthetic perceiver more as a multisensory, active agent than as the kind of perceiver one would be through the disengaged vision of the traditional position. Such inclusive accounts offer a more promising direction for environmental experience and design and have been developed in various forms, two of which I shall consider here.

Let me call the first the active model. Versions of this may be found in the aesthetics of pragmatism, especially in Dewey's *Art as Experience* and in the phenomenological aesthetics of Merleau-Ponty and others. What is common to the various forms of the active model is the recognition that the objective world of classical science is not the experiential world of the human perceiver. There is a sharp difference to be drawn between space, as it is presumably held to be actually and objectively, and the *perception* of that space. A theory of aesthetic experience thus must derive from the latter rather than from the former, from the manner in which we engage in spatial experience rather than from the way in which we objectify and conceptualize such experience. Dewey (1934), for example, maintains that art stirs into activity those inherent dispositions to an intimate relation to the surroundings that the human being has acquired through evolutionary and cultural development. In this portrayal of the experience of art, the organism is an activator of the environment. Perception is not purely visual, but somatic: It is the body that energizes space.

For Merleau-Ponty (1964), as well, perception starts with the body. The presence of the body as *here* is the primary reference point from which all spatial coordinates must be derived. Thus the perceived object is grasped in relation to the space of the perceiver: It does not stand as a discrete material object. Both perceiver and object inhabit the same space, a space continuous with life and its activity – "lived-space," as the phenomenologists term it. The notion of the lived-body further develops this sense of lived-space. The body is the vital center of our spatial experience from which we view existential space, determine its directional axes, and measure existential distance (Schrag, 1969). Thus the discernment of places, with their value and meanings, occurs in relation to the central position of the body.

Yet this is not enough. The environment is not wholly dependent on the perceiving subject; it also *imposes* itself in significant ways on the person, engaging one in a relationship of mutual influence. Not only is it impossible to objectify the environment, but the environment cannot be taken as a mere reflection of or response to the perceiver. Recognizing the influence of specific features in the environment makes it necessary to extend the active model of aesthetic experience to include such factors. The consciousness of self, of the lived-body, and of lived-space must be complemented by recognizing that the environment exerts influences on the body, that it contributes to shaping the body's spatial sense and mobility and, ultimately, to defining its lived-space. This leads us to a different conception of aesthetic experience. It is a view that understands the environment as a field of forces continuous with the organism, a field in which there is a reciprocal action of organism on environment and environment on organism and in which there is no real demarcation between them. Such a pattern is yet a third model of aesthetic experience, one I shall call the participatory model, and it leads us to an aesthetics of engagement.

It is perhaps easier to understand the forces that emanate from the body as it thrusts itself into the environment than it is to grasp the magnetism of environmental configurations as they exert a subtle influence on the body. We sense our own vitality more directly than we apprehend the action of spaces and masses. While the body and the environment extend interacting fields of force, what distinguishes the participatory model of aesthetic experience from the active model is its recognition of the way in which environmental features reach out to affect and respond to the perceiver. This phenomenon is not new: There have been artists and architects who have long utilized it. What has been missing, however, is a theoretical formulation of such environmental activity within the frame of an aesthetic theory. This essay offers the framework for such an account, for I believe that the participatory model is not a special case, an exception to the prevalent observational mode in aesthetics that is required by the unique conditions of environmental experience. Rather, I believe that it is a model that can be applied successfully to other – indeed, all – modes of art as well as to a general theory of aesthetic experience.[1]

My purpose here, however, is a more limited one. I should like to show how the participatory model applies most tellingly to certain environmental features, features that engage reciprocally with people in the aesthetic experience of environment. This reciprocity goes beyond the notion of invitational qualities suggested first by Lewin's (1936) field theory of psychological motivation. It urges us to recognize the ways in which environmental features impose themselves on the perceiver and the perceiver energizes the environment, both functioning in a reciprocal fashion.

Perhaps the most apparent of these features is the path. Paths, of course, are especially rich in significance. However, they are not experienced as cognitive symbols, but, if one insists on using that concept, as living symbols that embody their meaning, symbols that make us act, make us commit our bodies, our selves, to choices. What is most striking about paths is the way in which they act on us. In describing the hiking path, Bollnɋw (1961) comments that

the path does not shoot for a destination but rests in itself. It invites loitering. Here a man is *in* the landscape, taken up and dissolved into it, a part of it. He must have time when he abandons himself to such a path. He must stop to enjoy the view. (p. 36)

Roads, like paths, act on us in diverse ways, inviting us to move down them or putting us off. Thus routes are often unidirectional, more appealing in one direction than in the other. On a round trip that we make regularly, we are likely to follow one course going and a different one returning. Similarly, the habitual behavior with which we take a customary route may be explained as the largely unattended attraction of environmental cues that act on us to lead us routinely in the same direction.

Places, plazas, parks, and gardens may be inviting or discouraging in much the same manner. Participatory spaces encourage entry; they evoke our interest

Figure 8.3. Monumental space.

and draw us in. Instead of putting us off or offering a harmonious formal array that appeals through its orderliness when viewed from a distance, there may be comfortable irregularity and disorder. Great open spaces are divided into smaller protective ones, and enclosure replaces exposure, providing an easy habitation for the body. This is sharply different from the monumental forms and spaces of such places as the federal area in Washington, D.C., City Hall Plaza in Boston, and Brasilia – places that diminish and swallow the body (Steel, 1981).

Buildings may also offer opportunities for participation, and when they do, they contrast with the usual treatment of architectural structures as visual objects. Visual buildings may display a symmetrical structure. They may stand apart as monumental objects. They may be primarily façades whose third dimension is an incidental and unrelated appendage, or they may devolve into pure surface, as the curtain-wall skyscraper does (Yudell, 1977). Yet again, they may confront us with solid, opposing planes in which an insignificant opening for access is the only imperfection. In contrast, buildings that encourage participation possess human scale. They are not isolated objects that oppose the perceiver; instead, they are a part of the landscape that evokes our active interest by reaching out to us with embracing configurations that welcome our approach and invite access.

Nowhere is this invitation to participate more pronounced than in the case of entrances, doorways, and stairs. They can put one off or lead one on, and in ways that may be subtle or obvious. An effective entrance or doorway draws a person in instead of stopping movement through awe or confusion. It does not erect obstacles to be overcome or ambiguous shapes to be identified, nor does it present intimidating or insignificant ways of passing into a place or a building. Rather, a participatory entrance is easily and clearly recognized; it is appropriate to the body, inclusive in its perceptual character, welcoming in its affective qualities. So, too, can a staircase invite ascent, pulling the body upward through its own rising movement. A visual staircase becomes a pedestal to support an imposing structure; a participatory staircase beckons us to climb.

Recognizing participatory traits requires us to rethink what we mean not just by space, landscape, and the parts and features of landscape, but also by the environment as a whole. The etymology of the term *environment* notwithstanding, the perception of environment is not a perception of a foreign territory surrounding the self. It is, rather, the medium in which we exist, of which our being partakes and comes to identity. Within this environmental medium occur the activating forces of mind, eye, and hand, together with the perceptual features beyond the body that engage these forces and elicit their reaction. Every vestige of dualism must be cast off. There is no inside and outside, no human being and external world, even, in the final reckoning, no self and other. The conscious body moving within and as part of a spatiotemporal environmental medium becomes the domain of human experience, the human world, the ground of human reality within which discriminations and distinctions are made. We live, then, in a dynamic nexus of spaces that speak to us and to which we belong.

Marcel urges us to say not that I have a body, but that I *am* my body. So we can say, in like manner, not that I live in my environment, but that I *am* my environment. As the body can be considered an extrapolation from the unity of the self, the environment can be regarded in much the same way. Thus the concept of the environment must be deepened to assimilate the lived-body, on the one hand, and broadened to embrace the social, on the other. The social, however, not only is the institutions with which the self participates, but also includes the cultural meanings that infuse us. A full account of environment, then, needs to be a study of environmental perception and of the physical features that participate in a reciprocal fashion with the self. It also requires a correlative study of the semiotics of environment that would explore the meanings inseparable from such features.

An examination of both aesthetic attitude and aesthetic experience leads, then, to the idea of environmental engagement. Conversely, a participatory model of experience provides a key to environmental understanding. Aesthetic engagement enables us to grasp the environment as a setting of dynamic powers, a field of forces that engages both perceiver and perceived in an experiential unity. What is important are not physical traits but perceptual ones, not how things are but how they are experienced. In such a phenomenological field, the environment cannot be objectified; rather, it is a totality continuous with the participant. An environment can be designed to work in this mode, or it can be structured to oppose it. It can be shaped to encourage participation or to inhibit, intimidate, or oppress the person. When design becomes humane, it not only fits the shape, movements, and uses of the body, but also works with the conscious organism in an arc of expansion, development, and fulfillment. This is a goal that a consciously articulated aesthetic of the sort I have begun here can help the architect, designer, and planner to accomplish. Such an aesthetic can be a powerful force in the effort to transform the world we inhabit into a place for human dwelling.

It is undoubtedly true, as Goethe once remarked, that "people see what they know." The task, then, is to so enlarge our knowledge as to expand our vision. As we develop the conception of the participatory environment in greater detail through the various sensory modalities of environmental perception (the tactile, the kinesthetic, the auditory, and the olfactory, as well as the visual), of the physical characteristics that encourage reciprocity, and of the semiotics of environmental consciousness, we shall be better able to make intelligent and humane decisions in shaping the experiences and hence the lives of people through environmental design. What has been done in a groping and intuitive fashion can then take place more clearly and directly. The implications here are universal. As Yeats expressed it, "Does not 'the eye altering alter all'?" (Yeats, 1961, p. 159).

Note

1. I am undertaking this in a book in progress called *Art and Engagement*.

Section II

Empirical studies

Editor's introduction

The design of buildings, the cityscape, or the landscape has the potential to evoke favorable or unfavorable responses from the public. Without detracting from the role of a skilled designer, empirical researchers have attempted to advance knowledge of such environmental influences on affect. Ultimately, such information can be used to guide design decisions and to produce solutions that are pleasing to users.

Included in this section are studies of interior spaces (Locasso; Kasmar; Flynn), architectural exteriors (Hershberger; Hershberger and Cass; Oostendorp and Berlyne; Groat), central-business-district scenes (Nasar), housing scenes (Nasar; Talbot), retail-street scenes (Nasar), natural scenes (Fenton; Herzog; Nasar, Julian, Buchman, Humphreys, and Mrohaly), and rural scenes (Orland; Kaplan and Herbert), as well as methodological papers that apply to a variety of settings (Fenton and Reser; Russell). In addition, several of these studies (Nasar; Orland; Kaplan and Herbert) provide cross-cultural and intergroup comparisons of aesthetic preferences.

For convenience, the papers are grouped into five categories: (A) methodological comments; (B) architectural interiors; (C) architectural exteriors; (D) urban scenes; and (E) natural and rural scenes. Nevertheless, the studies show considerable agreement on some aspects of environmental preference. These conceptual links are briefly outlined here for dimensions of environmental affect (referring more to subjective feelings in response to the scene), dimensions of environmental perception and cognition (referring more to visual properties of the scene), and the relationship between these two sets of factors.

It is evident that aesthetic response includes both *formal* and *symbolic* components, the former involving structural properties of the object, such as size, shape, or complexity, and the latter involving situations where the object stands for something else, such as when people associate meanings with a particular style of building. Unfortunately, empirical study of symbolic factors has not progressed as much as the study of formal factors. As a result, this section has a greater emphasis on formal than on symbolic aesthetics.

101

The nature of environmental affect

Studies that examine the dimensions of emotional quality in the environment include those by Russell (diverse environments), by Oostendorp and Berlyne (architectural exteriors), by Hershberger and Cass (residential, commercial, and institutional buildings), and by Nasar (residential street scenes). In general, it seems that environmental evaluation consists of three components: pleasantness (like–dislike), excitement (arousing and pleasant or unarousing and unpleasant), and distress (arousing and unpleasant or unarousing and pleasant). Russell, for example, found that environmental affect for a wide variety of molar environments is the product of two primary orthogonal dimensions – pleasantness (an evaluative dimension) and arousal (a non-evaluative dimension). In combination, they produce two complementary evaluative dimensions – excitement and distress. Nasar et al. also found three evaluative components for natural urban settings – liking, excitement, and safety – comparable with Russell's three evaluative dimensions. The affective factors in Hershberger and Cass reveal primary factors – General evaluative and Activity – just like Russell. A third dimension (Utility evaluative), while similar to Russell's pleasantness, is perhaps a unique factor of particular relevance to buildings. Finally, an inspection of the Oostendorp and Berlyne dimensions reveals dimensions (Hedonic Tone/Arousal and Tension) that are similar to Russell's excitement and distress. The convergence of Hedonic Tone and Arousal may be an artifact of the mixture of collative and affective scales. Although further exploration of the dimensions of emotional meaning within various classes of environment is called for, the results across a variety of stimuli and methods have been surprisingly consistent and supportive of Russell's framework.

Salient perceptual and cognitive features

Three papers – Oostendorp and Berlyne (architectural exteriors), Nasar (residential street scenes), and Fenton (natural scenes) – present empirical analyses of the noticeable attributes of outdoor scenes. The studies share dimensions that might be labeled Obtrusiveness of the Built Elements (Monumentality in Oostendorp and Berlyne, building prominence in Nasar, and natural–man-influenced in Fenton), Visual Richness (Decorativeness in Oostendorp and Berlyne, diversity in Nasar, and walking paths in Fenton), and Clarity (clarity in both Oostendorp and Berlyne, and Nasar). The overall dimensional structures, however, differ with the kind of scene examined. Thus Enclosure (Fenton; Nasar) emerges as more prominent in studies of scenes with depth. As has been suggested by Michelson (1976), individuals may use different criteria for evaluating different kinds of scenes. Furthermore, the finding of a "use" dimension in Fenton (natural–man-influenced) suggests that cognitive coding by use may affect the perception of scenes.

Visual preferences

Of particular relevance here are the theoretical frameworks proposed by S. Kaplan, Appleton, and Nohl. Recall that Kaplan described environmental evaluation as a function of two processes: coherence, which enhances comprehension and aesthetic value; and complexity, which produces involvement and enhances aesthetic value up to a point. Appleton cited prospect and refuge as important aesthetic factors. And Nohl commented on the overdesign of urban open spaces. The studies here shed light on the aesthetic value of coherence, complexity, prospect, refuge, and the character of open space. Finally, several studies provide insight into the role of familiarity and novelty in aesthetic preference.

The studies demonstrate that environmental aesthetics is a complex phenomenon that may vary with the kinds of observers, kinds of scene, and associated activity studied. Nevertheless, it seems that aesthetic quality is generally enhanced by naturalness, coherence, order, or compatibility (Oostendorp and Berlyne; Groat; Nasar) and complexity (Oostendorp and Berlyne; Nasar for central business district; Nasar for housing; Nasar for signscape), and is reduced by man-made intrusion (Nasar; Talbot; Fenton; Herzog; Kaplan and Herbert). The findings for prospect and refuge are contradictory. In some cases, prospect received favorable ratings (Talbot; Nasar et al.); in other contexts, it had no influence or an unfavorable influence on preference (Talbot; Fenton; Nasar et al.). For refuge, the results indicate no effects (Fenton) or variable effects, depending on the sex of the respondent (Nasar et al.). The context (scene type, location, and respondent characteristics) may influence evaluative appraisals of prospect and refuge. Finally, the examinations of familiarity and novelty suggest that the kind of familiarity and type of scene being judged may influence response. Thus Orland found that a group of rural residents evaluated grassland scenes (which were similar to the landscape where they lived) more favorably than did a group of urban residents. Nasar found that Japanese and American observers each prefer novel (foreign) scenes to commonplace and familiar (native) ones, and Kaplan and Herbert discovered evidence for both favorable and unfavorable responses in relation to familiarity, depending on the kind of scene evaluated.

A. Methodological comments

Editor's introduction

In attempting to quantify emotional responses to visual attributes of the environment, investigators have employed a variety of methods. Studies vary in choice of subjects, scenes, modes of presentation, measures of environmental attributes, measures of affect, and analytic procedures. For purposes of application, ecological validity is desirable: The conditions of the study should approximate as closely as possible the real conditions to which the results are to apply. This inplies the use of a diverse or representative sample of respondents, and a diverse or representative sample of scene stimuli. In addition, the features of the environment and the kinds of responses obtained should be relevant to naturalistic experience.

Fenton and Reser focus on methodological choices. Three approaches to environmental assessment – objective, subjective, and phenomenological – are reviewed. The authors argue for an integrative approach, in which salient attributes of scenes are first derived and assessed, evaluative responses are then obtained, and the relationship between evaluation and the salient features are examined. Several studies (Flynn; Oostendorp and Berlyne; Nasar; Fenton) in this collection employ strategies similar to those proposed here.

Russell describes the results of a series of studies that derived dimensions of environmental affect. This paper is presented in this section rather than elsewhere because rather than focusing on one kind of setting, it examines a diverse array of molar environments (including interiors, urban scenes, and natural scenes). Using a variety of methodologies, Russell found environmental affect to consist of two primary components – pleasantness and arousal – and two additional components – excitement and distress – which represent mixtures of pleasantness and arousal.

9   The assessment of landscape quality: an integrative approach

D. Mark Fenton and Joseph P. Reser

Landscape-preference studies and the more general research area of human response to natural environments have been plagued by a number of unresolved theoretical and methodological issues, making any general statement of findings or of theoretical synthesis problematic. Researchers adopt either atheoretical, stimulus-defined methodologies, or theoretically prescribed response measures that may or may not bear any meaningful relation to stimulus material. This splintered effort has been compounded by the differing disciplinary orientations of those doing research in this area, with human geographers and landscape architects subscribing to a largely atheoretical and apsychological approach, and psychologists opting for a theoretically derived psychometric stance. There has been no successful attempt to date to develop an approach that would allow for a meaningful and working relationship between theory and method – and, indeed, among disciplines – although a number of individuals have suggested that this is a critical prerequisite for further development in the field of environmental perception and preference (e.g., Carlson, 1977; Wohlwill, 1976; Zube, Sell, and Taylor, 1982).

Approaches to the definition and measurement of landscape

The physical environment has typically been defined either in terms that are independent of the perceiving individual (Barker, 1965; Brunswik, 1956; Wohlwill, 1973) or in terms of individual perception and construct of the environment (Boulding, 1956; Ittelson, 1976; Klausner, 1971; Wapner, Kaplan, and Cohen, 1973). Research in environmental aesthetics, for example, which most often treats the physical setting as the sole predictor of preference, defines setting variables in either objective or perceptual terms, in which objective refers to the quantification of environmental variables through a yardstick that is other than the perceptual cognitive process. The objective approach has been referred to as an "instrumentalist" view and asserts "that nature's objects and events have inherent aesthetic value as causes of the aesthetic experiences of people, but that only those experiences themselves possess intrinsic aesthetic value" (Willard, 1980, p. 297). The perceptual approach, on the contrary, asserts that although there obviously exists an objective external environment, it is the individual's perception and construct of the environment that determines aesthetic value. The integrative approach that is outlined in this paper is basically an interactional perspective and holds

that beauty emerges as a result of the interaction between a human experiencer (or even a nonhuman one) and the natural objects and events experienced, but that beauty is not located only in, or attributable solely to, either the former or the latter. (Willard, 1980, p. 297)

Although theoretical problems in understanding the relationship between perceiver and perceived have been increasingly discussed in the environmental-perception literature (Krasner, 1980; Lévy-Leboyer, 1979/1982), this has been, for the most part, a rearticulation of a conceptual and methodological issue that has preoccupied scientific thinking for quite some time. The issue, very briefly and in general, is how to reasonably assess the relative contribution of perceiver and perceived in determining a complex behavioral product, such as a percept, a judgment, or an environmental preference. This general issue, in a behavioral-science context, has found most recent expression in the "interactionist" approaches that have been published in the past decade (Magnusson, 1981; Magnusson and Endler, 1977). Magnusson (1981), for example, in an introduction to an interactional perspective, identifies four fundamental problems confronting research: the analysis of "actual environments and situations"; "an analysis of perceived environments and situations"; the "description and classification of individuals in terms of situation perceptions and in terms of situation responses"; and the "analysis of person–environment interactions" (pp. 6, 7). Parallel theoretical and methodological issues are also evident in the research on semantics and on the relationship between the meaning of a word and its referent (Alston, 1964; Dale, 1972), and in the field of person perception, a precursor of the interactional approach, where characteristics of the perceiver and the perceived have been found to play an important role in interpersonal attraction (Milord, 1978). In the context of environmental-quality assessment, Craik (1981) distinguishes between observational and technical assessments, whereas Magnusson (1981) considers the world as it is – the "actual environment" – and the world as it is construed – the "perceived environment."

The general theoretical problems associated with understanding the relationship between perceiver and perceived are also evident in studies of landscape evaluation and preference, in which the landscape is exclusively defined in either objective or perceived terms. Kreimer (1977), in a critical analysis of environmental-preference methodologies, has emphasized that research in environmental preference often assumes a "direct" correlate between the individual's perception and construct of the environment and the "real" external environment. Such an isomorphism is, from all we know of perception and cognition, naïve at best. Similarly, Wapner, Cohen, and Kaplan (1976), in discussing approaches to research in environmental perception, emphasize a dichotomy in the definition of environment between two groups of researchers: Some researchers "conceive of the environment in terms of independent or quasi-independent physical features – objective properties of phenomena" (p. 235), while other researchers assume "that environments could not be characterized independent of either human perception or human action" (p. 235). Russell and Ward (1982) have also noted in a review of the field of environmental psychology the "continuing debate between environmental psychologists taking a cognitive approach and those emphasizing the study of the objective physical environment" (p. 664). The polarity of research approaches on this issue is also evident in the continuing and recent debate between those researchers emphasizing the "formal" and those stressing the "nonformal" attributes of the landscape underlying aesthetic response (e.g., Carlson, 1977, 1984; Ribe, 1982). Notwithstanding the fact that researchers have in many instances identified the problem, there has been little serious consideration given to how the two approaches might be reconciled. This paper attempts to directly address this problem and suggests that environmental perception, and, in particular, the perception of landscape quality, is an interactive phenomenon and simultaneously dependent on both objectively definable landscape variables and an individual's knowledge or cognitive representation of the landscape. There is an urgent need for an integrative theoretical and methodological approach to environmental perception, which would recognize that it is the perceived as well as the real world to which we respond.

   A review of the landscape-preference literature suggests that three principal approaches have been used to define physical-landscape variables thought to influence the perception of landscape quality: (1) the objective measurement of physical-setting variables; (2) the use of judges' ratings to define landscape variables with a clear environmental referent (Wohlwill, 1976, 1982); and (3) the description of landscape variables in phenomenological terms. Whether physical-setting variables are defined in objective or judgmental terms, a necessary assumption is required that these variables are salient to the perception of landscape quality. The phenomenological approach, in addition, assumes that the physical-landscape variables are indirectly perceived and construed by the individual making the preference judgment. A similar distinction is discussed by Gibson (1960), who differentiates between "potential" and "effective" stimuli, a distinction that, in turn, directly relates to whether the characteristics of the stimulus or environmental setting are defined independently of the perceiving individual – that is, in objective terms – or as a product of perception – that is, in subjective terms.

Objective quantification

One of the earliest landscape-preference techniques attempted, through the use of black-and-white photographs, to identify a number of objectively quantifiable landscape variables capable of predicting aesthetic quality (Shafer, Hamilton, and Schmidt, 1969). The selection of landscape variables was based on the identification of specific landscape zones, dimensions of such zones, tonal variables, and composite variables of sky, land, and water. Ten landscape zones were defined; they were sky, stream, waterfall, and lake zones, as well as zones of immediate, intermediate, and distant areas of vegetation and nonvegetation. Each zone was also described in terms of its perimeter, interior, area, and shape. Through factor analytic and multiple regression techniques, it was found that positive aesthetic response was best predicted by the perimeter of immediate and intermediate areas of vegetation and nonvegetation, respectively; negative aesthetic response, by the perimeter of immediate vegetation squared and area of water squared. However, these findings produce interpretive difficulties. For instance, it is not clear why perimeter of immediate vegetation should be associated with positive aesthetic response, while squaring this value produces a negative aesthetic response. As Shafer (1969) admits, the model is "difficult to interpret in terms that are meaningful to anyone except a mathematician" (p. 79). Perhaps one of the most important problems may be that the model lacks a theoretical rationale that a priori defines the relationship between objective measures and perceived quality.

A similar technique to that used by Shafer et al. (1969) was used by Zube, Pitt, and Anderson (1974) in the prediction of scenic resource values from fifty-six color photographs. On the basis of a review of the landscape-planning literature, twenty-three landscape dimensions were selected as possible predictor variables. These variables consisted, for example, of such measures as absolute relative relief, mean slope distribution, percent tree cover, and topographic texture. As an example of the measurement procedure used in quantifying these dimensions, a measure of topographic texture was "obtained by finding the one contour line with the greatest number of crenulations . . . the number of crenulations are counted and the sum is divided by the area of view (in square miles) to yield a topographic texture measurement" (Zube et al., 1974, p. 40). The prediction of scenic resource value from the twenty-three dimensions, comprising all fifty-six views, indicated that the best predictor of scenic resource value was the land-use-compatibility dimension, which accounted for 51.4% of the total predictive strength of the model.

The landscape-evaluation studies of Anderson, Zube, and MacConnell (1976), Guldmann (1979), Pitt (1976), Rabinowitz and Coughlin (1971), and to some extent Fines (1968) are also characteristic of those studies that assume a direct relationship between objective properties of the environment and perceived aesthetic quality. Fines (1968) and Guldmann (1979), for example, made use of objective physical indices to examine the impact of proposed developments on visual intervisibility and scenic values, where a measure of intervisibility was obtained through the measurement of distance and the height of the proposed development. Similarly, Rabinowitz and Coughlin (1971) used such objective measures as brightness – measured through the use of a photographic light meter – and grid square measures of area covered by large trees and water as predictors of landscape preference.

Those studies that attempt to relate perceived aesthetic quality to objective characteristics of the environment generally suffer from both the manner in which the physical variables are selected and the nature of the criteria used in selection (R. Kaplan, 1985). In some instances, variables are selected because they are considered by the investigator to be the most common attributes of the landscape on which people make judgments of scenic quality; in other instances, variables are selected because they are amenable to quantification, rather than being of any particular relevance to the assessment of landscape quality. As R. Kaplan (1975) has stated, "An overconcern with objectivity has tended to produce myopia; theoretical sense and even common sense are abandoned in an effort to squeeze prediction from unlikely but reliable variables" (p. 122). Appleton (1975b), in a discussion of the need for theory in landscape-evaluation research, has also questioned whether "a system of differentiation which has proved useful in geomorphological description is necessarily what we need for aesthetic evaluation" (p. 121). The use of only objectively quantifiable landscape attributes in the assessment of landscape preference has also led to a consideration of only those aspects of the landscape representing a group of "formal" properties, "such as shapes, colors, lines, and their patterns and combinations" (Carlson, 1977, p. 136). This is also the case in traditional experimental aesthetics, where research has tended to examine the structural characteristics of stimuli as they relate to aesthetic response, rather than a co-emphasis on the content characteristics (Berlyne, 1974a, p. 18).

Normative judgments

In moving toward a more subjective definition of environmental variables, it is possible to define the environment in quasi-objective terms through the use of normative judgments, for which "the investigator may need to resort to human judges to assess dimensions of the environment that have an objective referent in principle, but nevertheless are not susceptible to direct measurement via physical indices" (Wohlwill, 1976, p. 63). This approach is demonstrated in Wohlwill's (1968) research, which investigated the relationship between preference and complexity in scenes of the urban and natural environments. Complexity, a characteristic of the objective environment, was quantified by summing judges' ratings of "color, shape, direction of dominant lines, texture, and natural versus artificial" (Wohlwill, 1968, p. 308), which referred to "actual attributes of the environment, rather than purely subjective experience" (Wohlwill, 1976, p. 63). It is questionable whether the scale natural versus artificial could be conceived of as being a quasi-objective measure, since individual differences in experience or knowledge of the environment would surely determine whether the environment was perceived of as natural or not. However, the intent of Wohlwill's (1968) procedure was to obtain objective measures of the environment that could be used to predict preference; therefore, a quasi-objective procedure had to be used to quantify complexity, given that there were no other appropriate measuring instruments.

A procedure similar to that of defining important environmental determinants of preference has been used by S. Kaplan (1975), R. Kaplan (1975), and Ulrich (1977), wherein a small group of judges is used to define the environment on specific dimensions preselected by the experimenter. Although these approaches allow for a more subjective definition of the environment, as compared with the approaches of Shafer et al. (1969) and Zube et al. (1974), the aim is still to predict preference on the basis of characteristics of the environment rather than the subjective meaning of the environment to the individual. The physical-landscape variables of complexity, coherence, and mystery have been examined using this approach (R. Kaplan, 1973; Kaplan, Kaplan, and Wendt, 1972; Ulrich, 1977), these variables being selected not because they are amenable to objective specification and quantification, but because there are a priori theoretical reasons for assuming that relationships exist between these variables and the perception of landscape quality. In contrast, the landscape-feature checklist and the landscape-description and -evaluation scales characteristic of such studies as the Zube et al. (1974) research can be considered as descriptive of only the relationship between the variables measured and the perception of landscape quality, since there is no attempt to explain why the existence of such rated attributes of the landscape as hills, ridges, and slopes should be important to the perception of landscape quality. Also, of course, such adjectives themselves necessarily derive from imposed and somewhat arbitrary judgments on the part of the researcher.

In contrast to a number of landscape-preference studies, the landscape-evaluation research lacks a theoretical base that on a priori grounds can be used to select landscape attributes important in determining the landscape quality of an area. The landscape-evaluation studies that use a normative-judgment approach either rely on the sole judgment of the researcher (Linton, 1968) or use persons "trained and experienced in a discipline associated with landscape and expert in map interpretation'' (Fines, 1968, p. 54). For example, researchers such as Linton (1968) and Gilg (1974) found it useful to describe physical-landscape attributes in geomorphological terms, and have classified and assigned scores to aspects of landform (mountain, bold hills, hill country, plateau uplands, and lowlands) and land use (wild landscapes, rich varied farming landscapes, varied forest and moorland, treeless farmland, continuous forest, and urbanized and industrial landscapes). Using this approach, mountains, for example, are given a score of 6, while lowlands are given a score of zero in their contribution to the assessment of landscape quality. Such weightings are based on the perceptual judgment of the researcher developing the technique and are assumed to be "judgments to which we largely subscribe" (Linton, 1968, p. 230).

Phenomenological descriptions

Physical-landscape attributes can also be considered in phenomenological terms; through structured-interview techniques (Honikman, 1972; Leff and Deutsch, 1973), open-ended questions (Dunn, 1976), and verbal associations (Desbarats, 1976), individuals elicit from their cognitive domains those attributes of the landscape that are perceived as being important to the assessment of landscape quality. The phenomenological approach to defining physical variables has been emphasized by Heckhausen (1964) in a review of Berlyne's (1960) research on complexity and liking, in which he suggests that "what we need and still do not have is a multidimensional phenomenology of complexity, preferably in conjunction with operational definitions of its key variables and concepts" (p. 172).

Although the phenomenological approach to defining physical-landscape attributes has not been widely used in studies of landscape evaluation and preference, or, for that matter, within the area of experimental aesthetics generally (O'Hare, 1981), it is a particularly important approach to the description of physical variables that may have a "psychological impact" (R. Kaplan, 1975) on the individual's perception of landscape quality. The phenomenological approach is ultimately based on each individual perceiver's range and organization of experience, and whatever objective attributes of the landscape are assessed unquestionably and substantially colored by individual differences.

A study by Dunn (1976), which exemplifies this approach to understanding what variables are important in landscape preference, required respondents to give reasons for preferring a site ranked first and one ranked fifth. Through a content analysis of responses, such features as woodland, water, open space, and variety were perceived as being important reasons for ranking a site as first, whereas barrenness, lack of open space, and the existence of buildings were perceived by individuals as being important reasons for disliking the site ranked fifth. A central criticism of this study is that dimensions of the landscapes studied were not established beforehand in terms of any objective physical attributes, through the use of either physical indices or normative judgments, and therefore nothing meaningful could be said of the relationship between the objective and the perceived attributes of the landscape as they related to landscape preference. Additionally, it is quite possible that individuals doing the rankings simply supplied a convenient and concrete response to a difficult question.

A number of studies have adopted Kelly's (1955) personal construct theory, and grid techniques developed by Hinkle (1965) and Fransella and Bannister (1977), to obtain a phenomenological description of the environment. These studies have been concerned with specifying the underlying structures of images and cognitive representations of architectural space (Honikman, 1972, 1976), urban forms (Harrison and Sarre, 1975; Hudson, 1974), and natural environments (Fenton, 1985). Honikman (1972), in a development of Hinkle's (1965) grid methodology, applied the technique to describe the way in which living rooms are phenomenologically perceived and construed as similar and different. Through the use of the "implications" and "resistance-to-change" grids, aspects of the physical form of the living room were related hierarchically to a number of superordinate, or abstract, constructs, which were also elicited from subjects. The eliciting of superordinate and subordinate constructs is described by Honikman (1972) as follows:

If a trio of interiors resulted in the informant identifying a "friendly–hostile" construct and the living room being studied was identified as being "friendly," he was asked, "What evidence do you have for saying it is 'friendly'?" He would then give another construct, for example, "cosy"; then he would be asked, "What evidence do you have for calling this room 'cosy'?" He would then give another construct such as "It has a low ceiling." Eventually a series of three or four constructs would emerge, usually relating an abstract construct such as "friendly" to a series of "physical" or "tangible" constructs such as the "relationship between the two chairs," "the warm colors" or "the rough texture of the bricks." (p. 6-5-5)

Similarly, Leff and Deutsch (1973) have used a grid methodology to investigate the ways in which "environmental professionals" and "lay persons" construe imagined physical environments. Comparisons were made between the two groups through an examination of the organizational and content properties of elicited constructs. An important finding of the study was that when both groups were combined, the most commonly occurring dimensions of meaning were those that refer to physical–spatial aspects of environments. However, as was the case in studies by Honikman (1972) and Dunn (1976), no comparisons could be made between the objective or quasi-objective attributes of the environment and the construed representations of the environment.

An integrative approach to the definition and measurement of landscape

From the perspective of an integrative approach to the assessment of landscape quality, it is suggested that the initial stage of the assessment process be to dimensionalize the landscape along an objective or a quasi-objective continuum, which on a priori and theoretical grounds is assumed to be related to the perception of landscape quality. The objective attributes of the landscape may be assessed through objective quantification and/or through normative or psychophysical judgments. If the dimensions of the landscapes are established in these terms, subjective responses may be elicited from individuals in the form of constructs, verbal associations, and open-ended questions, representing the way in which each individual construes the physical landscape. Such a methodology may be usefully employed in examining the relationship between the perimeter measurements of immediate, intermediate, and distant zones of vegetation and nonvegetation (Shafer et al., 1969) and normative judgments of complexity. Evans and Day's (1971) early research in experimental aesthetics suggested, for example, an association between objective complexity and normative judgments of complexity for geometric stimuli. If a similar association could be found, for example, between perimeter measurements and normative judgments of complexity in natural settings, then this may better enable environmental designers to modify properties of the physical setting to suit the behavioral needs of users of the setting. Research by Fenton (1984) has found that while there appears to be a moderate association between perimeter measurements of immediate vegetation (Shafer et al., 1969) and normative judgments of complexity, normative judgments of "environmental" complexity tend to be associated more with judgments of meaningfulness – that is, the amount of perceptual information "afforded" (Gibson's [1982] terminology) by the setting – than with the objective–structural properties of the setting. Such a finding also supports Munsinger and Kessen's (1964) conclusion that high levels of complexity tend to have the greatest number of associations, and Ulrich's (1977) conceptual definition of complexity as being the number of independently perceived elements in the setting. Given these findings, it is suggested that environmental complexity, as distinct from stimulus complexity, comprises such properties as objective physical-setting properties (e.g., objective edge measurements) *in addition to* perceived setting properties (e.g., the amount of meaning associated with the physical setting).

In addition, and through a simultaneous examination of objective and perceived attributes of natural landscapes, Fenton (1984) has found that settings that "afford" perceptual involvement – that is, settings that are judged to be high in mystery, complexity, and meaningfulness (S. Kaplan, 1975) – are phenomenologically construed as more dynamic, active, and alive in terms of both the flora and the fauna of the setting. While normative judgments of mystery and complexity were related to aesthetic response, it appeared that the dimensions through which the individual construed the environment were the more salient determinants of aesthetic quality (Fenton, 1984). Again, these findings clearly demonstrate the usefulness of an integrative approach to the perception of landscape quality, by which the interrelationships between the objective and the perceived dimensions of the environment are simultaneously examined in terms of their contribution to perceived aesthetic quality.

In addition, the finding of substantial individual differences in judgments of aesthetic quality (Dearden, 1984; Fenton, 1985; Purcell and Lamb, 1984) does suggest the possibility that individuals may be basing their judgments of aesthetic quality on very different properties of the natural setting. For example, some individuals' judgments of aesthetic quality may be determined by the underlying objective properties of the setting, while other individuals may be basing their aesthetic judgments on the meaning they attribute to the setting. It is only through an integrative approach and the simultaneous and concurrent examination of both objective and perceived characteristics of the environment as they relate to judgments of aesthetic quality that future research will be able to examine the differential salience of these dimensions to judgments of aesthetic quality among individuals.

The need for such an integrative theoretical and methodological approach as that outlined has been emphasized by Pervin (1978), who states that

while definition of stimuli, situations, and environments in both objective and perceived terms is often recommended, and study of the relationship between the two would be a valuable endeavour, most studies define phenomena in either objective or perceived terms and gather data accordingly. (p. 82)

A summary of the general methodological approach that may be employed is given in what follows. It involves, in logical sequence, the following steps:

1. The selection of objective or quasi-objective attributes of the landscape that are theoretically, or at least intuitively, related to landscape quality. For example, the dimensions of landscapes may be described in terms of their collative (Wohlwill, 1976; Wohlwill and Kohn, 1976), informational (S. Kaplan, 1975), or invariant functional properties (Gibson, 1967/1982), or classified into landform and land-use dimensions (Linton, 1968).
2. The use of objective physical indices or normative judgments to quantify landscapes that vary in these attributes.
3. The eliciting of subjective descriptions of these landscapes, in terms of schemata (Neisser, 1976) or constructs (Kelly, 1955), representing the underlying structure of an individual's image (Boulding, 1956; Lynch, 1960) or cognitive representation (Moore and Golledge, 1976) of the landscape.
4. The examination, through, for example, property-fitting methods (Carroll and Chang, 1977; Fenton, 1985), of the relationship between the objectively defined

attribute dimensions of the landscape and the perceived or construed attributes
of the landscape as they relate to evaluative judgments of landscape quality.

The approach suggested here has important practical applications to the fields
of landscape architecture and environmental design, in that the physical attributes of the natural and built environment may be manipulated by the designer
with some knowledge of how the individual is going to experience the resulting
setting. For example, the design of trails in national parks can make use of such
features as complexity, coherence, and mystery in determining optimum levels
of involvement and perceptual response. An understanding of how parts of the
trail vary in such specified characteristics and with respect to associated meaning
and perceptual response allows designers and park officials to better plan or
modify existing natural settings to meet planning and recreational objectives.

An additional practical application has to do with matching particular settings
to particular needs, such that individual differences with respect to the screening
of information (Mehrabian and Russell, 1974), previously acquired adaptation
levels for various types of settings and stimuli (Helson, 1964), and simple
interest and information-processing requirements can be optimally met. Settings
designed to cater to particular educational and therapeutic problems can similarly
be planned with a realistic appreciation of how particular setting attributes interact with specific user requirements and needs.

Aesthetic considerations are also an important aspect of environmental-impact
assessment, where the interdependent effects of perceptions and values ultimately determine the desirability of interventions (e.g., Matthews, 1976). Indicators
of physical environmental quality emphasized by Matthews (1976) include such
aesthetic domains as scale, color, texture, pattern, and variety, with an implicit
criterion being that any such index "quantitatively summarizes or measures the
conditions of the . . . aesthetic environment" (p. 164). However, as Carlson
(1977) has emphasized, aesthetic appreciation of the environment "is a function
of various non-formal qualities *in addition to* formal qualities" (p. 159) of scale,
color, texture, pattern, and variety. It should be kept in mind, as well, that what
has been termed "aesthetic response" is really an organic component of perceived environmental quality and ultimately reflects a complex but hopefully
comprehensible human response to complex natural environments.

An adequate understanding of human response to natural settings and landscapes will ultimately depend on a meaningful synthesis of theory and method,
such that the perceiver and the perceived are given equal and concurrent consideration. What is suggested is that research not lavish "too much effort on
hypothetical models of the mind" (Neisser, 1976, p. 8), and neither should it
propose "a theory of perception in which mental events play no role at all" (p.
9). What is needed, and what the integrative approach attempts to do, is to

provide an approach in which an individual's construal of the environment is examined simultaneously with the information offered by the environment. The approach outlined allows for a simultaneous examination of formal, objective attributes of physical settings and nonformal, perceptually mediated attributes, such that the operational version of theoretical constructs more closely approximates the nature of environmental experience.

10   Affective appraisals of environments

James A. Russell

When we think about or encounter an environment, perhaps the most important judgment we make about it is whether it is interesting, gloomy, frightening, relaxing, or such. Whether we choose to go there, what we do there, and whether we return may be largely determined by such judgments, which I shall call *affective appraisals*.

Understanding affective appraisals requires knowledge in at least three related areas: (1) the nature of affective appraisals themselves, including some means of describing and measuring them; (2) the relationship of affective appraisals to physical or objective properties of environments; and (3) the relationship of affective appraisals to other psychological processes, including behavior, emotion, and cognition. In the rush to gather needed information on the second two topics, the first has been somewhat neglected, and it is the focus of this paper.

A characterization of affective appraisals

An affective appraisal occurs when a person judges something as having an affective quality, such as being pleasant, likable, exciting, and so on. Affective appraisals thus resemble both emotions and cognitions. They resemble emotions in that they concern affective feelings. They resemble cognitions in that they are one aspect of how someone interprets something. Still, we must distinguish affective appraisals from other phenomena that fall under the headings of emotion and cognition.

That, by the definition above, affective appraisals are judgments (and hence mental events) distinguishes them from the physiological and behavioral components of emotion. That, by the same definition, affective appraisals are always directed toward, are always about, something, distinguishes them from what I call moods – with mood defined as a subjective affective state not directed at any object, such as feeling melancholic or chipper, for no apparent reason (Russell and Snodgrass, in press). In short, I distinguish your judgment that Beirut is a frightening place – the affective appraisal – from any feelings of tension or any

120

shivering, trembling, or other behavioral or physiological signs of fear that you may or may not experience at the thought.

Distinguishing affective appraisals from other affective phenomena is not to deny that affective appraisals can be accompanied by changes in emotion or mood. But this distinction does have a direct implication for the measurement of an affective appraisal: The most direct measure possible for an affective appraisal is a verbal report. Because a person judges a particular place as distressing or relaxing does not necessarily imply that behavioral or physiological signs of stress or relaxation will be present. Their presence must be established empirically rather than assumed. Behavioral and physiological signs of mood and emotion could conceivably be used as less direct measures of an affective appraisal (just as height of mercury can be used as a measure of temperature), but only after we have learned under what conditions such signs accompany affective appraisals.

Affective appraisal is one aspect of how someone interprets an environment. To find a place pleasant, interesting, stressful, or the like is to attribute to that place an *affective quality* – a capacity to alter mood. In other words, to say of an environment that it is pleasant is to say that it can produce pleasure. The affective quality attributed to a place is a key component of the full meaning attributed to that place: In a series of studies on the meaning of large-scale environments, affective quality was repeatedly found to be a salient and important way in which environments are interpreted and compared with one another (Ward and Russell, 1981a, 1981b).

To distinguish affective appraisals from other aspects of the interpretation of an environment, we can divide the meaning of an environment into two components: affective and nonaffective. Some words, such as *pleasant, disgusting,* and *stressful,* describe the affective component. Other words, such as *green, tall,* and *old,* describe objective or physical components. But most words combine the two, as is shown by the vast literature on the semantic-differential technique (Snider and Osgood, 1969): *dangerous* describes places where harm is likely and that are frightening; *noisy* describes places with loud and unpleasant sounds. Any instrument that is claimed to measure affective appraisal must be carefully constructed to avoid confounding affective with nonaffective meaning. (For example, what is commonly called the semantic differential can, when used to assess places, produce an assessment in which it is difficult to disentangle affective from nonaffective meaning; see Craik, 1981b; Russell, Ward, and Pratt, 1981, for discussions of this issue.)

The structure of affective qualities

The environmental researcher looking for a simple means of assessing an individual's affective appraisal of a place may have a hard time of it. Which of the

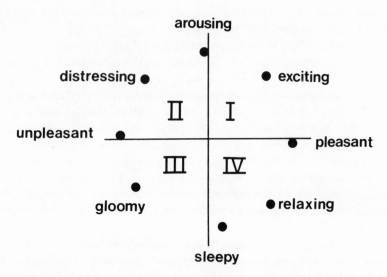

Figure 10.1. A spatial representation of descriptors of the affective quality of environments.

hundreds of terms potentially descriptive of affective qualities of environments should you use? The scientific literature complicates the picture by sometimes using dimensions, sometimes using categories, and rarely using the same measure twice. In this section, I propose a system that attempts to order the full set of relevant terms and that accounts for both dimensional and categorical approaches.

Terms for affective qualities of places are systematically interrelated, and the network of these interrelationships can be described through a spatial metaphor, as illustrated in Figure 10.1. Its base consists of two bipolar dimensions. The first, represented as the horizontal axis, ranges from extreme unpleasantness through a neutral point to extreme pleasantness. The second dimension, which is independent of the first and is represented as the vertical axis, concerns the arousing quality of the place and ranges from soporific to extremely arousing. A place appraised at the high end of this axis might be called frantically exciting, provided that "frantically exciting" is not thought of as necessarily positive or negative.

Categorical affective descriptors, such as *pleasant, gloomy, distressing,* and *relaxing,* are located at specific points in the two-dimensional space. The space can therefore be more fully described by examining the location of representative affective descriptors. A neutral appraisal is located at the center of the space, and the more extreme the appraisal, the more it falls toward the perimeter. When scales for eight representative descriptors were developed, they were found to

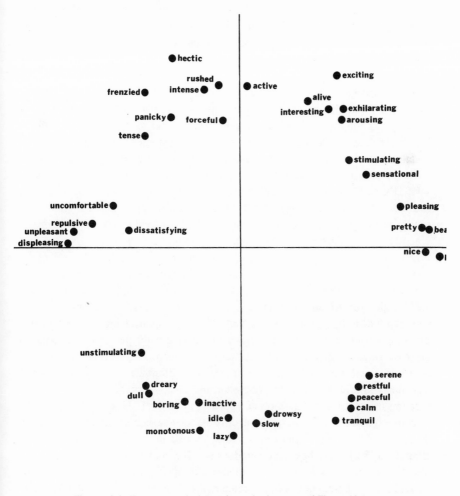

Figure 10.2. Forty categorical descriptors in the space of Figure 10.1.

fall in a circular order around the perimeter (Russell and Pratt, 1980), as shown in Figure 10.1. If each of these variables is thought of as representing a region of the space, the above description corresponds to a division of the space into octants. For some purposes, finer divisions may be required. For now, a division of the space into quadrants will be useful, and they are numbered in Figure 10.1.

The eight variables in Figure 10.1 are not the only possible descriptors. As is illustrated in Figure 10.2, Quadrant I contains descriptors of places appraised as highly arousing and highly pleasant. Examples are *exciting, interesting,* and *exhilarating.* Quadrant II contains descriptors of places appraised as highly arousing and highly unpleasant, such as *frenzied, tense,* and *hectic.* Quadrant III contains descriptors of places appraised as unarousing and unpleasant, such as *dull, unstimulating,* and *dreary.* And Quadrant IV contains descriptors of places appraised as unarousing and pleasant, such as *tranquil, serene, peaceful,* and *restful.*

The proposed structure of affective descriptors stems from several lines of evidence. First, the structure is consistent with what is known about affect in general. Although the English language contains up to 2,000 emotion-related words (Wallace and Carson, 1973), psychological studies are converging on the conclusion that they can be summarized as indicated in Figure 10.1. (Fisher, Heise, Bohrnstedt, and Lucke, 1985; Fromme and O'Brien, 1982; Plutchik, 1980; Russell, 1980; Watson and Tellegen, 1985). And similar evidence is now available from other languages and cultures (Russell, 1983) and from children as young as two years (Russell and Bullock, 1986). Second and more to the point, when adjectives descriptive of the affective qualities of *environments* were analyzed, the same two dimensions and circular ordering were found (Russell et al., 1981). This last study was of 323 actual environments, each rated on a relatively complete set of affective descriptors. Based on these results, highly reliable and valid scales for affective appraisal were developed (Russell and Pratt, 1980).

This evidence suggests, to me, that affective appraisal involves a two-step process. An environment is initially and automatically perceived in terms of pleasant versus unpleasant and arousing versus unarousing. Phenomenologically, these dimensions combine in a unitary perception: The environment seems, say, pleasant-and-arousing. In my model, this step is graphically represented as locating the place at an appropriate spot in Figure 10.1. Next, to articulate the affective quality of the environment, it is associated with an affective category. Thinking of both places and categories as located in the same space gives us a simple way to picture how a person does so: The applicability to a place of any affective category is proportional to the distance between the place and the category, which are now both located within the same space. An affective category, such as *pleasant* or *unpleasant* or *stressful,* is thus not so much a bin with sharp boundaries as it is an ideal point with which any place can be compared. Some places are prototypically pleasant, some less so, and some prototypically unpleasant; pleasantness – like all other affective descriptors – is a matter of degree rather than all-or-none.

The measurement of an affective appraisal must take into account that affective appraisal is not a tidy, scientific enterprise. To affectively appraise something is not to place it within one, and only one, carefully defined category; it would not be like assigning a name to a chemical element according to the periodic table of elements. Rather, each place will fit many affective categories, to some degree. Ratings on a small and unsystematic set of affective descriptors can result in ambiguity. To understand why a place is rated, say, *distressing,* one

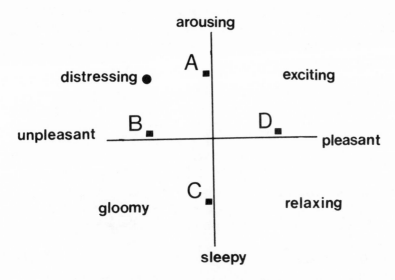

Figure 10.3. Four hypothetical places in affective space.

must look at where that place falls within the entire system of categories and dimensions – that is, where the place falls in the space of Figure 10.1. Affective appraisal is thus best assessed as a whole, rather than piecemeal.

This point is illustrated in Figure 10.3, where four places (A, B, C, and D) are plotted in the space of Figure 10.1. Suppose that subjects are asked to judge how *distressing* each place is. The meaning of their ratings is best understood by examining the location of each place in the space. Places A and B would be rated as equally distressing, and places C and D as equally unstressful – and yet the environmental researcher would achieve little by classifying A and B as distressing. Places A and B differ too much from each other: Place A is highly arousing, but neither pleasant nor unpleasant; place B is highly unpleasant, but neither arousing nor unarousing. Although both A and B would be rated as distressing, they may have little in common, and the reason that A is more distressing than C is different from the reason that B is more stressful than D. Imagine further that A, B, C, and D represent conditions in an experiment. A study of stress that compares A (experimental group) with C (control group) will not reach the same conclusion as a study that compares A with D, or B with C, or B with D – even though each study could legitimately be described as a study of stress.

The relativity of affective appraisals

An affective appraisal of a place is relative to the specific circumstances in which it is made, including peripheral and previously encountered places. Rather than fixed, the human standard of judgment adjusts to the range and distribution of

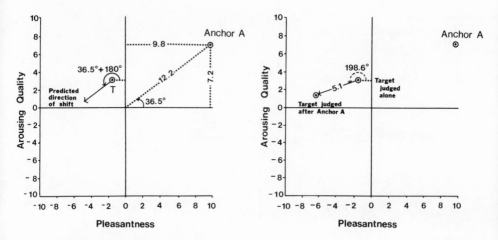

Figure 10.4. Predicted and observed displacement in the appraisal of the Target stimulus induced by prior exposure to Anchor A.

available experiences (Parducci, 1968). For example, migrants from rural areas judge their city as noisier and more polluted than do migrants from urban areas (Wohlwill and Kohn, 1973). Long-term residents of a smog-filled basin set a higher criterion for reporting the presence of smog than do short-term residents (Evans, Jacobs, and Frazer, 1982). The scenes that people find beautiful depend on the sorts of scenes they typically encounter in their everyday life (Sonnenfeld, 1969).

Combining the structure of affective appraisals seen in Figure 10.1 with Helson's (1964) adaptation-level approach to the relativity of judgments provides a precise characterization of how affective appraisals depend on previously encountered places. For example, if an extremely pleasant environment is made to be a salient aspect of the context of judgment, the next place encountered will seem less pleasant. This contrast effect has been demonstrated to occur for the judged pleasantness of stimuli (Parducci, 1968; Thayer, 1980), but theoretically, it should occur throughout the two-dimensional space of Figure 10.1. To illustrate, consider how an encounter with a pleasant and somewhat arousing scene (a place I shall call Anchor A) might influence the subsequent appraisal of a relatively neutral place (the Target scene). My hypothesis is illustrated in the left half of Figure 10.4. Assume that *without* an encounter with Anchor A, the appraisal of the Target would be point T. The encounter *with* Anchor A (which is to the right and slightly upward of point T) shifts the appraisal of the Target down and to the left, in a direction 180° different from the angle from the origin toward Anchor A. As a consequence, the Target should appear less pleasant and less arousing. Categorical descriptors in Quadrant I (such as *interesting*) should become less applicable, and categorical descriptors in Quadrant III (such as *boring*) should become more applicable.

This hypothesis was tested experimentally (Russell and Lanius, 1984), with the result shown in the right half of Figure 10.4. In the experiment, sixteen different "anchor" stimuli were used, of which Anchor A was one example. These were photographs of various environmental scenes, including a suburban street, a children's amusement park, a cement factory, and a highway with a burning car. A subject first saw and rated an anchor. Next followed the Target scene, which was the same in all conditions. The result shown in the right half of Figure 10.4 was typical of those from the other fifteen conditions: Consistently, the affective appraisal of the Target shifted, relative to its appraisal by a control group, in the way predicted.

Measurement of affective appraisals must take into account their relativity. Giving subjects a set of environments to rate will likely create an adaptation level different from their natural one. Giving subjects even one sample stimulus to rate may influence ratings on the Target stimulus. There is no vantage point from which an absolute affective appraisal may be made: An affective appraisal will always be relative to the particular circumstances in which it is made.

Environmental assessment

So far, I have discussed affective appraisals, which are individual matters. Different individuals may not affectively appraise the same environment in exactly the same way. Nor would the same individual at different times, nor different populations of individuals who have different backgrounds. Where does this leave the investigator who wants to assess the environment itself, to know how stressful, interesting, pleasant, and so on the place is? Affective quality is a key factor determining the human response to an environment and cannot be omitted from the assessment of that environment. Craik (1971) pointed out that environmental assessment and environment perception (which would include affective appraisal) are distinct but related topics, and bear a relation to each other similar to that of personality assessment and person perception.

In environmental assessment, as in personality assessment, consensus among a group of judges can be used to establish a judged property as a property of the object judged. We can therefore define an average or a consensual affective appraisal from a specified population of individuals for a particular environment. Of course, the amount of agreement among those individuals must be established empirically. Fortunately, evidence is available to show that there is fair consensus (e.g., Nasar, 1983, 1984). For example, various individuals were asked to affectively appraise various environments, and a mean judgment was used to define the average affective appraisal of each environment (Russell and Pratt,

1980). The split-half reliability of these averages for a group of twenty quite different environments was .97 for pleasantness and .97 for arousingness. The mean scores in each half-sample were based on between eight and twenty subjects. From an important series of analyses, Schroeder (1984) concluded that even fewer subjects may suffice and that results from this simple averaging technique are, for practical purposes, virtually interchangeable with those from more sophisticated ones. (Of course, a consensual affective appraisal is not necessarily identical with the actual affective quality – the mood-altering capacity – of the environment appraised. Although the two are likely to be closely related, I know of no systematic evidence on this issue.)

Concluding remarks

In a fascinating study in the history of science, Jammer (1957) followed the concept of force as it evolved from its unsystematic use in everyday thinking to its current use (and disappearance) in modern physics. Early writers thought of force as something residing within a moving object, pushing it along. The muscular strain that you feel while moving your arm seemed to be the prototype of force. Gradually, force came to be thought of as something outside the moving object, such as being struck by another moving object. For later writers, force sometimes meant energy, sometimes momentum, and sometimes any cause of motion. Slowly, these different but related concepts were distinguished and clarified.

I believe that psychological concepts, such as cognition, emotion, and affect, will similarly evolve. Different but related concepts will be distinguished and clarified. Such has been my goal in this paper. For the notion of affective appraisal, the ideas offered must therefore be taken as proposals and evaluated against the criterion of usefulness.

Affective appraisals can be distinguished from other aspects of the psychological response to an environment. An affective appraisal locates an environment within a network of interrelated categories and dimensions. Mapping out that network allows for a reliable and valid assessment of an individual's affective appraisal, which is a key aspect of that individual's response to the environment. Everyday human concepts involved in affective appraisal have limitations that must be taken into account when those concepts are pressed into scientific service. *Pleasant, stressful, peaceful,* and the rest are not formally defined and do not form a mutually exclusive set. Nevertheless, when their actual properties are recognized, these concepts are understandable and useful. An affective appraisal does not depend solely on the environment appraised, but like other human judgments, depends on prior experiences of the individual human judge. Still, there exists sufficient agreement among judges that, through averaging the separate affective appraisals from a representative sample, it is possible to include within the assessment of an environment a consensual affective appraisal, which is a key property of that environment.

Acknowledgments

This chapter is based on empirical and conceptual work carried out in collaboration with Larry Ward, Jackie Snodgrass, Gerry Pratt, and Merry Bullock, to all of whom I am grateful, and was supported by grant 410–85–1735 from the Social Sciences and Humanities Research Council of Canada.

B. Architectural interiors

Editor's introduction

Two landmark studies in both environmental aesthetics and environmental psychology are the Maslow and Mintz (1956) and Mintz (1956) comparisons of exposure to "beautiful," "neutral," and "ugly" rooms. The studies ostensibly found short-term and long-term effects related to aesthetics. In the first paper in this section, Locasso turns a critical eye to potential method artifacts in these studies. He replicated the "short-term" study with controls to overcome these artifacts, and he found little evidence of direct short-term effects of the aesthetic manipulation. Does this mean that aesthetic quality has no effect? No. Taken together, the findings of Maslow and Mintz and Locasso suggest that although short-term exposure to various aesthetic conditions per se may not affect behavior, the effects are likely to occur through the interaction of visual quality with social factors and long-term experience. Thus it is important to understand the nature of environmental aesthetics.

One line of research has aimed at deriving adjective scales relevant to the description of the physical surroundings. Such scales are of value to the researcher interested in the nature of aesthetic response, the designer interested in establishing user or client design preferences, or others interested in evaluating their own surroundings. Central to this line of research is the second paper in this section, in which Kasmar describes several studies that identify sixty-six bipolar adjectives that the public and designers commonly use in describing architectural interiors.

Lighting plays an important role in the perceived quality of interior space. In the final paper in this section, Flynn reviews the results of several studies on the relationship between interior lighting configurations and evaluative response. Flynn found three perceptual dimensions of interior lighting: uniform–nonuniform, overhead–peripheral, and bright–dim. When subjective impressions of these lighting conditions were obtained, Flynn found that clarity was enhanced through uniform, overhead, and bright lighting; spaciousness through uniform, peripheral, and bright lighting; relaxation through nonuniform, peripheral, and dim lighting; privateness through nonuniform, peripheral, and dim lighting; and pleasantness through nonuniform, peripheral, and bright lighting.

11 The influence of a beautiful versus an ugly room on ratings of photographs of human faces: a replication of Maslow and Mintz

Richard M. Locasso

Introduction

Although aesthetic concerns are central to many issues in architecture and environmental planning, there is little scientific evidence concerning the manner in which beautiful and ugly interior environments influence human behavior. Reviews of this work (Locasso, 1976) have indicated in general that few studies have been reported and that problems in experimental design, measurement, and methodology are common. There appears to be little solid empirical evidence demonstrating that attractive interior spaces exert some form of beneficial influence on human functioning and behavior.

Although not a cornerstone of the empirical literature, early work by Maslow and Mintz (1956) has received much exposure in environmental psychology and the environmental-design disciplines. They examined the effects of "beautiful" and "ugly" rooms on subjects' judgments of the amount of "energy" and "well-being" reflected in photographs of human faces. One of three experimentation rooms was decorated as a comfortable study and contained a mahogany desk and chair combination, a rug, drapes, paintings, sculptures, and other items. People who saw this room described it as "attractive," "pretty," "comfortable," and "pleasant" (Maslow and Mintz, 1956, p. 247). The "ugly" room was described as "horrible," "disgusting," "ugly," and "repulsive." It contained "battleship gray walls, an overhead bulb with a dirty, torn, ill-fitting lampshade, and 'furnishings' to give the impression of a janitor's storeroom in disheveled condition" (Maslow and Mintz, 1956, p. 248). Two Es alternated between the two rooms and showed Ss a series of ten negatives of photographs. Subjects made judgments concerning the amount of "energy/fatigue" and "well-being/displeasure" that they saw in the faces pictured in the photos. Results indicated that the Ss in the beautiful room gave significantly higher ratings on these two dimensions than Ss who were tested in the ugly room.

In a second research report, Mintz (1956) examined the behavior of the two Es

134

in the same study to determine the duration of any adverse environmental effects. Over a three-week period, the Es spent consistently less time in the ugly room than in the beautiful room, regardless of which E was in the room. Also, the Es made self-ratings using the energy and well-being scales at the end of each session of data collection (two sessions per week). They believed that the self-ratings were administered to check the reliability of the scales. Analyses of the self-ratings indicated that significantly lower scores occurred in the ugly room throughout the three-week testing period. These results were interpreted to mean that the adverse environmental effects were maintained over time and that the Es could not adapt to the different aesthetic experiences of the two rooms.

The Maslow and Mintz (1956) study has been widely quoted and discussed in the literature and generally has come to be regarded as a "classic" study in the area of environmental psychology. Their early effort is often cited as evidence that environmental-aesthetic impact is indeed important and that attractive and unattractive rooms will exert, respectively, a positive and a negative influence on human behavior. Because of this and also because no replications have been reported in the literature, the present study was conducted to attempt to replicate the Maslow and Mintz work.

Review of original methodology

The paradigm of the present research project attempts in part to address and resolve a number of methodological issues in the Maslow and Mintz work. One objective of the present study was to not replicate the earlier work per se, but to use an approximate replication design with altered methodology, which would allow a more precise interpretation of any experimental effects that might emerge. Thus a brief discussion of the more important methodological issues is warranted.

Both Es in the Maslow and Mintz study collected data in both experimentation rooms. They were thus exposed to both levels of the independent variable, environmental-aesthetic impact. It is plausible that any graduate student serving as an E might wonder why he or she would be asked to collect data in a beautiful and attractive context, only to be switched later and in counterbalanced fashion to what seemed to be a janitor's storeroom (or vice versa). A reasonable case can be made for the demand characteristics of the experiment influencing the two Es, who, in turn, influenced the Ss with whom they worked.

At the beginning of the research, Maslow and Mintz administered the energy and well-being scales to the Es in each of the two experimentation rooms. The Es were told that this was a practice period, so that they could learn the procedures of the experiment. During these sessions, it is plausible that the authors in one way or another could have unintentionally communicated the desired study outcome vis-à-vis the experimentation rooms. Is it possible that the effects reported by Maslow and Mintz are no more than experimenter bias carried by the Es to the Ss from their practice sessions with the authors?

The experimental design required the Es to alternate data-collection chores between two rooms of highly contrasting aesthetic impact. A body of research in psychology (Helson, 1964) holds that one's adaptation level, what one is "used to" and regards as normal, is pulled toward the anchor, or extreme element, in a series of stimuli. Switching back and forth between the rooms would buffet the adaptation level of the poor Es like a punching bag, heightening the impact each time a door was opened. Thus the operative factor accounting for the reported effects might not have been the room per se, but the continued contrast in aesthetic impact between the rooms.

To resolve some of these issues, the present study adopted a highly controlled and analytical approach of purposedly restricted domain, excluding from investigation the following factors. First, social-interaction effects, when two or more individuals co-occupy a space, were eliminated by having only one person, the S, in a room at a time. It proved impossible to develop a design that would utilize an E without some form of potential confound similar to those illustrated earlier. Second, symbolic impact – that is, associations or added meanings in addition to the stimulus-bound aesthetic impact – was low. Thus, for example, we did not say to an S about to use the ugly room, "Here, this room's kinda bad, but it's okay for you." Third, the present research examined task-oriented behavior, excluding such activities as play, recreation, social affiliation, and even daydreaming. Finally, this study focused on the first occasion, the initial exposure to a space, as did Maslow and Mintz, as opposed to long-term effects that could emerge over time and repeated visits to a space.

Methodology

Subjects

Two groups of freshmen and sophomore students, each containing six males and nine females, were recruited from introductory psychology classes for participation in the study.

Photographic stimuli

Thirty-two people were photographed individually by the E during one afternoon at a local shopping center using a 35-mm camera. The black-and-white photo was taken straight on at an approximate 4-foot distance, showing the person's head and upper chest, standing about 5 feet in front of the building's brick wall.

Facial expressions were assumed naturally by the Ss, and the sample contained "average everyday people" of differing sex, age, height, weight, physical attractiveness, social class, cleanliness, and so on.

Photos were sorted into five categories on a dimension of "pleasantness of facial expression" by twenty psychology graduate students working independently, and means and standard deviations were computed for each photo. Five photos were selected that spanned the range of positive to negative facial expression and had relatively small standard deviations – that is, high agreement by Ss. Three photos were selected that exhibited positive, neutral, and negative facial expression and had extremely large standard deviations – that is, low agreement by Ss.

The 3- by 5-inch photos were taped onto 5- by 7-inch pieces of white cardboard backing for presentation to the S, and the series of photos was numbered 1 to 8.

Experimentation rooms

Two rooms juxtaposed in the psychology department at The Pennsylvania State University functioned as the beautiful (B) and Ugly (U) rooms (Figure 11.1). Each room measured $6\frac{1}{2}$ feet by 15 feet, with 9-foot ceilings; was painted off-white and carpeted; and contained overhead lighting and acoustical-tile ceilings. The B room contained a window, and the U room was ventilated by forced noncooled outside air.

As depicted in Figure 11.1, the B room contained plants, curtains, framed photos, a Mexican serape, a reclining leather chair, and various personal touches, and looked like an attractive working office. The U room contained stacks of computer and research material, a box of electrical components, 10 feet of polygraph records taped to one wall, smudges and fingerprints on the walls, some missing ceiling acoustical tiles exposing building structure, and the like. The U room appeared as a working office, but a very ugly one. Lighting, ventilation, temperature, and noise level were similar between the rooms and fell in the normal range.

A room-assessment study using twenty-five male and twenty-five female students was conducted to measure the level of aesthetic impact of each room and to ensure that the only difference between the two rooms occurred on the aesthetic dimension. Subjects used a questionnaire containing eighteen bipolar environment-description adjectives (Collins, 1971; Hershberger, 1969; Kasmar, 1970) with seven response category format to individually rate the B, the U, and a third room using a design counterbalanced for room order. The aesthetic dimension was composed of five scales: beautiful–ugly, attractive–unattractive, good colors–bad colors, good style–bad style, and neat–messy. The mood dimension

Figure 11.1. The beautiful (B) and ugly (U) rooms.

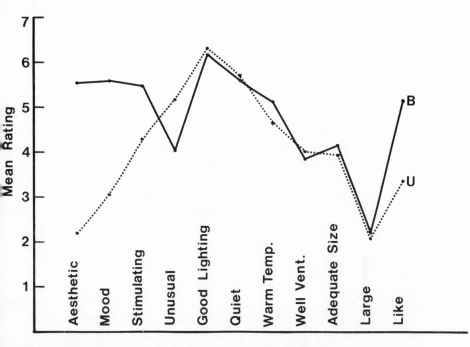

Figure 11.2. Environmental profiles of the beautiful and ugly rooms (listed adjectives were on point 7 of the scale).

included the scales of appealing–unappealing, pleasant–unpleasant, warm mood–cold mood, and friendly–unfriendly. The results (Figure 11.2) indicate that (1) the B and U rooms represent substantially different levels of the independent variable, (2) the rooms were highly similar along all other dimensions (the aesthetic and mood dimensions are highly intercorrelated when rating environments devoid of social stimulation (Locasso, 1971), and (3) all other dimensions fall in the normal range and are not so extreme that they could interact with potential aesthetic effects. Interestingly, the data indicated that if the B room was rated first, ratings of the U room were always more extreme than when the U room was rated first (and vice versa), indicating the operation in this study of the contrast effect described earlier.

Experimental design

The experimental design compared the dependent measures of one group of fifteen subjects who worked individually in a beautiful room with identical measures from another group of fifteen subjects who performed the same tasks in an ugly room. To control for the potential confounds described earlier, Ss worked alone in a room after receiving both verbal and a packet of typed instructions from an E who was blind to the room to be used by the S.

The following dependent measures were used in this study. First, photos were rated on the following bipolar adjective rating scales of seven categories each: physically attractive–physically unattractive; pleasant facial expression–unpleasant facial expression; energetic–fatigued; relaxed–tense; healthy–unhealthy; friendly–unfriendly; and secure–insecure. The energetic–fatigued dimension was included to parallel the "energy" dimension of Maslow and Mintz (1956), while their "well-being/displeasure" scale was judged multidimensional and replaced with healthy–unhealthy and secure–insecure. Second, the room was rated using the room-assessment questionnaire described earlier. Third, the experimental task was rated using eight bipolar adjective scales of seven categories each to assess task pleasantness, interest, frustration, and preference. Finally, the amount of time each S stayed in a room was estimated by recording when the S left and returned to the E's office.

Procedure

Each subject was told individually that this was an experiment in interpersonal perception and that after receiving an instruction packet, the S would go to a room, follow the instructions, complete the task alone, and then return to the E's office. The S was told that this procedure avoided the possibility of the E biasing the S's responses by being present.

After having any questions answered, the S left the office, opened the packet, and went to the room number given on the first page. On entering, the S seated herself at the desk with the stimulus materials located on a stand to the right about 3 feet from floor level. Following the instructions, the S reviewed the set of photos, spending five seconds on each, checked photo numbers for correct sequencing, wrote the first photo number on the first sheet of rating scales, viewed the photo for five additional seconds, and then completed the seven scales. This procedure was followed until all eight photos were rated. The task-evaluation scales were completed next, followed by the room-description scales and a brief biographical-data section. The S returned the eight photos to the viewing stand, checking to see that they were in the proper sequence, put the questionnaire back into the envelope, and returned the envelope to the E. At this time, the E stopped the timer that had been started when the S initially left the E's office.

Results

The results from the photograph-rating scales were evaluated by four different procedures. The first procedure was identical to that used by Maslow and Mintz

(1956). Each S rated each of eight photos on each of seven rating scales. The seven ratings were summed and averaged to provide each photo with one overall score. The scores of the eight photos were then summed and averaged to provide each S with one score. This score represented the average rating made by an S over all photos and all scales. Means were computed for the fifteen Ss in the B room and the fifteen Ss in the U room. This analysis tested the hypothesis that the mean ratings given by Ss in the B room (M = 4.28) would be more favorable than the average rating given by Ss in the U room (M = 4.34). A higher numerical value equals a more favorable rating. An independent t test between these two means was not significant (t (28) = .14, p > .05).

In the second analysis, the ratings for the eight photos were averaged by rating scale. Thus each S had seven average scores, one for each rating scale. Seven t tests, one for each scale, compared the average ratings of the Ss in the B room with the average ratings of the Ss in the U room. One test, on the "pleasantness of facial expression" dimension, was statistically significant (t (28) = 2.28, p < .05) but in the opposite direction, indicating that the facial expressions were rated as more pleasing to look at in the U room than in the B room.

A third analysis examined two orthogonal components of variation identified through a series of principle component and factor analyses of scale, subject, and photo matrices. Ratings of the eight photos were averaged for the following two clusters: (1) pleasant facial expression, relaxed, and friendly; and (2) physically attractive, energetic, and healthy. The t test for the "physically attractive" factor was not significant (t (28) = .11, p > .05), whereas the test approached significance for the "pleasantness of facial expression" factor (t (28) = 1.95, p = .06) but opposite the hypothesized direction. That is, the facial expressions depicted in the photos were judged to be more pleasant in the U room (M = 4.51) than in the B room (M = 4.25).

Finally, fifty-six t tests compared the scale means in the B room with the scale means in the U room. Only three of these comparisons were significant at the .05 level, and five were significant at the .10 level, results that are expected by chance. Twenty-three mean differences were in the predicted direction, and thirty-two mean differences occurred in the direction opposite the hypothesis, a finding that lends no evidence to either position. The "healthy" scale had mean differences in the hypothesized direction for six out of eight photos; the "pleasantness of facial expression" scale had mean differences in the opposite direction for six out of eight photos; and the remaining scales showed no trends.

A series of principle component and factor analyses with rotations were done on the nine task evaluation scales over the thirty Ss, identifying four orthogonal components. Factor scores were calculated for each S on the four rotated components, and comparisons of the mean task scores between rooms for each component were not significant (p > .05). Similarly, there was no difference in the average amount of time spent in the two rooms (p > .05).

Responses by Ss to the environment-description scales indicated profiles for the B and U rooms that were highly similar to those of Figure 11.2.

Discussion

The results of this study did not replicate the findings reported by Maslow and Mintz (1956). Photos of human faces were rated no differently when done in a beautiful or an ugly room. Subjects also did not spend more time in the B room than the U room while performing the experimental task, as reported in the original study. And although not addressed in that work, individuals in the present study indicated no differences in task enjoyment or other task-related characteristics that were accounted for by the B and U rooms. Thus the present research found no effects attributable to the different aesthetic impact of the two interior environments.

Methodological differences, which provided more control and eliminated some confounds in the present study, could also be responsible for the failure to replicate the earlier work. For example, aesthetic effects could have operated on or been mediated by social interaction between the E and the S. This represents a different phenomenon from aesthetic impact influencing the lone individual. Second, the same Es were used throughout the data-collection period, visiting the rooms on many occasions; in the present study, each S visited a room only once. Thus the variable of time or occasions is confounded with environmental-aesthetic impact on the S during the first occasion. Third, the Es alternated between rooms, evidently to counterbalance any possible experimenter effects. This procedure gives rise to the contrast effects that were described earlier. Finally, the potential symbolic aspect seems higher in the Maslow and Mintz study. The ugly room was so bad that it could not rationally have been regarded as a usual or typical environment by either the E or the S.

In the present study, all Ss visited the environment on the first occasion, no social interaction took place, no contrast effects occurred, and the aesthetics of the rooms offered little symbolic impact. That no significant effects occurred strongly suggests that the Maslow and Mintz effects were not due to the physical context per se, but occurred in interaction with one or more other variables. It is easy to speculate that the distaste held by the Es for the rooms (Mintz, 1956) was enhanced over time and by the contrast effects, and that this influenced both the social interaction between the E and the S, and the S's ultimate behavior when rating the photos. Perhaps it was the energy and well-being of the E's face that was being reflected in the ratings of each of the photo negatives!

Another difference between the two studies concerns the stimulus materials. Maslow and Mintz used negatives of photographs of human faces with no description of facial expressions, while the present study used black-and-white photos of individuals with natural facial expressions. Thus it might be suggested that the decreased stimulus structure of the negatives was more easily influenced, whereas the more clear-cut stimulus structure in the black-and-white photos could prove more resistant to environmental influences. However, three of the eight photos were selected precisely because of their ambiguity: Preassessment ratings indicated that Ss could not clearly agree on their descriptions of these photos. Data analyses indicated no substantial differences or trends in environmental effects when comparing the three low-structure photos as a group with the five photos of greater stimulus structure.

That the aesthetic impact of the physical environment is important goes without question. People go to great lengths and expense to surround themselves with that which appeals. To document with scientific rigor the behavioral influences attributable to these environments is a more difficult issue, but the present research suggests some possibilities. For example, using the B and U room paradigm, a good experimental bet might involve two subjects working together but without an E on a task of moderate attention or effort whose successful completion requires components of frustration resistance, creativity, and perhaps light-heartedness, as well as return to the working room for perhaps half a dozen occasions. Bundling social interaction over time with an appropriate "influenceable" task should provide a good chance of demonstrating the desired effects, particularly in light of the differences in both methodology and outcome of the present study and Maslow and Mintz.

12   The development of a usable lexicon of environmental descriptors

Joyce Vielhauer Kasmar

If people are expected to describe and distinguish among environments and architectural spaces, they will need tools to do so, and they will especially need a scale appropriate for the description of the physical environment.

It has been suggested that people are sensitive to and can respond to perceptual cues embedded within an environment – cues about the function of a given space, the type of people inhabiting an architectural structure, the behavior appropriate to the space, and so on. Any physical environment might be perceived as formal or informal, indicative of high or low status, warm or cold, public or private, and so on. References to this aspect of man's physical environment come from diverse sources. The architect Neutra (1954, 1956, 1958) refers to architecture not only as an instrument that caters to requirements and that shapes and conditions responses, but also as a reflector, or mirror, of conduct and living. After reviewing furniture arrangements found in the courtrooms of various countries, the law professor Hazard (1962) was led to conclude that it is not facetious to suggest that comparative-law scholars include the furniture arrangement in courtrooms within the scope of their study, for, as he noted, the arrangement will tell at a glance who has what authority. Redl and Wineman (1952) noted how sensitive otherwise defensive children are to the "atmosphere" suggested by architectural design, space distribution, furniture arrangement, and even the type of housekeeping. As the concept of milieu therapy has gained importance in the treatment of the mentally ill, psychiatrists and architects alike have begun to question the impression that both normal and deviant groups obtain as they approach a psychiatric hospital, as they step into the usually large and drab lobby, and so on. Their "introspective" conclusion is that the "message" generally is aweing, humbling, and even degrading (Baker, Davies, and Sivadon, 1959; Berger and Good, 1963; Osmond, 1959a, 1959b).

If people are indeed sensitive to and respond to the perceptual cues of architectural environments, then it behooves us to elicit their impressions. To do so, we

This paper was first published in *Environment and Behavior* 2 (1970): 153–8.

will need descriptive tools, and, especially, we will need to provide a scale appropriate for the description of the physical environment.

As Hall noted in 1966, there was still so much to learn about the emotional and meaningful aspects of space that the field was just in the process of being delimited. It should be obvious that if the physical and psychological aspects of architectural space or the physical environment could be identified, then it might be possible to construct a scale on which environments could be rated.

Although it may be tempting to borrow and use the semantic differential (Osgood, Suci, and Tannenbaum, 1957) as the descriptive tool, its use is at best a stopgap measure. The terms that compose it may or may not have relevance to the description of environments and may or may not have meaning for the user trying to describe architectural spaces. The results from a study that uses the semantic differential could be open to question, if not to meaningless, misleading, or erroneous interpretation.

If one ultimate aim of researchers is to ascertain the major dimensions of architectural space that are cogent for the description of space (Canter, 1969; Vielhauer, 1965) and finally to identify those architectural dimensions that are "central" and "peripheral" determinants of the perception of architectural space (Warr and Knapper, 1968), it will be necessary to identify the elements that might compose the various dimensions. Hence, it would seem helpful to begin with a relevant, meaningful, and unambiguous lexicon of spatial descriptors, rather than merely borrowing descriptors from other scales.

If the ultimate lexicon of architectural descriptors and terms making up an environment description scale are to be used by the public to describe environmental enclosures, then it seems most appropriate to elicit their architectural-descriptive language, in addition to the language of the professional, so that any final lexicon can have maximum meaning and relevance for the ultimate user – the public.

This research, then, had as its aim the beginning development of a lexicon of architectural descriptors that are relevant and meaningful and that can be used by nonarchitects to describe physical environments.

Stage 1. Initial search for architectural descriptors

Method

The first problem was to obtain a pool of adjectives thought to be descriptive of architectural space. There was no readily available source of such items. Therefore, questionnaires were used to elicit descriptive adjectives; additional descriptive terms came from architectural and interior-design magazines, as well as from previous research on the affective aspects of music, color and lines, art, and the theater (Farnsworth, 1954; Gordon, 1952; Hevner, 1935, 1936, 1937; Israeli, 1928; Lundholm, 1921; Poffenberger and Barrows, 1924; Ross, 1938; R. G. Smith, 1959, 1961; Springbett, 1960; van de Greer, Levelt, and Plomp, 1962; Wright and Rainwater, 1962).

Fifty-four undergraduate students (twenty-two males and thirty-two females) were asked to describe two rooms they liked and two rooms they disliked, listing the adjectives they believed to be descriptive of the four rooms. On completing this questionnaire, each student supplied the bipolar complement of each adjective that he or she had listed. Additionally, eleven fourth- and fifth-year architecture students, after completing the first questionnaire, completed a second questionnaire, which listed thirteen categories suggested by architects and designers as important in describing architectural space (size, volume, scale, mood, color, texture, function, illumination, aesthetic quality, climate, odor, acoustical quality, and miscellaneous). The architecture students listed descriptive adjectives appropriate to each of the thirteen categories and then supplied the bipolar complement of each adjective listed.

Results

Five hundred bipolar pairs of adjectives possibly relevant to the description of architectural space were obtained. These terms came from the free-response vocabulary of the nonarchitects as well as from the more technical vocabulary of architects, designers, and the like. To reduce the length of the adjective list to a more workable size, multiple linkages were eliminated. Any single word having a number of adjectives used as its complement was used only once. For example, the word *free* was linked by the respondents with *imprisoned, restrained, restricted, stifling,* and *confining.* The investigator and an additional judge arbitrarily determined the best of each multiple linkage pair to be used.

The original list was reduced in this fashion to the more workable number of 197 pairs of adjectives. These adjective pairs included terms referring to aesthetics, personal mood, and the physical characteristics of architectural space (Table 12.1).

Stage 2. Appropriateness of the adjective pairs to describe architectural space in general

Method

The 197 adjective pairs were rated on their appropriateness to describe architectural space in general. Twenty-three of the 197 adjective pairs were duplicated to assess the internal consistency of the ratings. Thus the total list presented for rating consisted of 220 pairs of adjectives. The entire list was divided into "words referring to esthetics and mood," 126 pairs, and "words referring to physical attributes," 94 pairs. Within the two sections of the list, the individual adjective pairs were printed in random order. Half of the Ss rated the adjective pairs referring to physical attributes and then rated the pairs referring to aesthetics and mood; for the other half of the Ss, the word order was reversed.

Ninety-two undergraduate students (forty-six males and forty-six females) served as raters. The Ss (1) had not participated in the earlier phases of the research, (2) were between the ages of seventeen and twenty-two, and (3) rated every adjective pair with four reasonable ratings (e.g., all ratings were not at the same point on the rating scale).

Each adjective pair (large–small, quiet–noisy, and so on) was rated on its appropriateness to describe architectural space in general. An 11-point rating scale, ranging from a rating of 1 – "extremely inappropriate" for the description of most physical environments – was used. The Ss also were given the option of using question marks to designate any word or pair of words whose meaning was unclear.

Results

Due to the constricted range of rankings for any one duplicated adjective pair, one reliability coeffecient was computed for the twenty-three duplicated pairs combined. The Pearson r of .81 for the total S sample is significant beyond the .01 level of confidence.

The average-ranked appropriateness of each pair of words, the amount of S agreement concerning the appropriateness of each adjective pair, and the clarity of meaning of the adjective pairs were computed and/or tabulated for each of the 220 pairs of adjectives, using respectively medians, interquartile ranges, and number of question marks.

The criteria for deleting adjective pairs were stringent. To be eliminated, an adjective pair (1) had a median at or below 6.00, the neutral point on the 11-point scale; (2) had an interquartile range of 5.00 or larger; or (3) had seven or more question-mark ratings. In addition, any adjective pair that had a median above 6.00 for all Ss combined, but for which the male or the female median was below 6.00 was eliminated. Most of the descriptive pairs had small interquartile ranges; thus the criterion of an interquartile range of 5.00 or larger was a *post hoc* empirical decision, signifying wide variability among Ss' ratings. The use of the 6.00 median criterion allowed the retention of any pair for which only 50% of the Ss gave ratings within the appropriate range.

Of the different appropriate pairs, 113 were retained. A total of 34 pairs of words were eliminated from the original list of 197 pairs of adjectives because they failed to meet one or more of the criteria for retention (Table 12.1). The majority of these pairs were deleted because of a median rating of less than 6.00 – that is, a median rating within the inappropriate range. Five pairs were deleted solely because the median rating of one sex was below 6.00, although the combined median was slightly above 6.00. Ambiguity or unclear meaning was the sole criterion for the elimination of only three pairs; it was, however, a concomitant criterion for the deletion of additional adjective pairs.

Stage 3. Appropriateness of the adjective pairs to describe specific architectural environments

Method

The 113 remaining adjective pairs now were rated on their appropriateness for describing specific environments. The appropriateness ratings progressed from the environment in general (Stage 2) to appropriateness ratings for a representative sample of environments (Stage 3), the logic being that an adjective pair might be appropriate for the architectural environment in general or for one type of environment and inappropriate for another type of environment.

Projected 35-mm colored slides of six interiors were used as stimuli for rating the appropriateness or inappropriateness of the adjective pairs. The slides, selected to present a representative sample of interiors in terms of function or use, included (1) Library, Watts Sherman House, Newport, Rhode Island; (2) Christ Church, Cambridge, Massachusetts; (3) Airport Terminal, St. Louis, Missouri; (4) Kitchen; (5) Executive Office, IBM, Pittsburgh, Pennsylvania; and (6) Bathroom.

A totally new sample of sixty-four undergraduate students (thirty-two male and thirty-two female) served as raters. The criteria for S selection were the same as the criteria in Stage 2.

Each adjective pair was rated on its appropriateness to describe each of the six specific environments presented pictorially. The format of the rating scale was identical to that used in Stage 2 – that is, an 11-point rating scale, with Ss again given the option to use question marks as a rating. The 113 appropriate adjective pairs plus 5 adjective pairs repeated to assess the internal consistency of the ratings were printed in random order on two pages. The rating booklet was made up of six complete two-page sets of the adjective pairs, or twelve pages. Within the six-set booklet, three of the two-page sets began with the first page of words followed by the second page of words, and in the other three of the six sets, the page order was reversed.

All Ss rated the appropriateness of the entire list of adjective pairs for one pictured environment before the next slide was projected. All Ss within the same testing session were presented with the same order of slides. A different order of slide presentation was used for each of the four experimental sessions required to collect the data from the sixty-four raters.

Results

The procedure for obtaining the internal-consistency reliability of the duplicated adjective pairs was the same as in Stage 2. Reliability coefficients were computed for the ratings on each specific environment as well as on the environments combined. All seven of the reliability coefficients were significant beyond the .01 level of confidence. The *r*s were as follows: environments combined, .91; library, .87; church, .95; airport, .90; kitchen, .92; office, .92; and bathroom, .92.

In Stage 3, the criteria for retention of adjective pairs were stringent. Only very appropriate, unambiguous adjective pairs were retained. Any adjective pair having 75% of all its ratings on appropriateness above 7.50, medians on all six of the environments at or above 7.00, little variability among Ss' ratings (interquartile ranges of less than 3.00), and no more than 3% question-mark ratings were retained.

It was interesting to note that irrespective of the room used as a stimulus, the medians, first quartiles, and interquartile ranges were fairly consistent, with little variation among the six pictured environments used. Perhaps this finding implies that these architectural descriptors are not environment-specific, but that they are appropriate or inappropriate for a diverse sample of architectural spaces.

Sixty-six appropriate adjective pairs remained after those failing to meet the criteria were eliminated. The sixty-six remaining adjective pairs are designated first in Table 12.1. The most stringent criterion for deletion of adjective pairs was that the first quartile be above 7.50. Forty-six of the forty-seven eliminated pairs included that as a criterion for rejection, and for *three* of the forty-six pairs, it was the sole reason for rejection of the pair. Only one adjective pair was discarded solely on the basis of too many question marks: decorated–stark.

Stage 4. Use of the sixty-six adjective pairs for environmental description

Method

The sixty-six remaining appropriate adjective pairs were put in scalar form (hereafter designated as the Environment Description Scale – EDS), and the EDS was used to obtain descriptions of "live" environments.

The sixty-six pairs of appropriate adjectives were set up as 7-point rating scales, with the bipolar terms (e.g., *large–small*) serving as the two ends of the continuum.

Three rooms were used as stimuli to be rated: the reserve reading room in the university library (RRm), a dining room in the student union (DR), and a large lecture hall (LH). Two of the three rooms were rated either by separate S samples (DR-students and DR-adults, LH students-1 and LH students-2, and LH students-2 and LH students-3) or by the same S sample at two different times (LH students-1 and LH students-3).

A new and independent sample of 100 Ss, 50 males and 50 females, rated each of the rooms. A total of 500 Ss served as raters. Since one S sample rated an environment on two separate occasions (LH students-1 and -3), a total of 600 room ratings were obtained, although only 500 Ss were used. The Ss for DR-adults were fifty male and fifty female adults, with a median age of thirty-nine and an age range from twenty-three to seventy years. The raters for the remaining rooms were university undergraduate students between the ages of seventeen and twenty-two years. The criteria for S selection were the same as the criteria in the earlier stages.

The sixty-six bipolarities were printed in random order on two pages. To counterbalance any possible order and fatigue effects, half of the Ss began with page 1 followed by page 2; for the other half of the Ss, the page order of the EDS booklet was reversed. While in the environment to be described, the Ss were given the test booklet, read the instructions, and completed their ratings of the environment using the sixty-six-item EDS.

Results

The stability coefficient of room ratings (LH students-1 and -3) was computed with the Pearson r. The retest followed after a three-week interval. Due to the constricted range of ratings for any one of the sixty-six adjective pairs, an overall stability coefficient was computed. The r obtained, .68, is significant at the .01 level of confidence, indicating stability in the Ss' ratings of the lecture hall over the three-week interval.

The average rating of a room on each adjective pair was calculated using medians. Sixty-six medians were computed for each room sample, a separate median for each bipolarity in the EDS. Since there were six room samples, six sets of sixty-six medians each were obtained.

The Mann-Whitney U test was used to compare the median ratings of the six room samples. It was found that the different rooms (RRm, DR, and LH) differed significantly from one another irrespective of the S sample rating the room. The reserve reading room was distinguished from both the dining room in the student union and the lecture hall, and the ratings of each of the latter rooms differed significantly from the other. All those differences noted among the three rooms were significant beyond the .01 level of confidence. The same room rated by different S samples (DR and LH) or rated by the same S sample at two different times (LH) did not differ significantly.

The room ratings were also compared using the Kolmogorov-Smirnov test. Here the cumulative frequency distribution of ratings for a bipolarity for one room sample was compared with the cumulative frequency distribution of ratings for the same bipolarity for every other room sample. It was expected that room samples that differed significantly on the Mann-Whitney U tests also would have a significantly greater than chance number of their sixty-six adjective pairs with significantly different cumulative frequency distributions. The results confirmed that expectation. The different rooms (RRm, DR, and LH) had a greater than chance number of the sixty-six bipolarities with significantly different cumulative frequency distributions. For different S samples rating a room or for the same S sample rating a room twice, a greater than chance number of the sixty-six bipolarities was not found to have significantly different cumulative frequency distributions.

Discussion

The results of this research represent the beginning of a workable and meaningful lexicon of architectural descriptors, a vocabulary that includes terms that are relevant and appropriate to describe architectural spaces, and terms that are understandable to the nonarchitects. The research also represents the beginning of an Environment Description Scale for use by the general public in describing architectural spaces. It is suggested neither that the lexicon of sixty-six adjective pairs is exhaustive nor that the EDS is a final scale. However, both the lexicon and the EDS represent beginnings. It should be possible for many researchers to use the sixty-six adjective pairs with confidence in their appropriateness.

Two words of caution appear to be in order, however. The Ss sampled in this research were fairly circumscribed and distinct. They were university students and adults associated with the university. The S sample probably was of above-average intelligence, and the results probably should not be generalized to any other population. Those presumably above-average-intelligence Ss in a free-response format produced many of the original descriptors, rated those descriptors on appropriateness, and used the sixty-six descriptive adjective pairs to distinguish among architectural spaces. Although the note of caution stands, the sixty-six adjective pairs have been used by other Ss of unknown intellectual level to differentiate among architectural enclosures (Kasmar, Griffin, and Mauritzen, 1968). Nonetheless, one might legitimately question if adults of average or even subnormal intelligence could understand all the descriptors in the final lexicon.

Of equal and perhaps even greater importance, the environment samples in

Stage 4 were public rooms – a room in a library, a public dining hall, and a lecture hall. To date, the scale has been neither used to assess, nor tested as a measuring device for, the description of private space. Its discriminant validity in this area is unknown. However, the research outlined suggests that the sixty-six adjective pairs, when used by Ss of above-average intelligence, will be relevant and appropriate for describing most architectural spaces. Nonetheless, any final definitive lexicon and scale cannot be considered complete until more environments are sampled. Only further research can answer the question if scales specific to classes of rooms are needed, or if the general EDS can be used meaningfully to distinguish among all different types of enclosures. Our next step would be to have a representative sample of Ss describe a representative sample of environments. Such a project might provide answers to the questions raised. With new and representative S samples, it might be advisable to reintroduce the question mark as a rating alternative. With the question-mark option, new S samples would have the opportunity to identify any descriptors that are unclear or ambiguous to them.

It certainly is possible that other researchers may not wish to be confined to just the sixty-six descriptive adjective pairs identified here. That there may be and probably are environment-specific descriptors that are not included on the final sixty-six-item list is accepted. Perhaps the results reported, however, can provide some guidelines for additional term selection. By judiciously selecting additional adjective pairs from the list provided, researchers could devise measuring tools that would have more relevance and appropriateness and greater clarity for the ultimate user than an armchair selection of terms would allow. This can be accomplished by selecting those terms having the least ambiguity and the most appropriateness – that is, from the beginning of Stage 3, in preference to those terms available at the beginning of Stage 2.

What are some of the implications of this research, and what are other possibilities for the future? First, given the beginning of a basic lexicon of environmental descriptors, perhaps ultimately it will be possible to refine that lexicon in order to identify the environmental dimensions or factors that are cogent and, also, those dimensions that are "central" and "peripheral" determinants of environmental perception. Ultimately, then, environments could be conceptualized in terms of N-dimensional models.

Second, an Environment Description Scale represents a usable tool. An EDS might be used as a tool to increase effective communication between architects and their clients. For example, the client using the EDS might specify the conceptual architectural qualities and characteristics that he desires. In addition, the client might use the EDS to rate the architect's physical models. Then, by correlating the client's conceptualizations with his ratings of the models, the architect would better understand the client's needs and desires. An EDS repre-

Table 12.1. *A lexicon of environmental descriptors (including descriptors retained and deleted)*

Descriptors retained

Adequate size–inadequate size	Huge–tiny
Appealing–unappealing	Impressive–unimpressive
Attractive–unattractive	Inviting–repelling
Beautiful–ugly	Large–small
Bright–dull	Light–dark
Bright colors–muted colors	Modern–old fashioned
Cheerful–gloomy	Multiple purpose–single purpose
Clean–dirty	Neat–messy
Colorful–drab	New–old
Comfortable–uncomfortable	Orderly–chaotic
Comfortable temperature–uncomfortable	Organized–disorganized
temperature	Ornate–plain
Complex–simple	Pleasant–unpleasant
Contemporary–traditional	Pleasant odor–unpleasant odor
Convenient–inconvenient	Private–public
Diffuse lighting–direct lighting	Quiet–noisy
Distinctive–ordinary	Roomy–cramped
Drafty–stuffy	Soft lighting–harsh lighting
Efficient–inefficient	Sparkling–dingy
Elegant–unadorned	Stylish–unstylish
Empty–full	Tasteful–tasteless
Expensive–cheap	Tidy–untidy
Fashionable–unfashionable	Uncluttered–cluttered
Flashy colors–subdued colors	Uncrowded–crowded
Free space–restricted space	Unusual–usual
Fresh odor–stale odor	Useful–useless
Functional–nonfunctional	Warm–cool
Gay–dreary	Well balanced–poorly balanced
Good acoustics–poor acoustics	Well kept–run down
Good colors–bad colors	Well organized–poorly organized
Good lighting–poor lighting	Well planned–poorly planned
Good lines–bad lines	Well scaled–poorly scaled
Good temperature–bad temperature	Wide–narrow
Good ventilation–poor ventilation	

Descriptors deleted at Stage 3

Busy–calm	e[a]	Interesting–uninteresting	e
Calming–upsetting	a,e	Livable–unlivable	a,b,e
Changeable–unchangeable	e	Lively–dull	e
Complete–incomplete	e	Long–short	e
Cozy–monumental	c,e	Natural–artificial	a,b,e
Cultured–uncultured	a,b,e	Nice–awful	e
Decorated–stark	c	Open–closed	e
Depressing–exhilarating	e	Pleasing–annoying	e
Dignified–undignified	a,e	Progressive–conservative	a,b,e
Dry–humid	e	Proportional–unproportional	e
Exciting–unexciting	a,e	Refined–unrefined	a,e

Table 12.1 (*cont.*)

Descriptors deleted at Stage 3 (cont.)

Expressive–unexpressive	e	Refreshing–wearying	e
Feminine–masculine	a,b,e	Restful–disturbing	e
Finished–unfinished	e	Romantic–unromantic	a,b,e
Formal–informal	e	Serene–disturbed	e
Friendly–unfriendly	a,e	Sociable–unsociable	a,b,e
Frilly–tailored	a,b,c,e	Soothing–distracting	e
Happy–sad	a,b,e	Sophisticated–unsophisticated	a,b,e
Hospitable–inhospitable	a,e	Stereotyped–unstereotyped	c,e
Hot–cold	b,e	Stimulating–unstimulating	e
Imaginative–unimaginative	b,e	Symmetrical–unsymmetrical	c,e
Impersonal–personal	e	Urban–rustic	a,b,c,e
Inspiring–discouraging	a,e	Versatile–nonversatile	e

Descriptors deleted at Stage 2

Active–passive	a	Meaningful–meaningless	a
Affected–unaffected	a	Mechanical space–nonmechanical space	a,c
Alive–dead	a,b	Mystic–nonmystic	a
Ascending color–receding color	a	No odor–strong odor	b
Coarse–smooth	a	Orthodox–unorthodox	b
Confused–clear	a	Plush–austere	c
Consonant–dissonant	a,c	Polished–unpolished	a
Content–discontent	a	Popular–unpopular	a
Coordinated–uncoordinated	b	Positive–negative	a
Dated–timeless	a,b	Pretentious–unpretentious	a,b,c
Deep–shallow	a	Real–phony	a
Defined space–undefined space	a	Rectilinear–curvilinear	a,b,c
Definite volume–indefinite volume	a	Regular–irregular	a
Directed–undirected	a,c	Related–unrelated	a
Downward scale–upward scale	a,c	Relaxed–tense	a,b
Dynamic space–static space	a,c	Reputable–disreputable	a
Encouraging–discouraging	a	Reserved–uninhibited	d
Euphonious–diseuphonious	c	Resonant–flat	a,c
Even texture–uneven texture	a	Restrained–unrestrained	a
Familiar–unfamiliar	d	Restricted–unrestricted	a
Fatiguing–invigorating	d	Reverent–irreverent	a
Flexible–rigid	a	Rhythmic–unrhythmic	a
Formed–formless	a	Rich–poor	b
Fragile–sturdy	d	Rickety–stable	a,b
Gentle–brutal	a	Scenic–unscenic	a,c
Glaring–unglaring	a	Sectionalized space–undifferentiated space	a,c
Good–bad	a	Sedate–flamboyant	c
Good odor–bad odor	d	Sensitive–insensitive	a
Graceful–clumsy	b	Sensual–prim	a,c
Hard–soft	a,b	Serious–humorous	a
Hard texture–soft texture	a	Shaped–shapeless	a
Harmonious–discordant	c	Sharp–blunt	a
Healthy–unhealthy	a		
Heavy–light	a		

Table 12.1 (*cont.*)

Descriptors deleted at Stage 3 (cont.)			
Heterogeneous–homogeneous	a,b,c	Sincere–insincere	a
High–low	a,b	Spiritual–nonspiritual	a
Honest–dishonest	a	Sterile–filthy	b
Horizontal volume–vertical volume	a,c	Strong–weak	a
Human scale–inhuman scale	a,c	Structured–unstructured	a
Inner-directed–outer-directed	a,c	Temporary–permanent	a
Lazy–energetic	a,b	Textured–untextured	a
		Threatening–unthreatening	a
		True–false	a
		Valuable–worthless	a
		Varied–repetitive	a

[a]Reason for elimination of descriptor: (a) low median, (b) wide interquartile range, (c) excessive question-mark ratings, (d) median sex difference – Stage 2, (e) low Q^1 – Stage 3 (see text for more complete explanation).

sents a potentially relevant and meaningful tool for assessing changes in the environment. Also, one could assess the interaction between environmental characteristics, as described on an EDS, and the behavior occurring within the rated environment. In addition, since Osmond (1959b) has suggested that schizophrenics differ from normals in their perception of distance and that they inhabit a space shaped very differently from the space that normal people perceive, an EDS could be used to delineate the physical environment as it is perceived by schizophrenics. It then might be possible to manipulate that environment to make it more a part of the therapeutic treatment program (Baker et al., 1959; Berger and Good, 1963; Kling, 1959; Osmond, 1959a, 1959b; C. W. Smith, 1959). Ultimately, using such a scale, it might be possible to assess children's perceptions of environments and the changes that occur developmentally. And, in addition, characterization of environments of different cultures and subcultures might be studied. These are only some of the implications. Only further research by many from diverse disciplines will elucidate other uses and problem areas.

13   Lighting-design decisions as interventions in human visual space

John E. Flynn

"Meanings" associated with lighting systems

One theme that seems to have recurred periodically in discussions of lighting is the idea that some lighting designs communicate impressions of meaning to the users of the space. Some of these ideas are traceable to Hesselgren's (1969) writing on this subject, and other researchers (Martyniuk, Flynn, Spencer, and Hendrick, 1973) have also stressed this theme.

The theory referred to here is the idea that some psychological aspects of lighted space can be recognized and perhaps documented if we are prepared to discuss and study lighting design as an exercise in visual communication. This theory suggests that we are dealing in part with a system of visual cues that tend to be recognized and interpreted in somewhat consistent ways by users who share cultural values and background. It further suggests that as designers change lighting modes (i.e., the character of patterns of light and color in the room), they change the composition and relative strength of visual signals and cues, and this, in turn, alters some shared impressions of meaning for the room occupants. While many lighting systems are intended to function in a "permissive" way (i.e., to permit performance or participation in some activity that involves vision, without attempting to influence user behavior or impressions), there is evidence that other lighting designs may function more actively as selective intervention in human visual experiences.

Concerning the latter category, several examples may be useful here.

The effect of light on user orientation and room comprehension

Some light patterns seem to affect personal orientation and user understanding of the room and its artifacts. For example, spot-lighting or shelf-lighting affects user attention and consciousness; wall-lighting or corner-lighting affects user

This paper was presented at the meeting of the International Commission on Illumination, Study Group A, Montreal, Canada, August 1, 1974.

156

understanding of room size and shape. Considered as a system, these elements establish a sense of visual limits or enclosure.

The effect of light on impressions of activity setting or mood

Other lighting patterns seem to involve the communication of ideas or impressions. In this sense, we are coming to recognize spatial lighting patterns as part of a visual language that can assist the designer in implementing such impressions as somberness, cheerfulness, playfulness, pleasantness, and tension. Similarly, the designer can use light patterns to affect such psychosocial impressions as intimacy, privacy, and warmth. I am noting that we can use lighting in one way to produce a carnival-like atmosphere, and we can use it in another way to produce a somber place for meditation. We can use lighting to produce a cold, impersonal, public space; or we can use light to help reinforce an impression of a warm, intimate place where users feel a greater sense of privacy.

Notice that we are discussing more than aesthetic amenities here because these impressions or moods are often fundamental in satisfying some experience and activity requirements in buildings. This discussion implies a very broad definition of the function of a lighting system. And in this sense, perhaps a major contribution is the perspective that functionalism (in lighting design, as in other things) cannot be limited to ideas associated only with concepts of utility, task, or physiological needs. Unless we are willing to write off the sensitivities and the psychosocial needs of the users, we should represent the idea that a truly functional lighting design must come to include the qualitative ideas of user attitude, well-being, and motivation.

Nonmathemetical guidelines for lighting designs

With these background ideas in mind, I will attempt to discuss in a preliminary way one category of study as an example. In this example, I must draw partially on work that my associates and I have been undertaking (with some assistance from the Illuminating Engineering Research Institute in New York).

Briefly, we have been testing the previously discussed lighting-cue theory by accumluating data through semantic differential scaling and multidimensional scaling techniques (Flynn, 1972–73, 1973; Flynn, Spencer, Martyniuk, and Hendrick, 1973, 1974; Martyniuk et al. 1973). To date, this work is suggesting that in our North American society and culture, there are at least six broad categories of human impression that can be influenced (cued) or modified by the lighting design:

1. Impressions of Perceptual Clarity
2. Impressions of Spaciousness

3. Impressions of Relaxation and Tension
4. Impressions of Public versus Private Space (Prominence or Anonymity)
5. Impressions of Pleasantness
6. Impressions of Spatial Complexity (sometimes Liveliness)

In addition, we have identified instances in which the lighting design may present cues that serve to guide some categories of behavior, such as attention, selection of path in circulation, selection of seating, and seating position. But space limitations require that we omit discussion of this aspect here, limiting our discussion to the six categories of impression.

We find the possibility that we may begin to "measure" some previously unmeasured lighting influences – at least with relative or comparative measurements. In attempting this, we may find that it is possible to identify specific design rules to guide decision making in these presently unmeasured aspects of lighting design.

I will demonstrate the current status of such "relative measurements" by using semantic differential scaling and multidimensional scaling data in several ways:

1. In an attempt to demonstrate the possibility of achieving some nonmathematical design guidelines for this aspect of lighting;
2. In an attempt to demonstrate the consistency of some lighting-cue effects; and
3. In an attempt to demonstrate some ideas concerning potential forms for presentation of nonmathematical performance information in this area.

Discussion under these three headings follows as a primary focus of this paper.

An attempt to demonstrate the possibility of achieving some nonmathematical design guidelines for this aspect of lighting

A basic cornerstone in the theory that is developing here is the idea that in discussing lighting design, we are dealing in part with a system of visual cues that can be recognized and interpreted in consistent ways by those who share a cultural background. I am further suggesting that semantic differential scaling and multidimensional scaling can provide useful patterns that can guide decision making in lighting design.

One set of patterned guidelines is demonstrated by plotting semantic differential data within a "volume" that represents the lighting. Clear-cut examples have evolved from our research (Flynn, 1972–73, 1973; Flynn et al., 1974), and they are shown in Figures 13.1 to 13.5.

Notice that various lighting systems are subjectively categorized by three major dimensions or modes of lighting. In this, our multidimensional scaling seems to consistently confirm a bright–dim dimension or mode, an overhead–peripheral dimension or mode, and a uniform–nonuniform dimension or mode.

Notice further that the composite multidimensional scaling and semantic differential plots in Figures 13.1 to 13.5, together with the multiple regression analysis shown, suggest a sense of priorities for the designer – indicating specific design decisions that are important for reinforcing various semantically defined objectives through illumination design.

When impressions of "relaxation" are of concern, for example, Figure 13.3 suggests that the designer stress (1) peripheral lighting and (2) a nonuniform mode. Of these, the multiple regression coefficients indicate that the peripheral effect is a particularly important design decision for reinforcing impressions of relaxation through spatial lighting. Similar design guidance can be gained from the other diagrams shown. There seems to be a fourth modifier – color tone of light – but in our research, it is not presently clear if it will turn out to be a fourth multidimensional scaling dimension with a magnitude of importance similar to the three shown in Figures 13.1 to 13.5. This question is presently under study.

The same data can be reassembled to produce a second set of patterned guidelines, demonstrated in Figures 13.6 to 13.8. This time, comparative semantic differential data are plotted in a way that provides insight regarding economic and energy-budgeting tradeoffs that confront the lighting designer. Several examples are shown.

An attempt to demonstrate the consistency of some lighting-cue effects

Once again, the theory developing here suggests that as designers change lighting modes (i.e., the patterns of light in the room), they change the composition of relative strength of visual signals and cues, which, in turn, alters some shared impressions of "meaning" for the room occupants.

In a further attempt to test this theory, three rooms were tested simultaneously, with independent groups of subjects using 7-point bipolar semantic differential rating scales. In this work, the three rooms varied significantly in size and shape, although the functions were similar because all were arranged as conference rooms. One room (A) was medium-sized and irregular in shape; the second room (B) was large and rectangular; and the third room (C) was medium-sized and rectangular. In each room, however, similar lighting modes were utilized:

Setting 1: overhead fluorescent (CW), 40 f. c. horizontal intensity on table tops;
Setting 2: only four walls illuminated, no overhead illumination;
Setting 3: combined setting 1 + setting 2;
Setting 4: overhead incandescent downlights, 8 f. c. horizontal intensity on table tops;
Setting 5: setting 4 + one wall only illuminated.

Each of the lighting modes was individually controlled by multiple switches (often dimmer controlled) so that intensities, color tone, and distribution of light were consistent among the rooms. Our objective was to explore the applicability of the previously shown nonmathematical guidelines (Figures 13.1–13.5) to test the idea that these lighting modes might communicate or cue impressions in a manner that is somewhat independent of the configuration and décor of the room. In this sense, we hoped to explore the possibility that these impressions might be broadly shared by a significant proportion of the users (at least by a significant proportion of transient users).

Figure 13.9 shows the generally consistent patterns of relative response among settings in the three rooms. Notice that the patterns of shift in averaged ratings among settings are generally consistent with one another and with the findings in Figures 13.1 to 13.5, except for a seemingly significant inconsistency in the ratings of "pleasantness" and "relaxation" for the downlighting (setting 4) in room "A". But study of the physical spaces involved suggests that even these seemingly erratic downlighting patterns are consistent with the nonmathematical guidelines shown in Figures 13.1 to 13.5. Because the room in question ("A") is a nonrectangular, the stray and reflected light from the centrally placed down-lights does not light the peripheral walls to the extent found in rectangular rooms "B" and "C." Since the walls in room "A" have become darker than those in rooms "B" and "C," the resulting shift toward more "unpleasant" and more "tense" ratings seems consistent with the guidelines in Figures 13.1 to 13.5. Note that Figures 13.1 to 13.5 indicate the major importance of peripheral-lighting patterns when these categories of impression are being considered. In fact, these apparent inconsistencies in Figures 13.6 to 13.8 actually reinforce the theory that the cued impressions are correlated with the spatial patterns of light, not with the lighting systems as such.

Figures 13.6 to 13.8 seem to document apparent parallels in visual experience by the different groups in the three rooms, thus tending to support the theory that lighting design deals in part with "cue patterns" that may be recognized and interpreted in consistent ways by the users. We have been testing this theory further, continuing to find evidence that supports the consistency of the general patterns. (In subsequent studies, we are also finding evidence of consistent modifying influences involving glare, color tone, and articulation of form [Flynn et al., 1974].)

An attempt to demonstrate some ideas concerning potential forms for presentation of nonmathematical performance information in this area

Ultimately, this type of study must consider the question of performance recommendations and standards. In this, we find that the semantic differential data may be replotted in still another form – as "polar diagrams." The diagrams in Figure 13.10 show shifts in composite user impression that are attributable to the lighting design, since the lighting was again the only variable in the situation that produced the data (Flynn et al., 1974).

In this study, we are finding some broadly distinctive patterns or profiles that might facilitate further nonmathematical description and recognition of various lighting-design conditions. Notice that each of the lighting systems exhibits a somewhat distinctive (and distinguishable) form and size, with the profile shape expanding, contracting, or shifting in position relative to a "neutral" diamond. Limited studies indicate similar shifts in relative shape, size, and position in different rooms when similar lighting systems are utilized.

It is possible that diagrams of this (or similar) type could ultimately be developed to assist a designer in selecting lighting systems or system combinations in order to provide an "appropriate" light setting or background for various categories of human activity.

The intention here is to suggest the possibility of nonmathematical design standards for some aspects of lighting design. In use, these might be similar in nature to Sound Transmission Class (STC) curves or Noise Criteria (NC) curves in acoustics, with which the designer attempts to approximate a performance curve in order to achieve specific design-performance objectives.

Conclusion

The work presented here is intended to articulate the need for lighting engineers and designers to be sensitive to more than limited task-oriented standards designed to permit reading, sewing, shaving, cooking, drafting, bookkeeping, and similar visual task activities. We should be sensitive to the psychological uses of focal emphasis, color tone, silhouette, sparkle, and other patterns of spatial light. And we should be sensitive to the fact that the correct use of these patterns is fundamental in satisfying some space-activity requirements, such as when we need to reinforce attraction or attention; when we need to strengthen attitudes of relaxation; when we need to stimulate sensations of spatial intimacy or warmth; or when we want to reinforce impressions of cheerfulness or playfulness.

Recognizing that we are discussing a sketchy and incomplete line of inquiry, I nevertheless believe that a pattern of data is beginning to emerge, suggesting that recently developed methodologies in the behavioral sciences may offer important insights into the implications of various nonquantitative decisions that are made in lighting design.

Acknowledgments

I would like to acknowledge the work of associates in the development of much of the work that is referred to in this paper – Dr. Terry Spencer and Dr. Clyde Hendrick in the Department of Psychology at Kent State University, and Mr. Osyp Martyniuk in the School of Architecture at Kent State University. I am also indebted to my associates and students in the Department of Architectural Engineering at The Pennsylvania State University – particularly Mr. Eui Il Lim, who prepared the illustrations for this paper.

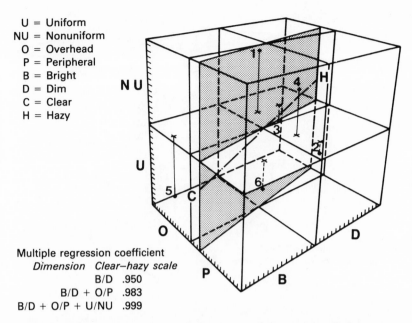

U = Uniform
NU = Nonuniform
O = Overhead
P = Peripheral
B = Bright
D = Dim
C = Clear
H = Hazy

Multiple regression coefficient
Dimension   Clear–hazy scale
B/D   .950
B/D + O/P   .983
B/D + O/P + U/NU   .999

Figure 13.1. Indicated lighting-design decisions for affecting impressions of Perceptual Clarity.

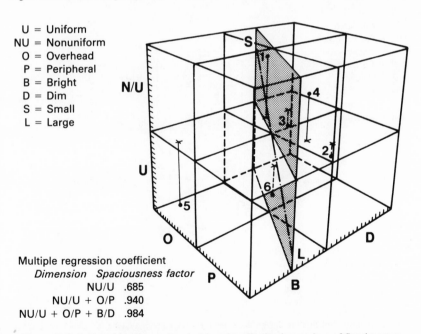

U = Uniform
NU = Nonuniform
O = Overhead
P = Peripheral
B = Bright
D = Dim
S = Small
L = Large

Multiple regression coefficient
Dimension   Spaciousness factor
NU/U   .685
NU/U + O/P   .940
NU/U + O/P + B/D   .984

Figure 13.2. Indicated lighting-design decisions for affecting impressions of Spaciousness.

164   JOHN E. FLYNN

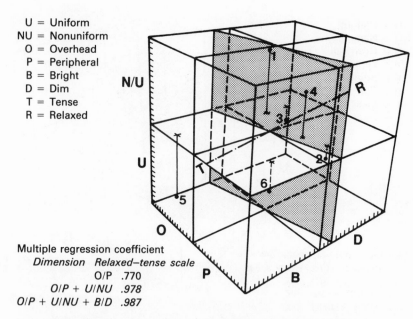

U = Uniform
NU = Nonuniform
O = Overhead
P = Peripheral
B = Bright
D = Dim
T = Tense
R = Relaxed

Multiple regression coefficient
Dimension Relaxed–tense scale
           O/P   .770
      O/P + U/NU   .978
  O/P + U/NU + B/D   .987

Figure 13.3. Indicated lighting-design decisions for affecting impressions of Relaxation.

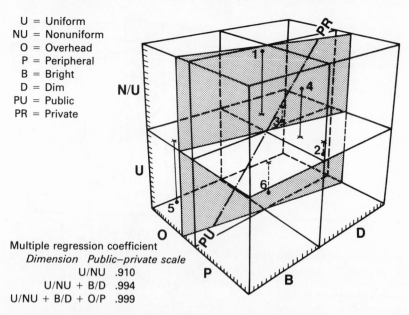

U = Uniform
NU = Nonuniform
O = Overhead
P = Peripheral
B = Bright
D = Dim
PU = Public
PR = Private

Multiple regression coefficient
Dimension Public–private scale
          U/NU   .910
      U/NU + B/D   .994
  U/NU + B/D + O/P   .999

Figure 13.4. Indicated lighting-design decisions for affecting impressions of Public versus Private Space.

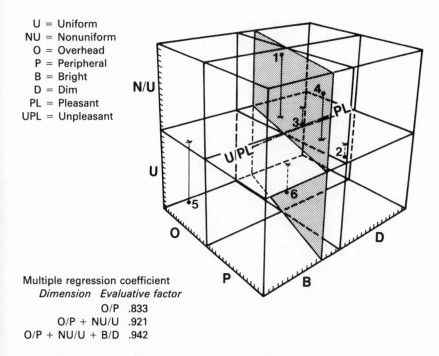

U = Uniform
NU = Nonuniform
O = Overhead
P = Peripheral
B = Bright
D = Dim
PL = Pleasant
UPL = Unpleasant

Multiple regression coefficient

Dimension	Evaluative factor
O/P	.833
O/P + NU/U	.921
O/P + NU/U + B/D	.942

Figure 13.5. Indicated lighting-design decisions for affecting impressions of Pleasantness.

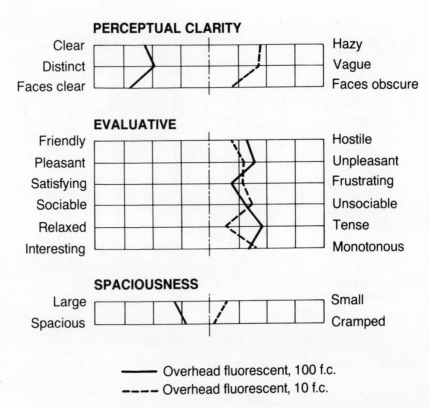

- Overhead fluorescent, 100 f.c.
- Overhead fluorescent, 10 f.c.

Figure 13.6. Study of variations in intensity of horizontal illumination. Overhead fluorescent lighting is compared, with all room factors identical except the intensity of light emitted from the luminaires. The systems vary significantly for Perceptual Clarity and Spaciousness. Differences on Evaluation are marginal.

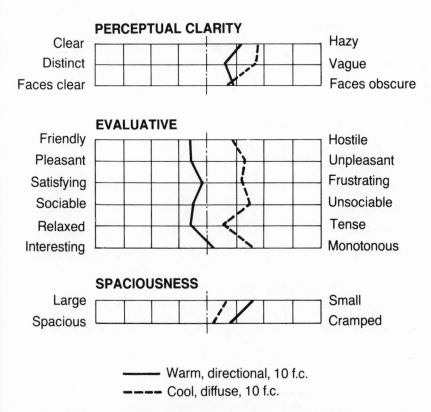

Figure 13.7. Study of variation in distribution of illumination from low-intensity systems. Warm-toned directional lighting and cool-toned diffuse lighting are studied for their suitability for low-intensity installations. The systems show marginally significant differences in Perceptual Clarity and Spaciousness, and significant differences on Evaluation, suggesting that the distribution of light may be critical in affecting acceptance of energy-budgeted systems.

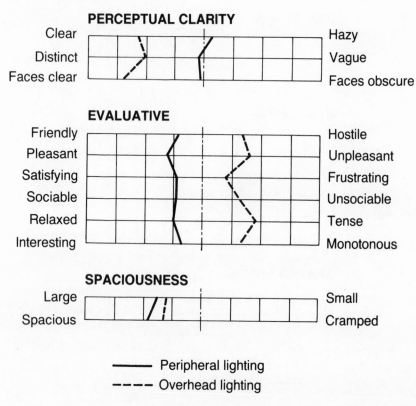

Figure 13.8. Study of equal-wattage design alternatives. Overhead fluorescent and peripheral fluorescent systems are compared, with all room factors identical except the distribution of lighting watts. (The two systems consume approximately equal wattage.) The overhead system produces the better impression of Perceptual Clarity; but the peripheral system produces the better Evaluative impressions. Thus there may be correct and incorrect ways to accomplish lighting energy budget, depending on the precise needs of the space and on the activity involved.

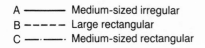

A ——————— Medium-sized irregular
B — — — — — Large rectangular
C —·——·— Medium-sized rectangular

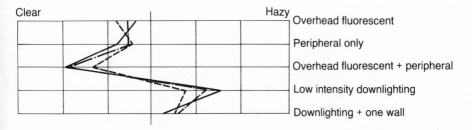

Clear ... Hazy

Overhead fluorescent

Peripheral only

Overhead fluorescent + peripheral

Low intensity downlighting

Downlighting + one wall

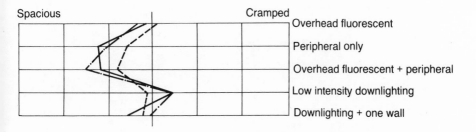

Spacious ... Cramped

Overhead fluorescent

Peripheral only

Overhead fluorescent + peripheral

Low intensity downlighting

Downlighting + one wall

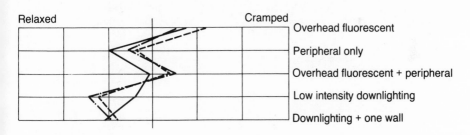

Relaxed ... Cramped

Overhead fluorescent

Peripheral only

Overhead fluorescent + peripheral

Low intensity downlighting

Downlighting + one wall

Figure 13.9. Patterns of response among settings in the three rooms.

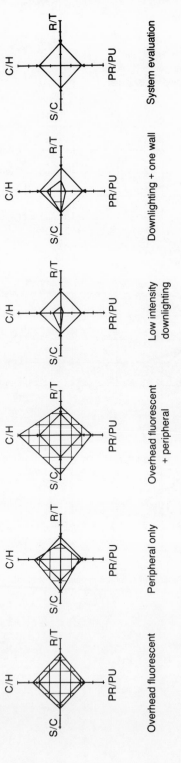

Figure 13.10. Semantic differential replotted as ''polar diagrams.''
Key: C/H = Clear/Hazy; R/T = Relaxed/Tense; PR/PU = Private/Public; S/C = Spacious/Cramped.

C. Architectural exteriors

Editor's introduction

The four papers in this section are concerned with the study of the visual quality of architectural exteriors seen alone or in relation to their immediate surroundings. If architects and the public differ in their interpretation of buildings, then design decisions by architects may fail to communicate desirable meanings to the public. Thus the question of similarities and differences between architect and public responses to buildings is an important one. In an early and throrough examination of this question, Hershberger (first paper in this section) documents several important differences in the meanings attributed to buildings by architects (graduating thesis students, undergraduate pre-architecture majors, and undergraduate architecture majors) and laypersons. Notably, the groups differed in assessments of aesthetic quality and Novelty–Excitement. The evident differences between the groups suggest a need for research aimed at informing architects about how the public interprets buildings. Central to such research is the development of an instrument for assessing architectural meaning.

In the second paper, Hershberger and Cass extend Kasmar's lexicon for use in assessing building exteriors. With separate investigations of response to houses, commercial buildings, and institutional buildings, the authors derive a set of ten factors and related adjective scales (ten primary and ten secondary) for the assessment of architectural exteriors. One can select adjectives representing these factors with some confidence that the adjectives are relevant to the assessment of many exteriors.

Because of potential complications in transporting respondents to buildings in order to obtain evaluations, it is useful to know what simulations are likely to produce valid data. Hershberger and Cass also examined this question. Through examining responses to buildings on site and to various media presentations of buildings, they confirmed earlier findings that responses to color film and color slides of buildings are good predictors of responses on site.

In the third paper, Oostendorp and Berlyne report the results of a series of studies on the relationship between visual features of architectural exteriors and subjective appraisals of those exteriors. Using similarity ratings of color pho-

173

tographs of architecture from around the world and multidimensional scaling, they found four perceptual dimensions that subsequent comparisons to stylistic and technical ratings revealed as Flamboyancy (straight vs. curved), Monumentality (verticality and functional expression), Conventionality (aesthetic vs. nonaesthetic norms), and Decorativeness (composition and color). Then factor analysis of affective and collative ratings of photographs yielded three affective and collative factors of response: Hedonic Tone/Arousal, Order, and Tension. When the authors examined the relationship between the dimensions of style and form and the affective and collative ratings, they found that Hedonic Tone/Arousal correlated with Monumentality and that Order correlated inversely with Flamboyancy.

In the last paper, Groat examines one aspect of architectural order: the compatibility (or contextual fit) of buildings to their surroundings. Compatibility can be seen as an ordering factor. Groat both identified architectural influences on perceived compatability of buildings and examined preference in relation to compatibility. The patterning of façade elements was a major determinant of perceived compatibility; and not unexpectedly, preference increased with compatibility.

14   A study of meaning and architecture

Robert G. Hershberger

There are few forms in architecture to which people do not attach some meaning by way of convention, use, purpose, or value. This includes the very mundane realization that a wood panel approximately 3 feet wide by 7 feet high is a *door* (object that one opens to pass through), the more subtle feelings of warmth and protection at the entrance to some buildings, and some of the most profound experiences of beauty and art. Indeed, the transmission of meaning through the architectural medium is essential to both the use and the enjoyment of architecture. Meaning is of considerable importance in perception (Creelman, 1966), "one of the most important determinants of human behavior" (Osgood, Suci, and Tannenbaum, 1957, p. 10), and unquestionably involved with human feelings. Furthermore, it has been argued by architects and planners alike that in an increasing number of situations, the underload, overload, or confusion of meaning in architecture has become such that both the use and the enjoyment of architecture are seriously jeopardized (Brown, undated; McHarg, 1962). In consequence, it would seem appropriate at this time to undertake serious studies of the nature of architectural meaning to learn what is needed to create physical environments that can be satisfactorily perceived, felt, and used.

The research reported here was addressed to this problem, taking the point of view that the forms, colors, spaces, and other qualities of architecture are media through which architects communicate to the users of their buildings, and focusing on the "fidelity" of this communication. Is there a close correspondence between the meanings that architects intend for buildings and the meanings that laymen attribute to them? Do architects and laymen share similar representations when they experience architecture? Are they affected in the same way by their representations? Are their resultant evaluations and behavior similar?[1]

It was decided that the most straightforward and effective way to approach this problem was to conduct an experiment in which architects and laymen would be

This paper was first published in H. Sanoff and S. Cohn (Eds.), *EDRA 1: Proceedings of the First Annual Environmental Design Research Association Conference* (Raleigh: North Carolina State University, 1969), pp. 86–100.

directly compared in their attribution of meaning to buildings. The primary objectives of the experiment were (1) to determine if the physical attributes of buildings can be considered to constitute a "code" capable of communicating architects' "intentions" to the users of their buildings, and (2) to determine if the areas of disagreement, if any, should be attributed to the professional education of the architects. These objectives were translated into three experimental hypotheses: (1) Architects and laymen will not differ greatly in the underlying dimensions of meaning used to judge architecture; (2) differences in specific judgments of meaning will occur most often on the affective and evaluative dimensions and least often on the representational dimensions;[2] and (3) the differences that are found (dimensional and specific) will be attributable to the professional education of the architects.

Method and procedure

The experimental design required each of four respondent groups to rate the connotative meanings of twenty-five building aspects (represented by colored slides) on thirty semantic scales.

Respondent groups

The experiment utilized three groups of twenty-six students each from the University of Pennsylvania as respondents: (1) the graduating thesis students in architecture, (2) a group of pre-architects, and (3) a random sample of non-architects. It also utilized a group of twenty-one architectural students from Drexel Institute of Technology.

There were several reasons for selecting the initial three groups over all other groups that might have been selected: (1) The choice of three student groups from the same institution allowed the testing of the hypothesis that differences that might be found between architects and non-architects could be attributed largely to the professional education of the architects. Thus the pre-architect group served as a control group whose ratings, if similar to the non-architects, would tend to confirm the hypothesis or, if similar to the architects, would tend to discredit the hypothesis. (2) The selection of the student groups from the same institution tended to ensure a reasonable similarity in social background, pre-professional education, intelligence, and age. This was essential not only to allow the hypothesis to be tested, but also to ensure, aside from professional education, that there would be no marked difference among the three groups in their understanding of the various words used in the instructions and rating scales. (3) It was also possible, by selecting three groups from the same university, to control a number of other factors that might have confounded the results. For instance, it was possible to obtain approximately equal prior exposure to the "building aspects" used in the experiment and to conduct the experiment at one time and place with equal conditions for all three groups. In this way, outside factors – such as differential instructions, time of day or week, weather, and the like – could not enter to cause differences among the responses of the three groups. The architectural students from Drexel Institute of Technology were included in an attempt to determine if differences between the Penn Architects and the other two groups should be attributed specifically to the education of architects at the University of Pennsylvania or, rather, if they could be attributed to more general characteristics of the architectural education.

Building aspects

In order to select a group of building aspects that would ensure a full exploration of the dimensionality of connotative meaning applicable to architecture, the following steps were taken: (1) Criteria were formulated for the selection of buildings and building aspects; (2) a relatively large number of buildings that seemed to meet the criteria were selected, and several aspects of each were photographed; and (3) the resulting slides were then compared with one another and with the criteria in an effort to reduce the number of buildings and aspects to the minimum that would meet the criteria.

The criteria were as follows: (1) The aspects selected must represent both humanly used and unused parts of buildings.[3] (2) a wide range of building types, sizes, styles, aspects, and qualities must be represented; (3) members of the respondent groups should have seen or used a majority of the buildings prior to the experiment (to ensure that the dimensionality of meaning uncovered would relate to the everyday use and experience of architecture); and (4) the prior use and experience of the buildings chosen should be approximately equal for the various respondent groups (to ensure the differences in factor structure or concept meaning could not be attributed to differential familiarity with the selected buildings). The majority of buildings selected were from the campus of the University of Pennsylvania. Other buildings were selected only when it was felt that the campus buildings could not fulfill the criteria. In such cases, buildings from Philadelphia, well-known and -visited international buildings, or equally unknown buildings were substituted.

The measuring instrument

The semantic differential, a general measuring technique developed by Osgood et al. (1957) to measure connotative meaning, was selected to obtain judgments of meaning from the various respondent groups on the architectural material. Briefly, the semantic differential utilizes a number of scales consisting of polar adjectives, such as good–bad, strong–weak, and active–passive, to differentiate the meaning of concepts (in this case, aspects of buildings). The scales are divided into seven steps, as follows:

strong____: ____: ____: ____: ____: ____: ____: weak

Each subject is asked to consider the concept (building aspect) and place a check in the blank in which he or she feels the meaning of the concept (building aspect) lies. From left to right on the example, a check in the blank would indicate: extremely strong, quite strong, slightly strong, neither strong nor weak (or not applicable), slightly weak, quite weak, extremely weak. Each concept is judged on several such scales, the scales being varied to suit the subject material.

Semantic scales

For this experiment, the attempt was to generate a "logically comprehensive" list of adjective pairs applicable to architecture, as was done by Vielhauer (1965) for the environment. To accomplish this, reference was first made to the extensive lists of adjective pairs generated for previous studies that utilized the semantic differential (Osgood et al., 1957). Additional adjectives and adjective pairs were obtained by referring to an assortment of architectural books and periodicals, to *Roget's Thesaurus*, to an adjective checklist developed by Craik (1966), and to adjective lists developed in studies by Canter (1969) and Collins (1969). Finally, the list generated was given to seven of the Ph.D. candidates in architecture at the University of Pennsylvania in an attempt to generate polar opposites for a group of adjectives for which no appropriate opposites had been provided and to add any additional adjective pairs that were thought to be appropriate to architecture, but were absent from the list. The extensive list of adjective pairs generated in this manner was reduced to the total of thirty pairs used in the experiment by grouping them into ten major categories of meaning and selecting the three most diverse scales from each of the categories. In this way, enough scales were provided from each major category to create a separate dimension in the factor analytic study *if* the meaning of the three adjective pairs selected from that category was, in fact, used by the respondent groups to have nearly the same meaning for architectural subject material. At the same time, the selection of the adjective pairs with the most diverse meanings from each category tended to ensure enough dissimilarity of meaning to prevent the automatic occurrence of all ten categories as dimensions of meaning.[4]

The experiment

The judgments of meaning over all aspects were obtained from the three respondent groups from the University of Pennsylvania during a ninety-minute session in a large lecture room in the Fine Arts Building. The three respondent groups were seated systematically in the room to ensure that effects of sound or vision would not differentially influence the results. The judgments of the Drexel Architects were obtained in a similar manner during two scheduled design studios.[5]

Phase I: Dimensionality of judgments

The first concern in comparing the four respondent groups was with respect to similarities or differences in the dimension used to rate the architectural subject material. The basic procedure was to subject the data obtained from each respondent group to a separate factor analysis, and then to compare the four sets of factors obtained.[6] A factor analysis, in effect, "groups" scales that are consistently used in the same way by members of a respondent group into "factors" or "dimensions." For example, if building aspects rated "complex" were consistently rated "interesting," these two scales would be highly correlated. If several such scales were highly correlated, factor analysis would group them around a common factor or dimension.

The comparison of the four respondent groups at this level was of fundamental importance because differences in the dimensions of meaning used by architects and laymen would involve total disagreement – the absence of a "common frame of reference" with respect to architectural subject material. In this event, further comparisons among the groups would become spurious, since the groups, in effect, would not have a common language.

Comparison of factors

Three common factors, together accounting for approximately 50% of the total variance, were obtained for each respondent group. Factor loadings over .40, by factor and respondent group over all aspects, are shown in Table 14.1. The first factor for each respondent group was clearly a Space–Evaluation factor, displaying high and generally restricted loadings on such scales as spacious–confined, open–closed, and loose–tight, and on such evaluative scales as cheerful–gloomy, pleasing–annoying, and welcoming–forbidding.

The second factor was considered to be an Organization dimension for the three groups from the University of Pennsylvania, because it had high and restricted loadings on such scales as ordered–chaotic, clear–ambiguous, and rational–intuitive. The same group of scales also loaded highly on the second factor for the Drexel Architects. In addition, such scales as beautiful–ugly, comfortable–uncomfortable, and good–bad loaded highly and positively on this factor, indicating that the Drexel students consistently attributed beauty, com-

Table 14.1. *Varimax factor loadings over .40, by factors and respondent groups over all aspects*

Penn Architects I Space–Eval		Pre-Architects Space–Eval		Non-Architects Space–Eval		Drexel Architects Space–Eval	
Cheerful	.80	Cheerful	.82	Cheerful	.85	Open	.73
Welcoming	.69	Delightful	.82	Pleasing	.82	Cheerful	.71
Spacious	.69	Pleasing	.79	Delightful	.81	Spacious	.68
Delightful	.68	Good	.77	Good	.80	Welcoming	.57
Open	.67	Welcoming	.76	Interesting	.78	Delightful	.56
Good	.64	Beautiful	.75	Beautiful	.75	Loose	.55
Beautiful	.64	Open	.69	Welcoming	.73	Pleasing	.48*
Pleasing	.64	Spacious	.68	Comfortable	.67	Comfortable	.47*
Comfortable	.64	Interesting	.67	Open	.66	Beautiful	.41*
Interesting	.53*	Comfortable	.63	Spacious	.66		
Revolutionary	.44	Loose	.58	Exciting	.54		
Profound	.43*	Revolutionary	.56	Unique (c)	.54		
Loose	.40	Exciting	.52	Loose	.54		
		Unique	.50	Revolutionary	.51		
		Active	.42*	Active	.49		
II Organization		Organization		Organization		Organization	
Ordered	.76	Ordered	.85	Ordered	.72	Ordered	.74
Straightfwd	.65	Clear	.69	Simple	.69	Straightfwd	.64
Clear	.62	Simple	.69	Clear	.65	Simple	.63
Rational	.57	Straightfwd	.68	Straightfwd	.64	Considered	.61
Controlled	.56	Controlled	.67	Rational	.52	Beautiful	.60
Considered	.48	Rational	.65	Controlled	.51	Clear	.55
Simple (c)	.43*	Considered	.58	Continuous	.50	Rational	.52
		Continuous	.57	Common (u)	.47*	Comfortable	.51
		Plain	.55	Considered	.43	Continuous	.51
				Plain	.41	Controlled	.50
						Good	.49*
III Poten–Aesth		Poten–Excite		Potency		Poten–Aesth	
Bold	.74	Strong	.69	Strong	.63	Interesting	.73
Strong	.70	Bold	.67	Permanent	.61	Unique	.72
Interesting	.68	Permanent	.57	Profound	.59	Bold	.69
Unique	.66	Profound	.54	Rugged	.48	Strong	.67
Activity	.64	Rugged	.49	Bold	.41	Good	.62
Exciting	.62	Active	.48	Controlled	.40*	Exciting	.62
Profound	.61	Unique	.48*			Pleasing	.56
Rugged	.58	Interesting	.48*			Active	.55
Specific	.58	Exciting	.46*			Profound	.52
Complex (s)	.57					Specialized	.51
Good	.56*					Permanent	.50
Permanent	.54					Delightful	.48*
Pleasing	.51*						
Ornate	.48						
Delightful	.42*						

Note: Scales marked with an asterisk load more highly on one of the other factors. Parentheses indicate that the loadings on the opposite pole of the scale are indicated under another factor.

fort, and goodness to those buildings that they considered to be rational and clearly organized.

The third factor for all four respondent groups was composed of scales indicating potency: strong–weak, bold–timid, permanent–temporary. However, beyond this common characteristic, there was considerable difference among the groups. The Non-Architects, for example, showed no other type of scale loadings, making this a "pure" Potency dimension. The Pre-Architects, however, had moderately high, but not restricted, loadings of such scales as unique–common, interesting–boring, and exciting–calming. For the Pre-Architects, then, buildings perceived to be potent were more often than not thought also to be somewhat novel and exciting. For the two groups of architects, the scales indicating novelty and excitement loaded highly and exclusively on this factor. In addition, there were moderately high, although secondary, loadings of such scales as good–bad, pleasing–annoying, and delightful–dreadful. It appeared, therefore, for these two groups, that this factor could correctly be labeled an Aesthetic dimension. The architects were "pleasurably moved" by buildings that they considered to be potent and unique.

The results of the comparisons of factor structure thus seemed to indicate that the spaciousness, organization, and potency parts of the factors were both stable and orthogonal for all four respondent groups. The affective and evaluative portions of the factors, on the contrary, appeared to be unstable and non-orthogonal, loading on one factor for one respondent group and on another factor for another respondent group, and in some cases loading on two or even all three factors at one time. Indeed, it appeared that the affective and evaluative aspects of connotative meaning in architecture were not "independent" dimensions of judgment at all, but "dependent" on one or more of the three stable and orthogonal subfactors: Spaciousness, Organization, and Potency.[7] Therefore, although there was probably a sufficient amount of agreement among the factor structures of the four groups to assume that a basis exists for connotative communication between architects and laymen on those portions of the factors relating most closely to the perceptual characteristics of buildings – spaciousness, organization, and potency – the shifting and breaking down of the affective and evaluative portions of the factors left considerable doubt as to whether buildings perceived in the same way by architects and laymen would be equally appreciated.

The results were ambiguous with regard to the hypothesis that differences between the architects and the laymen would be attributable to the professional education of the architect. Inspection of the scale loadings on each factor indicated that the Pre-Architects were midway between the Penn Architects and the Non-Architects on each of the dimensions of meaning. The Drexel Architects seemed to have a somewhat independent structure, being closest to the Penn Architects on the Space–Evaluation and the Potency–Aesthetic dimensions, but farthest from the Penn Architects on the Organization–Evaluation dimension.

Table 14.2. *Percentage of common variance accounted for by each factor for each respondent group, over all aspect groups*

Respondent group	Factors		
	I	II	III
Penn Architects	38.3	22.2	(39.4)
Pre-Architects	(47.6)	30.9	21.4
Non-Architects	(53.8)	29.1	17.1
Drexel Architects	28.6	34.2	(37.2)

Note: The dominant factor for each group is indicated by parentheses. The factors were labeled as follows:

Factor I   Space–Evaluation for all four groups
Factor II Organization–Evaluation for the Drexel Architects
           Organization for the other three groups
Factor III Potency for the Non-Architects
           Potency–Excitement for the Pre-Architects
           Potency–Aesthetic for the Penn and Drexel Architects

Relative factor salience

It has already been seen from inspection of the factor matrices that some differences existed in the relative dominance of the three factors obtained for each of the four respondent groups. Comparisons at this level are rather important because strong emphasis on different dimensions by architects and laymen could cause disruption in communication fidelity. That is, if architects place primary importance on the potency attributes of a building while laymen place primary importance on the spatial attributes, then *even* if both attributes were seen to be approximately the same by the two groups, there would be a dissimilar overall impression of the building. Accordingly, the groups were compared in this respect using the percent of common variance accounted for by each factor as the index of factor salience.

Table 14.2 shows the results of comparisons of the common variance accounted for by the three factors. It is apparent that some differences existed among the four respondent groups. The Potency–Aesthetic dimension was of primary importance for the two groups of architects, but was least dominant for the Non-Architects and Pre-Architects. Conversely, the Space–Evaluation dimension was dominant for the Non-Architects and Pre-Architects, but of only secondary importance for the Penn Architects, and of least importance for the Drexel Architects. The Organization–Evaluation dimension was not dominant for any of the groups. However, it was of secondary importance for the Drexel Architects, Non-Architects, and Pre-Architects.

Table 14.3. *Homogeneity of judgments*

Respondent group	Mean standard deviations	Significance of difference			
		Penn Architects	Pre-Architects	Non-Architects	Drexel Architects
Penn Architects	1.333	—	.001	.000001	.27
Pre-Architects	1.415		—	.001	.008
Non-Architects	1.509			—	.000001
Drexel Architects	1.303				—

Having already observed the shifting of the affective and evaluative scales from one to another of the more stable and independent, perceptually related portions of the factors, it seems that these shifts probably accounted for the changing dominance of the overall factors. Nevertheless, it is significant that the two groups of architects were more concerned with the aesthetic nature of the buildings viewed – their potency, interest, excitement, and the like – whereas the Non-Architects and Pre-Architects were more concerned with the pleasantness of the buildings – their spaciousness, comfort, cheerfulness, and the like. With regard to the hypothesis concerning the architect's professional education, there was strong evidence here that it has an appreciable effect.

Homogeneity of judgments

Another important comparison among the four respondent groups concerned their internal agreement, or homogeneity of judgments. Ordinarily, it has been found that members of an organization or a profession are more homogeneous in their ouflook than the population in general. Likewise, it has been found that homogeneity of judgment increases during the college years for groups following the same course of study (Tannenbaum and McLeod, 1967). Therefore, it would be expected that the two architect groups would display the most homogeneity of judgments, that the Pre-Architects would be next because of their similarity of interests, and that the Non-Architects would be last because of their diversity of interests and education. Table 14.3 presents the mean standard deviations for each respondent group over all building aspects as indices of their homogeneity of judgment. The table also presents the results of comparisons among the respondent groups utilizing the Mann-Whitney U test. Scores for this comparison were the mean standard deviations computed over all thirty scales for each building aspect, with the twenty-five building aspects as replications.

The results tended to confirm expectations, showing the Drexel Architects and Penn Architects to be significantly more homogeneous in their judgments than the Pre-Architects and Non-Architects. Similarly, the Pre-Architects were significantly more homogeneous in their judgments than the Non-Architects. The highly significant differences between the Architects and the Non-Architects can perhaps be accounted for by a combination of the professional education of the architects and the exceptional diversity among the Non-Architects, whose random selection provided twenty-three major study areas out of twenty-six respondents. The greater, but not significant, homogeneity of judgments of the Drexel Architects over the Penn Architects can perhaps be accounted for by the fact that most of the Drexel Architects had been involved in a professional program of architecture longer than the graduating thesis students in architecture at Pennsylvania or, perhaps, that most of the Drexel Architects had been engaged daily in similar experiences in architectural offices. It could also, of course, have been purely by chance.

Meaningfulness of judgments

Another characteristic on which the four respondent groups were compared was their extremity, or meaningfulness, of judgments. Since the midpoint of semantic differential scales is taken as an index of meaninglessness, departure from the midpoint may be referred to as meaningfulness. The appropriate measure here is the generalized distance function, D, computed as the multidimensional distance of each concept from the origin (Osgood et al., 1957). In this study, mean scale values for each respondent group were used in computing the D (or distance) among the groups. The four groups were compared in pairs utilizing the Mann-Whitney U test, with the twenty-five building aspects serving as replications. It was believed, although not hypothesized, that the Penn Architects and Drexel Architects would find the architectural material to be most meaningful, that they would be followed by the Pre-Architects, and that the Non-Architects would find the architectural material to be least meaningful.

Table 14.4 presents the mean D (or distance) from the origin of the "semantic space" for each respondent group taken over all twenty-five building aspects, and the results of the comparisons utilizing the Mann-Whitney U test. As was expected, the Penn Architects found the building aspects to be most meaningful. They were followed by the Pre-Architects and Non-Architects from the University of Pennsylvania. Surprisingly, the Drexel Architects found the building aspects presented to be even less meaningful than did the Non-Architects. In fact, the only difference approaching significance was that between the Penn Architects and the Drexel Architects. It was significant at the .02 level when considered alone; however, when Ryan's correction for experiment error rate was applied, even this comparison fell below the .05 level of significance (Ryan, 1960).

Table 14.4. *Meaningfulness of judgments*

Respondent group	Mean D score	Significance of difference			
		Penn Architects	Pre-Architects	Non-Architects	Drexel Architects
Penn Architects	6.523	—	.48	.21	.08
Pre-Architects	5.892		—	.98	.22
Non-Architects	5.744			—	.21
Drexel Architects	5.296				—

The lack of significant difference in the meaningfulness of the architectural material for the architects and the laymen was quite surprising, since it was believed that the architects would consider architectural material to be much more meaningful than would the Non-Architects. The result, however, can be considered encouraging because the differences were probably not great enough to create a serious problem of communication among any of the respondent groups. The even more surprising difference between the Penn Architects and the Drexel Architects can, perhaps, be accounted for by the different experience with architecture of the two groups. Most of the Drexel Architects had been employed full-time in architects' offices. Some, in fact, had been so employed before beginning their professional training. In this capacity, they were more likely to have considered buildings primarily as *objects* to be designed, detailed, and constructed, rather than as a potential method of communication with the users of the buildings. The Penn Architects and Pre-Architects, on the contrary, owing to their educational and social backgrounds as well as to their lack of practical office experience, might have tended to be more concerned about how buildings are perceived and experienced by their human occupants – that is, with their communication possibilities.

Another possible explanation for differences in the meaningfulness attributed to the architectural material for all groups is the differences in previous experience with the building aspects judged. It may or may not be coincidental that previous exposure to the aspects judged was ordered in the same manner as the judged meaningfulness of the aspects: The Penn Architects had seen or used the largest number of building aspects (83%); the Pre-Architects were next (75%), then the Non-Architects (58%), and finally the Drexel Architects (44%).

Phase II: Judgments on specific building aspects

The objectives of this phase of the analysis were essentially the same as for the first phase: (1) to determine if there is a sufficient basis to assume that buildings

Table 14.5. *Significant differences among respondent groups over all buildings and factors (Duncan multiple range tests: .05 level of significance)*

Over all factors; total: 200 (67)

	Penn Architects	Pre-Architects	Non-Architects	Drexel Architects
Penn Architects	—	38 (12)	53 (21)	21 (3)
Pre-Architects		—	30 (9)	23 (8)
Non-Architects			—	35 (14)
Drexel Architects				—

Pleasantness; total: 59 (28)

	Penn Architects	Pre-Architects	Non-Architects	Drexel Architects
Penn Architects	—	14 (6)	14 (7)	6
Pre-Architects		—	9 (4)	5 (5)
Non-Architects			—	11 (6)
Drexel Architects				—

Organization; total: 40 (12)

	Penn Architects	Pre-Architects	Non-Architects	Drexel Architects
Penn Architects	—	11 (5)	8 (2)	4 (1)
Pre-Architects		—	2 (1)	8 (1)
Non-Architects			—	7 (2)
Drexel Architects				—

Potency; total: 33 (3)

	Penn Architects	Pre-Architects	Non-Architects	Drexel Architects
Penn Architects	—	5	11 (2)	3
Pre-Architects		—	6	2
Non-Architects			—	6 (1)
Drexel Architects				—

Novelty–Excitement; total: 42 (14)

	Penn Architects	Pre-Architects	Non-Architects	Drexel Architects
Penn Architects	—	5	11 (5)	5 (1)
Pre-Architects		—	8 (2)	4 (2)
Non-Architects			—	9 (4)
Drexel Architects				—

Table 14.5 (*cont.*)

Spaciousness; total: 26 (10)				
	Penn Archi-tects	Pre-Archi-tects	Non-Archi-tects	Drexel Archi-tects
Penn Architects	—	3 (1)	9 (5)	3 (1)
Pre-Architects		—	5 (2)	4
Non-Architects			—	2 (1)
Drexel Architects				—

Note: The first value in the matrix indicates the number of significant differences. The second value, in parentheses, indicates differences in opposite directions.

can be used effectively by architects to communicate their "intentions" to laymen, and (2) to determine if the areas of difference can be attributed to the professional education of the architect; that is, to determine if the Pre-Architects, having had no professional architectural education, were closer to the Non-Architects than to the Architects in their attribution of meaning to buildings. In this regard, the difference between this phase and the first phase was not in objectives or hypotheses, but in the level and methods of comparison. It was an important phase of the analysis, however, because respondent groups that use the same dimensions of meaning might, nevertheless, be quite different in their judgments of meaning for specific buildings. The converse could also be true; groups that use different dimensions of meaning might be quite similar in their judgments of meaning for specific buildings. In any event, it was only by considering these two levels of comparison in relation to each other that a comprehensive picture could be obtained of the similarities and differences in the judgments of meaning of the four respondent groups.

The basic procedure was to use one-way analysis of variance and Duncan multiple range tests to compare the factor scores of the four respondent groups on each building aspect. A separate analysis was conducted for each of the five common factors extracted in the oblique factor analyses: Pleasantness, Organization, Potency, Novelty–Excitement, and Spaciousness (cf., Canter, 1969). The results of these analyses over all factors, and for each factor, are shown in Table 14.5. Looking first at the summary over all buildings and all factors at the top of the table, it can be seen that the greatest number of significantly different factor scores (at the .05 level) were between the Penn Architects and the Non-Architects, with 53 significant differences out of 125 comparisons, and with 21 of the significant differences in opposite directions. The smallest number of significantly different factor scores were between the Penn Architects and the Drexel Architects, with 21 significant differences out of 125 comparisons, and with only

3 in opposite directions. It should be emphasized that both the number and the direction of significant differences are extremely useful in revealing the *nature* of the differences among the specific judgments of meaning of the respondent groups. Considering again the Penn Architects and Drexel Architects, it can be seen that in only three instances on the five salient dimensions of meaning were the judgments of the two groups on opposite sides of the neutral point of the semantic differential. They disagreed once as to whether a building was basically organized or unorganized, once as to whether a building was exciting or *un*exciting, and once as to whether a building was spacious or *un*spacious. Most of their comparatively small number of differences were, therefore, only in magnitude, not in direction. If the Drexel Architects considered a building aspect to be potent, the Penn Architects also saw it as potent, but much more potent, or vice versa. In eighteen instances, a difference between the two groups in magnitude of judgment, but not in direction, was large enough to attain significance at the .05 level. This, of course, did not indicate any fundamental differences in their attribution of meaning to the individual buildings, but only a difference in intensity of the attributed meanings. Indeed, it is likely in the case of the Penn Architects and Drexel Architects that most of the differences were due to the general tendency of the Penn Architects to consider architectural material to be more meaningful than did the Drexel Architects and, therefore, probably had little to do with the meaning attributed to the specific building aspects as such.

When the Penn Architects and Non-Architects were compared with respect to the sign of differences, a rather different picture emerged. In this case, almost half (twenty-one) of the sizable number of overall differences (fifty-three) were in opposite directions. This pointed to a rather fundamental difference between the two groups in their attribution of meaning to specific buildings' aspects. The differences between the Drexel Architects and the Non-Architects were similar in this respect, although the number of significant differences were not as pronounced, with a total of thirty-five significant differences and with fourteen in opposite directions. The Penn Architects and Pre-Architects, while second in the overall number of significant differences, with a total of thirty-eight, had a somewhat lower proportion of differences in sign, with twelve. Most of the differences in this case were only differences in intensity of meaning and, therefore, did not indicate a communication difficulty of the magnitude of that between the Penn Architects and the Non-Architects.

Comparisons among factors

An advantage of the comparisons utilizing analysis of variance and Duncan multiple range tests was the ability to determine on which of the dimensions of meaning the greatest differences between the architects and the laymen would occur. In this regard, it was hypothesized that very few differences, particularly in the direction of judgment, would be found for what were considered to be the perceptually related dimensions of meaning: Organization, Potency, and Spaciousness. Conversely, it was hypothesized that there would be a substantial number of differences in both magnitude and direction on the affective and evaluative dimensions of meaning: Novelty–Excitement and Pleasantness. As usual, it was hypothesized that the professional education of the architects would account for the differences.

By inspecting the remainder of the matrices on Table 14.5, it can be seen that the expectations were generally confirmed. The greatest number of differences between the architects and the Non-Architects were on the Pleasantness dimension. The second greatest number were on the Novelty–Excitement dimension. Differences on the Organization, Potency, and Spaciousness dimensions followed in that order.

On the Pleasantness dimension, there were only six significant differences between the Penn Architects and the Drexel Architects and *none* in opposite directions. In comparing the Penn Architects with the Non-Architects, however, the results were quite different; there were fourteen significant differences out of a possible twenty-five (over one-half) and seven (nearly one-third) were in opposite directions. Furthermore, an almost identical relationship held with the Pre-Architects, with fourteen significant differences and six in opposite directions. Similar, but less pronounced results can be seen comparing the Drexel Architects with the Pre-Architects and Non-Architects.

On the Novelty–Excitement dimension, the differences were similar, with only five significant differences between the Penn Architects and the Drexel Architects and one in opposite directions, but eleven significant differences and five in opposite directions between the Penn Architects and the Non-Architects. On this dimension, the judgments of the Pre-Architects were more like those of the Architects than those of the Non-Architects.

The most surprising finding was that the differences among the architects, Pre-Architects, and Non-Architects on the Organization dimension were nearly as great as those on the Novelty–Excitement dimension in both magnitude and direction. The differences between the Penn Architects and the Pre-Architects were particularly puzzling. The greatest differences in both magnitude and direction on the Organization dimension were between the Penn Architects and the Pre-Architects, at eleven and five, rather than between the Penn Architects and the Non-Architects, at eight and two. This can, perhaps, be accounted for by the fact that several of the scales making up the Organization dimension really did not relate closely to the perceptual qualities of the building, but to inferences regarding the processes used in planning the building: rational–intuitive, considered–arbitrary, controlled–accidental. The Pre-Architects, being visually oriented, might have made their inferences solely on the basis of the perceived *surface* organization, while the architects, knowing more about design, could have made their judgments on functional criteria. In any event, it appeared that the professional education of the architects actually caused their judgments of meaning relative to organization to move from one side of the Non-Architects to the other.

The judgments on the Potency and the Spaciousness dimensions were as expected, with a relatively small number of differences in magnitude and direction among the groups. Even where there were a relatively large number of differences, as between the Penn Architects and the Non-Architects on the Potency dimension, at eleven and two, they were primarily with regard to magnitude and not to direction.

Conclusions

Architectural communication

Since the same three independent (orthogonal) and two dependent (nonorthogonal) dimensions of connotative meaning were extracted for all four respondent groups, it is felt that there is no *fundamental* reason why the physical attributes of buildings should not be considered to constitute a "code" capable of use to communicate architects' "intentions" to the users of their buildings (laymen and other architects). This would seem to be especially true, even now, for the Potency and the Spaciousness dimensions of meaning. On the Organization dimension, however, the relation to nonperceptual aspects of architecture (and the Drexel Architects' exclusive tendency to attribute goodness and beauty to buildings considered well organized) would create some doubt if the architect's intended meaning would coincide with the user's attributed meaning – at least in any uniform way.

The gravest doubts concerning communication fidelity, however, related to the affective (Novelty–Excitement) and evaluative (Pleasantness) dimensions of meaning. Here there were rather substantial differences in the scalar composition of the dimensions, in their loadings with the three independent dimensions, and in the judgments of meaning on specific building aspects. First, the Novelty–Excitement dimension of the Non-Architects was generally composed of a minimum number of scales and exhibited low factor salience. The equivalent dimension of the Penn Architects and Drexel Architects included a greater number of scales – several of an evaluative nature – to form a strong Potency–Aesthetic dimension. Since the Non-Architects appeared not to consistently value (find good and pleasing) architectural objects that they found interesting, exciting, unique, and so on, there is a strong possibility that forms intended by architects to be both exciting and good would not be interpreted in this way by laymen. Similarly, since the Aesthetic subfactor of the architects loaded with the Potency dimension, while the equivalent subfactors of the Non-Architects loaded with the Spaciousness dimension, it would seem very likely that here, too, some communication difficulties would arise. The most serious difficulties on these dimensions, however, appeared to relate to judgments on specific building aspects. On both the Pleasantness and the Novelty–Excitement dimensions, approximately half of the judgments by the Penn Architects were *significantly different* from those of the Non-Architects, and half of these differences were in *opposite* directions! The combination of significant and non-significant differences in opposite directions was, of course, even more numerous, to the point that it could be expected that approximately 30% of the time when the Penn Architects would judge a building to be good, pleasing, beautiful, interesting, exciting, and unique, the Non-Architects would judge it to be bad, annoying, ugly, boring, calming, and common. Such a large number of differences between the two groups would, of course, seriously affect the success of architects in communicating their intentions to laymen.

The architects' professional education

The professional education of the architects in this study appeared not to have a great effect on the basic dimensions of meaning used to describe or judge architecture. The Penn Architects, Drexel Architects, Pre-Architects, and Non-Architects employed the same basic dimensions or factors.

The professional education did, however, appear to have minor effects on the scales loading on some of the dimensions, the way in which dependent dimensions loaded with the independent dimensions, and the salience of the dimensions. With regard to factor salience, the effect was quite substantial, the professional education of the architects causing them to attach greater affective and evaluative importance to building aspects perceived to be potent. The Pre-Architects and Non-Architects, on the contrary, attached their judgments in the affective and evaluative areas primarily to the perceived spaciousness of the building.

The meaningfulness of architectural material appeared not to be related at all to the architects' professional education. Rather, it appeared to relate either to personality type or to previous experience with the buildings judged. The Penn Architects and Pre-Architects, who were probably quite similar in personality, background, and the like, and who had seen or used more of the aspects judged in the experiment, also had the highest overall scores of meaningfulness. The Non-Architects and Drexel Architects, who had seen or used fewer of the aspects used in the experiment, had the lowest scores. The homogeneity of judgments, on the contrary, appeared to be influenced quite markedly by the professional education of the architects, the differences between the architects and the other two groups being highly significant. Again, it was on judgments of the meaning of specific building aspects that the professional education of the architects appeared to have the greatest influence. While the two groups of architects were almost identical in the direction of their judgments on all dimensions, they were quite different from both the Pre-Architects and the Non-Architects in their judgments on the affective, evaluative, and Organization dimensions. Indeed, on the Organization dimension, it would appear that the professional education of the Penn Architects actually moved them from a position on one side of the layman as a pre-architect to a position on the other side as an architect.

Implications

Assuming that the results of the experiment could be generalized beyond the immediate respondent groups, the following implications would be apparent. If architects hope to utilize their medium (architecture) to communicate intentions to laymen, they must (1) reorient the architectural education such that it does not change architects' way of experiencing architecture from that which they had as pre-architects, (2) reorient the architectural education such that architects are taught how forms, spaces, and the like are interpreted by laymen, as well as by architects, so that they can consciously manipulate them in such a way as to successfully communicate with both groups, or (3) make greater efforts to educate the general public to see and appreciate architecture in the same way as architects. It is felt that the first alternative is neither desirable nor possible without abandoning the architectural education almost in its entirety. A combination of the second and third alternatives would seem appropriate, along with greater efforts to teach architects to empathize with what is important to themselves, their instructors, their peers, or those who select their buildings for publication.

Some will argue, of course, that most schools of architecture have already come to this conclusion. Nevertheless, the fact remains that there was a wide disparity in the specific judgments by the architects and laymen in the present study, such that if this course of action is now being followed, it is not being done successfully. More important, however, is the fact that some of the specific problems of communication have been isolated in this study. We need not be concerned about bringing the basic dimensions of meaning closer together; they are already similar. We need not be concerned about the perceived spaciousness of a building; there are few disagreements in this regard. We *do* need to concentrate on what it is that makes nonarchitects fail to appreciate some "potent" buildings that architects appreciate. We should also try to determine which characteristics of buildings cause them to appear organized to laymen and pre-architects but not to architects, and vice versa. Similar efforts should be directed toward the extreme differences between architects and laymen with respect to the affective and evaluative areas of meaning. Indeed, to discover what it is in architecture that "pleasurably moves" some people but not others should be at the very core of our concern as architects. If we hope to communicate our intentions to those who use our buildings, we must know the meanings that they will attribute to the forms, spaces, colors, and other qualities that we choose to employ. The curtains are hardly drawn on the vast amount of research that could be done in this area.

Notes

This paper is based on the experimental portion of research conducted by the author in partial fulfillment of the requirements for a Ph.D. in architecture at the University of Pennsylvania. The dissertation, by the same title, includes detailed descriptions and explanations of all aspects of the experiment as well as an extensive theoretical study into the nature, types, and levels of architectural meaning. The dissertation study was supported in part by grants from the University of Pennsylvania's Computer Center and Institute for Environmental Studies. Copies of the dissertation can be obtained from University Microfilms, 300 N. Zeeb Road, Ann Arbor, Michigan 48106.

1. Implicit in research that takes such a focus is the belief that meaning is not *contained in* the elements of architecture, but is something that is intended for or attributed to them by human beings; that such meanings may or may not be held in common by those who experience architecture; indeed, that fundamental differences in human experience will cause fundamental differences in the meanings that people attribute to their environments.

2. In the theoretical portion of the dissertation, a model of meaning was formulated that indicated that there are essentially two kinds of meaning related to architecture. The first kind, representational, or objective, meaning, was considered to embrace all meaning dealing with objects, events, ideas, and the like that are external to the person having the meaning. This meaning dealt with such phenomena as percepts, concepts, and thoughts. The second kind of meaning, responsive, or subjective, meaning, embraced all meaning relating solely to the person having the meaning. This meaning dealt with such phenomena as feelings, emotions, attitudes, and evaluations.

3. One aspect of the experiment not reported here was to determine if there are fundamental differences in the dimensions of meaning relative to the use or nonuse of building aspects.

4. The rather elaborate procedure used to obtain and select the scales was considered to be extremely important, because the truth regarding work with the semantic differential and particularly with factor analysis is that you get out only what you put in. An incomplete list of adjective pairs will necessarily yield an incomplete list of factors!

5. Each student from the University of Pennsylvania was paid $2 for his or her efforts from a grant provided by the Institute for Environmental Studies.

6. Factor analysis was by the principal factor method, using the Pearson product-moment correlation matrices as input. The diagonal elements, initially the squared multiple correlations, were iterated by refactoring until the maximum change in communality estimates was less than .001. Rotations were based on the oblimin criteria, utilizing Kaiser normalization. In the first series of analyses, the factors were kept orthogonal to yield varimax rotations. The minimum eigenvalue for this series of analyses was unity. This value determined both the number of principal components computed and the number of rotations. In the second series of analyses, the oblimin criteria were applied to the primary factor loadings, allowing the factors to be oblique and yielding simple loading rotations. For this series, extraction and rotation of six factors were required. Both series of calculations were performed utilizing the BMD X72 program on the IBM 360 computer at the University of Pennsylvania Computer Center.

7. Subsequent factor analyses utilizing oblique rotational criteria and requiring extraction of six factors confirmed the existence of the evaluative and affective subfactors. It also provided a quantitative comparison of the correlation of the various factors, which revealed that the Organization, Potency, and Spaciousness dimensions were essentially uncorrelated (orthogonal), but that the affective (Novelty–Excitement) and evaluative (Pleasantness) dimensions, while strong, generally correlated with one or more of the "independent" dimensions as well as with each other. The factors obtained in the oblique analyses were Organization, Potency, Spaciousness, Pleasantness (evaluative), and Novelty–Excitement (affective).

15   Predicting user responses to buildings

Robert G. Hershberger and Robert C. Cass

An approach is discussed whereby architects can learn to predict user responses
to the buildings they design. An argument for the importance of prediction in
architecture is presented, initial research efforts are discussed, specific scale- and
media-development experiments are reported, and two professional applications
of the resulting instrument are described.

Background

The recent history of architecture has been marked by an increasing involvement
of architects with client-user groups with which they had had little or no contact.
Commissions are obtained by architects not only in their own communities, but
also throughout the country and, for some firms, throughout the world. They are
obtained not only from clients from the same socioeconomic class or even the
ruling elite, as was the case in previous centuries, but also from client groups
having widely diverse socioeconomic and ethnic backgrounds (Appleyard,
1969). Often the clients represent user groups with special age, health, or mobility problems (Carp, 1970). Occasionally, user groups or potential user groups are
so large or ill-defined as to be virtually unobservable in any primary way. And
almost invariably, because of pressures brought on by rapidly increasing construction costs, architects are expected to perform their services in the shortest
conceivable period of time, "fast-tract" becoming the commonplace rather than
the exception.

From a technological viewpoint, architects appear to be managing quite well
under these circumstances. New buildings for all clients and users incorporate
the finest materials and systems to provide physical conveniences far beyond
those offered in previous times. The buildings are sturdy, durable, and often
quite attractive, at least from the architect's point of view. There is an increasing
awareness that such buildings may be excessively consumptive of energy and
resources, but there is now evidence even here that architects will be able to
make the necessary adjustments (Balchen, 1974).

195

Where the system often seems to break down is in terms of user satisfaction (Michelson, 1966). Architects seem to have neither the time nor the ability to understand the pluralistic user groups for which they are designing. In consequence, they seem prone to design environments that compromise the aspirations of these groups and, at worst, are intolerable for them, as was the Pruitt-Igo Housing Development in St. Louis (Yancey, 1971).

The problem is prediction

When architects and their client-users come from the same socioeconomic strata or share environmental and architectural beliefs, the architects' intuition may serve them well. Architects may be able to predict how client-users will respond to the specific designs that they propose. But when architects and client-users have very little in common, the problem of prediction becomes acute. If architects attempt to empathize with such groups, they are likely to err. They are likely to attribute to such groups environmental values, needs, and interests that in fact they do not have. They are likely to make erroneous predictions about how such groups will comprehend and use the buildings that they design (Hershberger, 1974). Users may not even perceive the same design features as do architects. They may have no knowledge of the referents that architects intend, and so may respond with uncertainty or even distaste to buildings that architects feel would be ideally suited to users' needs.

Many architects understandably feel caught in a bind. On the one hand, architects are being pushed toward the role of social scientist, forced to develop ''clinical'' skills and research abilities so that design solutions might reflect the needs and preferences of diverse client-user groups. On the other hand, under the increasing constraints of design time and budget that characterize the profession, architects find themselves hard pressed to do the job. They yearn for the social scientist who will step in and tell them the environmental needs and preferences of each client-user group, or hope for quick and inexpensive approaches to discover what the social scientist is unwilling or unable to reveal. Unfortunately, social scientists rarely have information on the specific user groups for which architects are designing and are understandably reluctant to make generalizations to groups they have not studied. Any quick and inexpensive research approaches that architects can use and analyze during the design process to obtain reliable and valid results about user-group needs and preferences are difficult to come by. If time and budget permit, wise architects may turn for help to one of the new environmental-analysis and programming firms. If the project is small, or time is of the essence, they more often must turn to their own resources: intuition, casual observation, and limited interviews followed by simple tabulation of results.

Quite often, this approach is satisfactory. Occasionally, it is not and results in the need for extensive redesign and drawing in order to satisfy client-user demands, or worse, the project gets built and does not work.

Learning to predict

This paper discusses the efforts of the authors to develop practical research and training procedures that architectural designers can use to improve their ability to make correct and consistent predictions about how specific client-user groups will perceive and respond to the buildings that they design. Our efforts to date have focused on predictions of user comprehension (or meaning) of the designed environment rather than prediction of specific use. We have been concerned with such affective and evaluative judgments as perceived satisfaction, pleasantness, usefulness, safety, comfort, excitement, and beauty as well as with the physical attributes of the designed environment to which such judgments relate, such as spaciousness, permanence, potency, complexity, temperature, and lighting.

We do not wish to deemphasize the importance of actual use, but to recognize, on the one hand, that is it not possible to use a building in a physical sense during the design process because it does not yet exist, and, on the other hand, that the actual building cannot be used, except in the most primitive sense, in the absence of comprehension or meaning (Hershberger, 1970; Osgood, Suci, and Tannenbaum, 1957), such comprehension or meaning having more direct expression in verbal than in overt physical behavior. And such verbal responses are important in their own right. From the day children first learn to talk, verbal responses become one of the primary ways for them to manipulate their environment, to express satisfaction or distaste, and to cause others to manipulate the environment for them. This is true not just for children, but also for poor or otherwise powerless adults, such as the aged or mentally disabled, who find themselves unable physically to influence the types of environment in which they are placed. Even for normal adults, it should not be assumed that physical behavior is all that is important or that a person's attitudes or beliefs can be inferred from physical behavior. Just because a person cheerfully eats breakfast every morning at a table in a dark, windowless room, it should not be assumed that he is satisfied with the arrangement. He may be a basically cheerful person who enjoys his wife's cooking and tolerates the arrangement, confident that he will someday be able to move into much more satisfactory accommodations – and will tell you so if you ask. It is important for the architect to learn to predict how people will respond to their physical environment. Learning to predict their verbal responses is an important first step.

Previous research

The first solid research evidence to support the notion that architects view the world in a substantially different way from laymen was obtained in a doctoral dissertation at the University of Pennsylvania (Hershberger, 1969). In this study, semantic differential scales and single-color slide presentations of familiar buildings were employed to obtain judgments from architects, pre-architects, and non-architects (1) to determine if representational, affective, and evaluative judgments by the three groups contained important differences, and (2) to determine if the differences could be attributed to the professional education of the architects. Pronounced differences were found between the architect and the non-architect groups, and because the pre-architect group was similar to the non-architect group in nearly all comparisons, it was evident that the judgment of the architects had been influenced by their professional education. The magnitude of the differences suggested to the author that architectural education may actually decrease the ability of architects to predict user responses. Furthermore, because the groups compared at the University of Pennsylvania were very similar in nearly all other respects – age, race, intelligence, economic, geographical, and such – it was felt that difficulties of prediction might become pronounced if the architect is different from the user in these other respects as well.

Developing the research instrument

Subsequent research efforts by the authors have been directed toward the development of research instruments that could be used to attack the prediction issue directly (Hershberger, 1971). Our first effort was to establish a short set of semantic differential scales (Table 15.1) that would fully cover the presentational, affective, and evaluative areas of architectural meaning. A group of twenty scales was identified by reviewing a number of research efforts that had utilized semantic scales (Canter, 1969; Collins, 1969; Craik, 1966; Hershberger. 1969; Vielhauer, 1965) and by selecting lead scales for the factors that had previously appeared. This research was presented and reported at EDRA 3 (Hershberger, 1972). A further analysis of the proposed set of scales was completed in 1972 and presented in a workshop at EDRA 4 (Cass and Hershberger, 1972). In this study, the twenty scales presented at EDRA 3 and ten additional scales of interest to the authors were used to obtain responses from students to twelve buildings selected from the campus at Arizona State University and the surrounding community. Factor analysis of the results and least squares comparisons revealed nine district factors or dimensions of meaning having eigenvalues over unity, as well as one superordinate evaluative factor. These ten factors are listed in Table 15.1, along with ten primary and ten alternative scales

Table 15.1. *Semantic scales to measure the meaning of designed environments: Hershberger-Cass Base Set*

Factor	Primary scales	Alternative scales
1. General evaluative	Good–bad	Pleasing–annoying
2. Utility evaluative	Useful–useless	Friendly–hostile
3. Aesthetic evaluative	Unique–common	Interesting–boring
4. Activity	Active–passive	Complex–simple
5. Space	Cozy–roomy	Private–public
6. Potency	Rugged–delicate	Rough–smooth
7. Tidiness	Clean–dirty	Tidy–messy
8. Organization	Ordered–chaotic	Formal–casual
9. Temperature	Warm–cool	Hot–cold
10. Lighting	Light–dark	Bright–dull
		Alternative
	Secondary scales	secondary scales
	Old–new	Traditional–contemporary
	Expensive–inexpensive	Frugal–generous
	Large–small	Huge–tiny
	Exciting–calming	Beautiful–ugly
	Clear–ambiguous	Unified–diversified
	Colorful–colorless	Vibrant–subdued
	Safe–dangerous	Protected–exposed
	Quiet–noisy	Distracting–facilitating
	Stuffy–drafty	Musty–fresh
	Rigid–flexible	Permanent–temporary

that in terms of high correlations and communalities seem best suited to represent them. The primary set of scales should be considered the absolute minimum essential for coverage of the range of independent meanings attributable to designate environments. The alternative set of scales can be used in whole or in part in place of the primary scales for subject matter or respondent groups to which their use would seem more appropriate. The alternative scales might also be used in addition to the primary scales to ensure coverage of various nuances of meaning within each dimension, to permit more robust factor analytic comparisons, to further test the orthogonality of these ten semantic dimensions, and to create a larger data base to improve reliability of the measuring instrument.

There are also ten secondary scales and ten alternative secondary scales included in Table 15.1. The secondary scales were not accounted for clearly in the factor analytic results. They (1) did not load heavily on any of the identified dimensions, (2) loaded contrary to previous studies on one of the evaluative dimensions, or (3) behaved somewhat erratically across media. Where time and resources permit, it would be advisable to include one or both sets of alternative scales to ensure more comprehensive coverage of the range of meaning attributable to buildings. We are currently engaged in two research studies in which these scales have been included in order to ascertain if their absence as separate dimensions was an artifact of the particular buildings used in the initial study or was in fact a result of their not having essentially different meaning from the ten identified factors.

Evaluating representational media

We have also been engaged in efforts to discover which, if any, media of representation might serve as adequate substitutes for actual environments. Specifically, the authors have sought to discover the degree to which responses to designed environments represented by various media might serve as useful indices of responses to the actual environments. This is an important question because of the time, expense, and control problems associated with transporting people to the environments one wishes to have evaluated. In the case of built environments, it is especially relevant because the ones being designed do not yet exist.

Review of the literature in simulation (Howard, Mlynarski, and Sauer, 1972; Seaton and Collins, 1971; Woods, 1971) indicates that photographic media would probably be the most satisfactory, for improving the architect's ability to predict user responses during the early stages of design. The usual architect's sketches and line drawings as well as finished renderings appear to be the least promising because of the inability of most laymen to read them (Seaton and Collins, 1972). Architect's models also seem difficult to employ because of the time and expense involved in preparation as well as the difficulty in representing interiors (Woods, 1971). Furthermore, our interest is to improve architects' understanding, and hence prediction, of client-users' environmental perceptions and evaluations prior to heavy investment of time and energy by the architects. It is of little satisfaction to architects to find that they have misread the client after they have expended their entire design budget. Architects really need to have an understanding of client-users' environmental perceptions and attitudes at the first stages of the design process. Our experimental efforts were directed, therefore, to finding appropriate media to fulfill this need.

In our first experiment, views of 12 prototypal housing examples were judged on 30 semantic differential scales by 120 architecture students randomly assigned to 6 equally sized respondent groups. One respondent group was bused to the selected examples of housing in order to make its judgments; each of the other five respondent groups based its judgments on one of the following five representational media: (1) 35-mm single-color slides, (2) 35-mm multiple-color slides,

(3) super-8-mm color film, (4) super-8-mm black-and-white film, and (5) black-and-white videotape. Factor analysis, least squares factor comparison, and analysis of variance were used to analyze the results.

Very similar factor structures were found for the real environments and each medium type, with closeness to the real as follows: color film, black-and-white film, videotape, single-color slides, and multiple-color slides. The principal factors for each medium and the real visits based on an eigenvalue of 1.00 were Aesthetic (Evaluative), Pleasantness (Evaluative), Organization, Ruggedness, and Spaciousness. The single- and multiple-color slides varied from the real and from the other media in that the otherwise independent Space and Pleasantness dimensions collapsed into one dimension. On all other dimensions, the single- and multiple-color slides were actually very similar to the real.

Multivariate analysis of variance indicated that there was a significant difference among judgments of the buildings on the six media groups (including the real). Univariate one-way analysis of variance on each of the thirty scales revealed ten significant differences. Dunnett's Test was employed to compare judgments based on real visits with judgments based on each of the representational media for these ten scales. Results of this analysis indicate that there were significant differences between judgments based on real visits and those based on media for only five scales. As can be seen in Figure 15.1, there were no significant differences in mean judgments comparing real visits with color film, and only one difference with color slides. There were two each comparing real with multiple-color slides and videotape, and three with black-and-white film. Interestingly all the differences were found for scales loading on the Organization or evaluative dimensions: simple–complex, ambiguous–clear, accidental–controlled, generalized–specialized, and rational–intuitive. In each case, judgments based on real visits indicated a greater degree of organization than those based on the media. In only one case was there a significant difference in opposite directions – for videotape on the generalized–specialized scale.

Scale-by-scale correlations among judgments on the real and on each of the media were often very high, going from .72 on the average for color film up to .79 on the average for multiple-color slides. Most scales with primary loadings on factors were in the .85 to .95 range. This suggests that even where significant differences occur, it may be possible with careful scale selection to use linear regression equations to predict judgments of real environments using judgments based on color film or color slides.

The second experiment was a replication of the first, with three variations. The views of the twelve prototypal housing examples were replaced by views of twelve institutional and commercial buildings. The original thirty semantic scales were replaced by the thirty factor representative scales (Hershberger, 1972). Only the two most promising media representations, color slides and

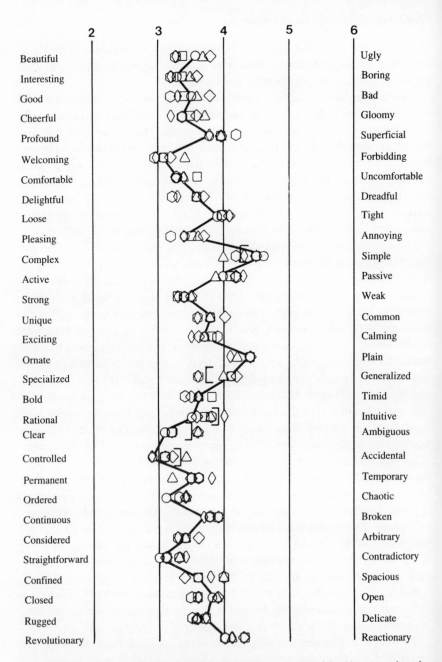

Figure 15.1. Experiment 1 Media Profile. The mean responses on each of the thirty semantic scales are plotted for judgments made during real visits and with respect to each of five representational media for twelve prototypal housing examples: o real visits; □ super-8-mm color film; △ super-8-mm black-and-white film; ◇ 35-mm single-color slides; ○ 35-mm multiple-color slides; ◇ videotape.

color film, were included in this investigation. Three randomly selected groups of pre-architecture students of twenty-seven subjects each responded to all twelve buildings either by visits to the real building or by viewing super-8-mm color film or 35-mm single-color slides.

A separate factor analysis of the thirty scales was performed for each respondent group. Eight factors with eigenvalues greater than 1.00 emerged for each groups. Comparison of the factor structures again revealed remarkable similarities and only minor differences. Major dimensions, such as Aesthetic evaluation, Activity, Potency, and Space, were very similar over all media. A Warmth dimension emerged from the judgments made during the media presentations. Conversely, a Lighting dimension emerged from the judgments made during real visits to the buildings. Most differences, like these, however, involved dimensions accounting for less than 5% of the trace variance: Lighting, Temperature, and Tidiness. A synthesis of the factor structures forming a total of ten factors is shown at the top of Table 15.1.

Results of a multivariate analysis of variance indicated that there were significant differences among the judgments of the three groups. When mean judgments for each of the two media groups were compared with mean judgments for the real-visit group, significant differences in judgment were found on twelve of the thirty scales (Figure 15.2). The buildings viewed in person were judged significantly more good, beautiful, pleasing, friendly, and unique than were the buildings judged on the basis of color slides or color film. All these scales formed part of the two evaluative dimensions: Aesthetic and Pleasantness. In addition, the buildings were judged as more quiet and safe during real visits, whereas the media seemed to enhance the size and publicness of the buildings. Not surprisingly, significant differences were found on the Lighting and Temperature scales, the media apparently not being able to represent these dimensions with any degree of accuracy.

Preliminary evaluation of the results of the two experiments reveals that media such as color film and color slides can be used to simulate actual designed environments. Evidence for this conclusion was provided by the factor analytic studies and least squares factor comparison, which indicated that very similar and easily identifiable dimensions were operating for both the real visits and the media representations even for dissimilar sets of scale and widely diverse building types. The very small number of significant differences found in the analysis of variance comparisons gives similar promise. The uniform suppressing tendency of the two representational media on evaluative judgments found in the second study again suggests that regression equations might be applied to allow closer prediction of evaluative judgments. The generally high scale by scale correlations between each of the media judgments and the real judgments further supports this possibility. Indeed, the fact that there were no significant differences in

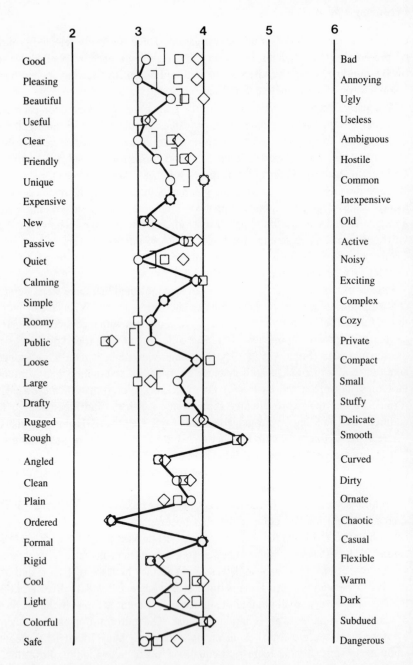

Figure 15.2. Experiment 2 Media Profile. The mean responses on each of the thirty semantic scales are plotted for judgments made during real visits and with respect to each of two representational media for twelve institutional and commercial buildings: o real visits; □ super-8-mm color film; ◇ 35-mm single-color slides.

opposite directions for color film and color slides when compared with real visits for either study suggests that errors that occur in prediction based on these media should be only in degree rather than in direction of judgment – that is, more or less complex, not simple versus complex; or more or less good, not bad versus good. This is an extremely important consideration when evaluating the adequacy of media to represent the real. Neither medium distorts the judgments enough to cause a reversal of meaning; hence the use of either medium will tend to ensure, over a group of judgments, that the qualitative aspects of bipolar judgments will not be misrepresented.

Applications to architectural practice

We are currently utilizing the Hershberger-Cass Base Set of semantic differential scales in a postconstruction evaluation study of Federal Aviation Administration regional office buildings in Los Angeles and Seattle. This Governmental Services Administration–sponsored research is being conducted by the People Space Architecture Company of Spokane, Washington, Sam A. Sloan, principal, along with Robert Sommer, Ph.D., social psychologist; Walter Kleeman, ergonomist; and Robert Hershberger, Ph.D., architect, as research consultants. The entire base set of semantic scales is being employed in this study as part of a larger questionnaire dealing with user satisfaction relative to the new office facilities. The semantic differential portion is used to elicit responses to (1) the employees' own FAA regional office building and (2) their own work stations, as well as responses to three photographs each of (3) a standard GSA office arrangement, (4) a typical office arrangement in the other FAA regional office building, and (5) a typical office arrangement in their own regional office building. The thrust of this research is essentially comparative. The primary objective is to determine if the office arrangement in Seattle is more or less satisfactory to the users than the office arrangement in Los Angeles. The choice of semantic differential questions was made according to the following rationale. The first two questions, in which the respondents rate the actual offices in which they work, are included to obtain basic information about user satisfaction as it relates to the representational, affective, and evaluative attributes of the offices. The three photographic displays of various office environments are used to obtain comparative attitudes or preferences relative to office arrangement. Taken together, the two questions and three photo displays will make it possible to determine not only which office, Seattle or Los Angeles, is more satisfactory in terms of actual user evaluations, but also which is considered more preferable by both user groups. It may also be possible to determine which physical attributes recognized by the users are most highly correlated with satisfaction. In uncovering these relationships, we will, of course, be giving architects information that will make it possible for them to predict more accurately how office workers will respond to their environment (Craik, 1968). By utilizing all forty scales on the Hershberger-Cass Base Set, we will also be able to employ factor analysis to further refine our understanding of the essential dimensions of architectural meaning and, possibly, to reduce the number of scales required to elicit the information that the architect requires for prediction.

Developing the prediction process

Having established in previous research that architects' representations and responses to the designed environment become increasingly different from those of laymen as a result of architects' professional education (Hershberger, 1969), and noting that architects' ability to predict user responses becomes notably weaker as their distance from the user increases in terms of place, race, language, sex, age, and so on (Hershberger, 1971), we have become increasingly convinced that a research instrument is needed that architects can employ directly in practice to learn to predict how specific client-users will respond to the buildings that they design. A set of semantic differential scales has been established and refined that appears to cover the architectural subject material with which most practicing architects are concerned (Cass and Hershberger, 1972; Hershberger, 1972). Other studies have been undertaken to determine which media of presentation are acceptable alternatives to direct experience of designed environments (Hershberger, 1971; Hershberger, 1974).

It now remains to utilize the scales and presentation media to develop a research instrument specifically geared to application: to help architects learn to predict user responses to their buildings before the buildings are designed and, if possible, even before initial design studies begin. This is being attempted by the authors with respect to a commission of Par 3 Planning, Architecture, and Research Studio to design a new facility for the First Southern Baptist Church of Tempe, Arizona.

The procedure we are testing is based on concept formation (Bourne, 1966; Bruner, Goodnow, and Austin, 1956; Haywood, 1966) and on traditional interrater reliability procedures developed in the social sciences. We first obtain from the members of the user group, or a random sample of the user group, their assessments of a representative sample of buildings presented by color slides and using the Hershberger-Cass Base Set of semantic scales. The sample of buildings is selected by the architect to include (1) a variety of building types and styles that represent a range of approach to the solution of the particular design problem, as well as (2) a number of unrelated buildings that might reveal the extremities of judgment of the user group. It is felt at this time that a standardized set of buildings could not be selected that would adequately reflect the diversity of architectural commissions possible, and, furthermore, to be most effective as a practical research tool, the building selection ought to reflect the particular architect's preferences and nonpreferences in order to convey the most useful information to him. In this regard, it should be pointed out that individual architects might wish to add one or more bipolar scales to the Hershberger-Cass Base Set in order to obtain ratings of particular interest to them. Factor analysis of the results will, of course, reveal if they are actually tapping new dimensions of meaning in so doing, or if the scales established in the previous studies actually represent these areas of meaning as well. A sample of the semantic differential response sheet utilizing the full set of forty scales used in this research is shown in Table 15.2.

The mean and the standard deviation of the user responses to each building for each scale are then calculated and plotted in the manner shown in Figure 15.3. For simplicity of illustration, only the ten primary scales of the Hershberger-Cass Base Set are shown. The architect, meanwhile (before the above analysis is revealed to him), responds to the set of buildings and scales in two ways: (1) He indicates what he personally feels about the buildings; and (2) he predicts the mean responses of the client-user group. His judgments are then plotted on the same diagram as for the user group so that the direction and magnitude of the differences among the architect, his prediction of the user, and the user responses are revealed. This is also shown in Figure 15.3.

At this point, an associate in the office (not the design architect) carefully reviews the results, paying particular attention to (1) differences in opposite directions between the user responses and the architect's predictions, as can be seen for the useful–useless scale in Figure 15.3, and (2) very large differences in the same direction, as shown on the cozy–roomy scale. The associate then calculates the magnitude of the absolute distance between the mean user responses and the architect's predictions scale by scale over all the buildings assessed, noting the total number of differences in opposite directions for each scale, and the total number of differences greater than 1 standard deviation. A hypothetical sample of the results of such a tabulation using twelve buildings and ten scales is shown in Table 15.3.

The design architect is then apprised of the calculated results, but not shown any of the profiles for specific buildings. If the architect's predictions are very close to the responses of the user group, there is no reason to proceed further. The architect has essentially confirmed that his intuition about the user group is sufficient, that his ability to predict their responses is probably quite adequate. We plan to confirm this by comparing the architect's predictions with actual user responses (1) to the architect's design presentations, and (2) to the completed, occupied buildings.

If the architect's predictions about the user are quite different from the actual

Table 15.2. *Bipolar adjective scales*

Cozy	—— —— —— —— —— —— ——	Roomy
Common	—— —— —— —— —— —— ——	Unique
Clean	—— —— —— —— —— —— ——	Dirty
Dark	—— —— —— —— —— —— ——	Light
Useful	—— —— —— —— —— —— ——	Useless
Delicate	—— —— —— —— —— —— ——	Rugged
Active	—— —— —— —— —— —— ——	Passive
Bad	—— —— —— —— —— —— ——	Good
Ordered	—— —— —— —— —— —— ——	Chaotic
Cool	—— —— —— —— —— —— ——	Warm
Old	—— —— —— —— —— —— ——	New
Colorless	—— —— —— —— —— —— ——	Colorful
Stuffy	—— —— —— —— —— —— ——	Drafty
Flexible	—— —— —— —— —— —— ——	Rigid
Expensive	—— —— —— —— —— —— ——	Inexpensive
Calming	—— —— —— —— —— —— ——	Exciting
Noisy	—— —— —— —— —— —— ——	Quiet
Safe	—— —— —— —— —— —— ——	Dangerous
Small	—— —— —— —— —— —— ——	Large
Hot	—— —— —— —— —— —— ——	Cold
Simple	—— —— —— —— —— —— ——	Complex
Pleasing	—— —— —— —— —— —— ——	Annoying
Smooth	—— —— —— —— —— —— ——	Rough
Clear	—— —— —— —— —— —— ——	Ambiguous
Formal	—— —— —— —— —— —— ——	Casual
Dull	—— —— —— —— —— —— ——	Bright
Friendly	—— —— —— —— —— —— ——	Hostile
Messy	—— —— —— —— —— —— ——	Tidy
Private	—— —— —— —— —— —— ——	Public
Boring	—— —— —— —— —— —— ——	Interesting
Traditional	—— —— —— —— —— —— ——	Contemporary
Generous	—— —— —— —— —— —— ——	Frugal
Huge	—— —— —— —— —— —— ——	Tiny
Beautiful	—— —— —— —— —— —— ——	Ugly
Unified	—— —— —— —— —— —— ——	Diversified
Subdued	—— —— —— —— —— —— ——	Vibrant
Protected	—— —— —— —— —— —— ——	Exposed
Facilitating	—— —— —— —— —— —— ——	Distracting
Musty	—— —— —— —— —— —— ——	Fresh
Temporary	—— —— —— —— —— —— ——	Permanent

user responses, but his personal responses are quite similar, there is also no need to proceed further. In this case. the architect will have seen that the user is really quite a lot like himself and that he need only let his own judgment be the guide. He can essentially design the building for himself.

Table 15.3 presents a hypothetical problem in which errors made by the

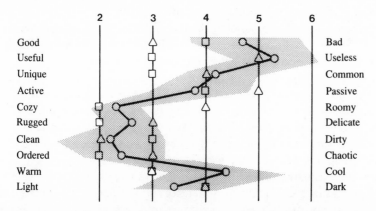

Figure 15.3. Sample building prediction profile. This is an example of a hypothetical prediction problem in which the architect has personally evaluated a given building (□) and has predicted how a group of client-users would assess the building (△). The hypothetical mean judgment of the client-users is indicated by (○), and the standard deviation of the judgments is indicated by the shaded area.

Table 15.3. *Architect prediction summary table*

Scale	Sum of absolute distances (1)	Mean absolute distance (2)	Opposite judgments (3)	Standard deviation differences (4)
Good–bad	25.3	2.1	6	6
Useful–useless	10.8	0.9	0	2
Unique–common	9.6	0.8	2	3
Active–passive	16.8	1.4	3	1
Cozy–roomy	8.4	0.7	3	0
Rugged–delicate	15.6	1.3	7	5
Clean–dirty	4.8	0.4	0	0
Ordered–chaotic	7.2	0.6	3	1
Warm–cool	9.6	0.8	0	1
Light–dark	4.8	0.4	2	0

architect in predicting the client-user responses are calculated. Absolute distances between the architect's prediction and the client-user responses are summed across all buildings assessed and entered in column 1. These distances are divided by the number of buildings assessed – in this example, twelve – and entered in column 2. The number of predictions made by the architect that are opposite to the client-user responses are entered in column 3. The number of predictions that differ from the client-user responses by more than 1 standard deviation are entered in column 4.

If the architect's predictions about the user are quite different from the actual user responses for some or all of the buildings, then a systematic prediction and learning process is in order. Otherwise, the architect is likely to design the building with erroneous ideas about how the user will respond to it.

At least three approaches to the prediction and learning process are possible. The first, most obvious, approach is to show the architect the results of each building comparison so that he can observe the magnitude and direction of the differences between his predictions and the mean user responses. In this way, he can gain a better understanding of the user group, himself, and his ability to predict. If some differences are particularly disturbing to the architect, he can contact representatives of the client-user group to begin an interactive educational process to discover why the differences occur, to try to establish a mutual understanding of the differences, and to try to determine if they are important enough to cause problems.

The second approach is more of a behavior-modification process. In this case, the architect actually trains himself to assess buildings in the same way as his client-users. The authors are utilizing the following procedure for this commission. The architect will first be shown the prediction-user profile for the building in which the overall magnitude of differences were greatest. After studying the results, he will be asked to reevaluate his predictions for one of the other buildings. The reevaluated predictions will then be charted over the earlier user responses and architect predictions, and will be shown to the architect. He will again study the results, noting on which scales he has come closer to the client-user mean and those on which he has gone farther away. The same procedure is continued through the remainder of the buildings, or until the architect learns to predict the user responses with considerable consistency and accuracy. On completion of this procedure, it is, of course, possible to contact representatives of the client-user group, as in the first procedure, in order to establish a mutual understanding of the differences.

A third approach to learning how to predict user responses, more closely related to concept-formation theory, which we hope to explore in a subsequent commission, would be conducted somewhat as follows. An associate of the design architect would analyze the results of the several building-scale predictions and develop a dimension-by-dimension learning experience. The architect would be presented with slide representations of two buildings that had been assessed by the client-users as differing widely on one factor analytic dimension (i.e., Potency), while being more similar on all other dimensions. The architect would assess the two buildings on the scales most representative of that dimension (i.e., rugged–delicate) and be given immediate feedback on the user's responses to that scale. The procedure would be continued on the same dimension for various scales until the architect were successful in making sufficiently fine, consistent, and accurate discriminations among the buildings. After the criteria of successful prediction (i.e., within .5 standard deviation from the user mean) have been achieved on one dimension, the process would be repeated for the remaining dimensions until the architect performed satisfactorily on each. Having obtained success with unidimensional discriminations, the architect would be asked to assess each building on all the scales with an expected large degree of success. The criteria of successful prediction will, of course, depend on the architect's determination and those traditional limitations of architect–client interaction: time and money. Again with the third procedure, either one or both of the previously described procedures could be applied to further refine the architect's ability to predict correctly and accurately. And comparisons of user responses could be made with the architect's predictions of user responses to his design solution and to the building itself in order to confirm his understanding of the client-user.

Conclusion

We believe that this prediction and learning approach will afford substantial benefits to architects and client-users with a minimum of investment on the part of the architects. It does not eliminate the need for careful architectural programming to make certain that the functional and physical needs of client-users will be properly accommodated. It does eliminate the need to generalize the results of other research on presumed similar client-user groups to the current commission and, quite possibly, will allow architects to avoid extensive clinical or experimental work with client-user groups, trying to determine their specific environmental preferences. Architects are not, of course, bound to design buildings that will be immediately satisfactory to clients or users along all or any of the dimensions covered in the set of semantic scales. If they choose to produce designs that will not be immediately satisfactory on some dimension, because they believe that it will be more satisfactory in the long run or for whatever other reason they might have, including aesthetic preference, they will at least be able to do so with full understanding of the likely response of the client-users. They will also have the opportunity from very early stages of design to "educate" client-users to appreciate the type of architecture that they value, with full knowledge of where the client-users' values do not correspond to their own. Most important, architects will be able to design satisfactory buildings for client-users while avoiding some of the pitfalls of time and expense inherent in the slower, traditional design exchange.

16   Dimensions in the perception of architecture: identification and interpretation of dimensions of similarity

Anke Oostendorp and Daniel E. Berlyne

Introduction

Environmental and architectural psychology have expanded and made great headway in recent years. But almost all the research that has so far been done in these areas suffers from two limitations. First, investigators have relied more or less exclusively on verbal judgments of one kind or another. Verbal responses are of interest in themselves, and they are frequently assumed to provide information about attitudes on which nonverbal forms of behavior depend. However, work in such diverse fields as social psychology, market research, and experimental aesthetics (see Berlyne, 1974a) shows that what people do cannot always be predicted from what they say, in accordance with assumptions that are often adopted. So there is an urgent need to supplement verbal measures with nonverbal measures.

Second, investigators have generally requested subjects to rate environments, buildings, or other stimuli for particular attributes that were chosen because the investigators thought them likely to be important and revealing. They thought so partly because of their own intuitions and everyday-life experience and partly for theoretical reasons. But there is, of course, the danger that the attributes chosen, which must in any case represent only a small part of the variables that can distinguish stimulus patterns, do not actually include the most important ones and will not therefore provide guidance toward fundamental dimensions of perception or emotional reactions.

A few attempts have been made to mitigate the latter problem. For example, Kasmar (1970) and Küller (1972) have systematically enumerated and sampled terms that are applicable to the description of architectural spaces. Through elimination procedures, they derived a set of descriptors that could function as relevant tools with a minimum of ambiguity and a maximum of appropriateness.

This paper was first published in *Scandinavian Journal of Psychology* 19 (1978): 73–82.

Factor analysis of such carefully selected attributes may point to basic dimensions in classification of and reactions to environments.

Recently, researchers have turned more and more to new research techniques that seem to offer prospects for overcoming the above-mentioned limitations. First, the technique of multidimensional scaling of similarity judgments has already frequently proved to be of great value whenever a classification of more complex stimuli was attempted. In the field of experimental aesthetics (e.g., Berlyne and Ogilvie, 1974) as well as in environmental psychology (e.g., Gärling, 1976a), this technique has been found worthwhile. It allows for a classification of stimuli that is not biased by the researcher's a priori conceptions of dimensions or of qualities of the stimuli (see Shepard, Romney, and Nerlove, 1972). Furthermore, the problem of interpretation of dimensions revealed by multidimensional scaling analysis can be overcome to a great extent by comparing the mean ratings on a battery of scales with the results of the multidimensional scaling solution (e.g., Berlyne, 1976). And the newly developed technique of canonical correlations and redundancy indices (Stewart and Love, 1968) helps one to find out how much of the variance in the similarity judgments is covered by a battery of scales (and emerging factors).

The major purpose of the present project is to test the value of the above-mentioned new research techniques. But it is also an attempt to demonstrate the overlap between the fields of experimental aesthetics and architectural psychology. The problems concerning lack of theoretical framework and the search for relevant independent and appropriate dependent variables are similar in these two fields. Therefore, in this project, we will attempt to find relevant variables in the perception of the (aesthetic) architectural environment.

In the first experiment, similarity–dissimilarity ratings of twenty buildings will be subjected to the multidimensional scaling procedure INDSCAL (Carroll and Chang, 1970). The INDSCAL procedure has found frequent application in experimental aesthetics (e.g., Berlyne and Ogilvie, 1974; Hare, 1975; O'Hare and Gordon, 1977). This is because the INDSCAL takes account of individual differences in the perception of stimuli. The analysis is based on the assumption that individuals perceive the stimuli in terms of a common set of dimensions, but the dimensions may be of different degrees of importance to the subjects. The results, therefore, point to those dimensions that are predominant in the perception of the given sample of stimuli by the employed group of subjects. It must then be clear that this project alone will not solve the large problems of identifying the basic dimensions in the perception of the architectural environment. Nevertheless, the results to be reported here will give some preliminary indication of the basic dimensions.

Interpretation of the dimensions, which is the major aim of the other two experiments in this report, will be possible by comparison with ratings of buildings on two batteries of scales: (1) collative and affective bipolar scales, and (2) judgments of the importance of stylistic and technical characteristics.

Stimulus information

Twenty color slides of buildings were prepared. As in most experiments of this kind, the stimuli do not represent a random or systematically constructed sample of architecture. Instead, on the basis of architectural literature, a list of twenty major architectural styles, belonging to a particular historical period or geographical region, was compiled. Subsequently, slides of twenty buildings, each representing one of the major styles, were made from pictures in readily available books or were made available through the Library of the University of Toronto School of Architecture. Therefore, the stimuli constitute a sample set of buildings, which is as varied as architectural material might be. The twenty stimuli and the sources of the pictures are listed in Table 16.1.

Experiment 1: Pairwise similarity–dissimilarity ratings

Subjects

Ten unpaid volunteers (four females and six males) were recruited from a subject pool, consisting of students taking an introductory psychology course.

Procedures

The subjects were individually tested, since the 190 possible pairs of the 20 stimuli were presented to each subject in a different, computer-generated random sequence.

The stimulus pairs were projected side by side onto a screen, 1.5 meters in front of the subject. The image of the slide was 36 × 54 centimeters. The two Kodak Carousel (Model 860) projectors were loaded manually, and each pair of slides was presented for five seconds. The subjects were instructed to examine the paired stimuli as long as they were on the screen and then to rate the degree of similarity for the pair on a scale with numbers 1 through 7. It was explained to the subjects that 1 represented extreme similarity and 7 extreme dissimilarity between the stimuli. The interval between the presentation of successive pairs was approximately five seconds. The subject was given six practice trials with slides of buildings that were not part of the experimental stimulus set. After this, the subject went through all 190 pairs without any interruption.

Table 16.1. *Stimulus information*

1 Egyptian: Pyramid of Cheops, Gizeh (twenty-eighth century B.C.). C. Rambert, *Architecture des origines à nos jours* (Paris: Deux Coqs d'Or, 1968), p. 36.

2 Greek: Temple of Poseidon, Paestum (ca. 500 B.C.). C. Rambert, *Architecture des origines à nos jours* (Paris: Deux Coqs D'Or, 1968), p. 44.

3 Roman: Temple of Vesta, Tivoli (first century B.C.). M. Wheeler, *Roman art and architecture* (New York: Praeger, 1964), plate 79.

4 Early Christian: Cathedral Apse, Palermo (ca. 500). Department of Architecture, University of Toronto.

5 Turkish Byzantine: Church of St. Saviour, Istanbul (ca. 550). Department of Architecture, University of Toronto.

6 Pre-Columbian: The Castillo, Chichen Itza, Yucatán (eighth century). C. Rambert, *Architecture des origines à nos jours* (Paris: Deux Coqs d'Or, 1968), p. 60.

7 Japanese: Temple of Daigoji, Kyoto (ca. 1000). Department of Architecture, University of Toronto.

8 Indian: Kandariya Temple, Khajuraho (ca. 1000). H. R. Hitchcock (Ed.), *World architecture* (London: Hamlyn, 1963), plate XVIII.

9 Russian Byzantine: Crypt Monastery, Kiev (ca. 1100). Department of Architecture, University of Toronto.

10 Romanesque: Abbey Church, Marmoutier in Alsace (twelfth century). Department of Architecture, University of Toronto.

11 Gothic: Notre Dame, Paris (1163–1225). C. Rambert, *Architecture des origines à nos jours* (Paris: Deux Coqs d'Or, 1968), p. 133.

12 Chinese: Temple of Heaven, Peking (fifteenth century). H. R. Hitchcock (Ed.), *World architecture* (London: Hamlyn, 1963), plate XI.

13 Islamic: Herbala Mosque, Iran (ca. 1500). Department of Architecture, University of Toronto.

14 Renaissance: Hampton Court Palace (south front), London (1689–1702). Department of Architecture, University of Toronto.

15 Baroque: La Superga, Turin (1717–1731). Architect: Filippo Juvarra. R. F. Jordan, *European architecture in colour* (London: Thames & Hudson, 1970), plate 95.

16 Rococo: Nymphenburg Castle, Munich (ca. 1755). Architect: François Cuvilliès. Department of Architecture, University of Toronto.

17 Neo-Classicism: The British Museum, London (ca. 1833). Architect: Sir Robert Smirke. R. F. Jordan, *European architecture in colour* (London: Thames & Hudson, 1970), plate 109.

18 Neo-Realism: EAG Turbine Factory, Berlin (1909). Architect: Peter Behrens. W. Hofmann and U. Kultermann, *Modern architecture in colour* (London: Thames & Hudson, 1972), plate 40.

19 Art Nouveau: Sagrada Familia, Barcelona (1909–1926). Architect: Antonio Gaudi. R. Barilli, *Art nouveau* (London: Hamlyn, 1966), plate 12.

20 Modern: Pilgrimage Church of Notre-Dame-du-Haut, Ronchamp (1950–1954). Architect: Le Corbusier. Department of Architecture, University of Toronto.

Results

Analysis of variance of the similarity judgments for all 190 pairs showed a significant main effort for pairs ($F = 5.01$, with $df = 198, 1701$ and $p < .001$). The corresponding reliability coefficient (Winer, 1971, pp. 283ff) is .80. It can therefore be said that there was a considerable degree of intersubject consistency.

Table 16.2. Experiments 1, 2, and 3: INDSCAL coordinates and estimated factor scores

	Experiment 1				Experiment 2			Experiment 3			
	Four-dimensional solution				I Hedonic tone/ arousal	II Order	III Tension	I Flamboy- ancy	II Monumen- tality	III Convention- ality	IV Decora- tiveness
	(1)	(2)	(3)	(4)							
Architectural style											
1 Egyptian	.02	-.16	-.29	-.42	.21	.79	.01	-.94	1.01	.24	-.24
2 Greek	.38	-.15	-.29	.01	-.26	.24	-.40	-.89	-.09	.26	.59
3 Roman	.37	-.12	-.08	.12	-.19	-.44	-.31	.26	-.61	.13	-.10
4 Early Christian	.11	.24	.18	.16	.01	-.08	.48	.77	-.51	.05	.07
5 Turkish Byzantine	.07	.49	.08	-.05	-.89	-.95	.18	.78	-.78	-.67	-.46
6 Pre-Columbian	.12	-.25	-.33	-.26	.08	.26	.12	-.72	1.07	.39	-.27
7 Japanese	-.39	-.21	-.17	.01	.21	.36	-.13	.15	-.58	.36	.76
8 Indian	.04	-.09	.18	-.26	.97	-.76	.10	.54	-.34	.22	.23
9 Russian Byzantine	-.15	.17	.33	-.05	-.31	-.16	-.46	.44	-.23	-.17	-.44
10 Romanesque	-.14	-.13	.24	.22	-.10	.95	.43	.06	.38	.11	-.30
11 Gothic	.17	-.11	.25	-.05	.58	-.32	-.21	-.00	.74	.46	-.29
12 Chinese	-.35	-.07	.04	-.18	.23	.42	-.05	.51	-.01	.31	.39
13 Islamic	-.25	.35	.10	-.17	-.13	-.77	-.43	.93	.31	.55	.32
14 Renaissance	.06	-.09	-.08	.41	-.15	.45	.03	-.60	-.61	.30	-.06
15 Baroque	.15	.13	.30	.18	-.07	-.36	.08	.48	-.34	.35	-.15
16 Rococo	-.19	-.14	-.07	.41	-.07	.69	-.01	-.75	-.56	.10	.37
17 Neo-Classicism	.19	-.18	-.17	.28	-.02	-.02	-.55	-.61	-.58	.22	-.52
18 Neo-Realism	.15	.39	-.28	.03	-1.30	.26	.44	-.92	-.23	-.98	-.92
19 Art Nouveau	.03	-.29	.32	-.25	.94	-.17	.64	.89	.91	-.65	-.16
20 Modern	-.40	.20	-.26	-.15	.27	-.39	.06	-.22	.36	-1.59	1.17

The data were then subjected to an INDSCAL analysis. The four-dimensional solution accounted for 40.88% of the total variance (between and within subjects), and the correlations for all subjects showed considerable improvement from the three- to the four-dimensional solution. The variances associated with each dimension were 9.39%, 9.56%, 10.68%, and 11.25%, respectively. The coordinates of the twenty buildings are designated by the names of the architectural styles they were meant to represent (Table 16.2).

Inspection of the stimulus coordinates on the four dimensions allows us to form some hypotheses concerning the attributes determining the similarity judgments. Clustering of the Modern, Japanese, and Chinese buildings along dimension 1 seems to point to color and uncommon shapes. On dimension 2, a cluster is formed by buildings with very angular shapes or steeples (e.g., Art Nouveau and Pre-Columbian), as opposed to rounded forms (Islamic, Neo-Realism, and Turkish Byzantine). Highly decorative buildings form a cluster along dimension 3. On dimension 4, buildings from the more western and northern parts of the world (Rococo and Renaissance) are separated from the more eastern and southern styles (Pre-Columbian, Egyptian, and Indian). This hypothesis was somewhat supported by the finding that the latitude of the locations of the buildings had a significant positive correlation with the coordinates on the fourth dimension ($.54, p < .05$). The stimuli with the higher dimensional coordinates were further north. None of the other dimensions showed significant correlations with longitude, latitude, or estimated year of construction.

Further guidance toward the attributes underlying the dimensions will be forthcoming from the results of the following experiments.

Experiment 2: Collative and affective ratings

In this experiment, the buildings were rated on thirteen scales. The battery of scales was the same as the one used by Berlyne and Ogilvie (1974), except for the addition of Unusual–Commonplace. This was included to test the hypothesis that the fourth INDSCAL dimension was related to the familiarity of a building's style.

As pointed out by Berlyne (1972), the rating scales fall into three classes. First, there are judgments that are indicative of the subject's evaluations of the stimulus pattern. Other judgments refer to his or her internal state while presented with the stimulus. The third category consists of descriptive judgments, concerned with collative (see Berlyne, 1960) or informational (Berlyne, 1974a) properties.

Subjects

Twenty subjects (twelve females and eight males) were recruited from the same pool as those in Experiment 1.

Procedures

Every subject rated each of the stimuli on the thirteen scales. A 20 × 20 balanced Latin square was constructed, such that each stimulus occupied every one of the twenty temporal positions once. Each subject was randomly assigned to one of the rows of the square. The scales were also presented in orders derived from a balanced Latin square, each order being randomly assigned to one or two temporal positions occupied by stimuli.

The scales were typed in the format popularized by Osgood (see Osgood, Suci, and Tannenbaum, 1957) for the semantic differential technique: Seven compartments appeared between the two labels. Scales descriptive of the subject's internal state were indicated by asterisks. One of the practice slides was used to acquaint the subject with the procedure.

The subject rated the stimuli on all the scales in whatever order they appeared on the computer print-out. The subject was instructed to go on to the next slide, pressing a button to cause it to appear, whenever he or she was ready. The experimenter recorded how long it took the subject to rate each stimulus. This measure (Rating Time) constituted the fourteenth variable.

Results

Analysis of variance and factor analysis. Analysis of variance showed the main effect of buildings to be significant at the .05 level for every scale as well as for Rating Time.

Since all the scales showed significant intersubject agreement, the mean scores on the thirteen scales were subjected to a principal-component analysis with help of the FACTOR program (Veldman, 1967). The factor loadings that emerged out of the Varimax rotation are shown in Table 16.3. Loadings with an absolute value exceeding 0.50 are printed in italics. The percentage of variance accounted for by the three factors was 84.6%, Factor I accounting for more than half. The mean estimated factor scores of the twenty architectural styles are presented in Table 16.2.

The factor structure is clear; that is, no scale has a high loading (above .50) on more than one factor. The evaluative scales load on Factor I, together with some of the judgments relating to informational properties (collative variables). In previous studies in experimental aesthetics, these two classes of scales have usually been found to load on different factors. Displeasing–Pleasing, Ugly–Beautiful, and related scales generally define the factor labeled Hedonic Tone (see Berlyne, 1974d), whereas such scales as Weak–Powerful, Drowsy–Alert, and Uninteresting–Interesting have defined a factor labeled Arousal (see

Table 16.3. *Experiment 2: collative and affective scales (factor loadings)*

Scale	Factor			
	I	II	III	h^2
1 Unusual–commonplace	−.77	.36	−.16	74.4
2 Unbalanced–balanced	.13	.90	.02	81.9
3 Simple–complex	.75	−.39	.27	75.5
4 Clear–indefinite	−.15	−.83	.20	74.8
5 Disorderly–orderly	−.04	.97	−.11	94.9
6 Displeasing–pleasing	.90	.22	.32	96.7
7 Weak–powerful	.80	.30	.36	85.9
8 Ugly–beautiful	.87	.22	−.30	89.6
9 Uninteresting–interesting	.94	.01	.07	88.9
10 No discomfort–extreme discomfort	−.25	.44	.73	79.0
11 No pleasure–extreme pleasure	.95	.21	−.10	95.5
12 Drowsy–alert	.72	.30	.32	71.3
13 Relaxed–tense	.22	−.05	.92	89.2
% variance	44.69	24.78	15.16	

Berlyne, 1975). However, in studies using visual material representing a limited range of differences in information content, the Hedonic Tone and Arousal factors have sometimes been found to merge (Berlyne, 1974b). Consequently, Factor 1 will be tentatively labeled Hedonic Tone/Arousal.

Factor II is dominated by three scales: Clear–Indefinite, Disorderly–Orderly, and Unbalanced–Balanced. These ratings represent the extent to which a stimulus pattern calls for information-processing effort. The factor appears to correspond to those found in previous studies that were labeled Uncertainty. However, there is one major difference: In contrast with earlier findings, the Simple–Complex scale, judgments of which are indicative of subjective uncertainty, does not show a high loading on this factor (see Berlyne, 1974d). Therefore, the label Order is suggested to indicate the difference. It should be pointed out that this Order factor is very similar to factors found in other factor analyses of ratings of buildings. Canter (1969) termed this factor Coherence; Küller (1972), Unity; and Hershberger (1972), Organization.

High loadings on the third factor were found on only two scales: Relaxed–Tense and No Discomfort–Extreme Discomfort. It seems that this factor is connected with aversive levels of arousal evoked by the stimuli (Berlyne, 1967, 1971). However, the constitution of this factor differs to a great extent from the Arousal factor that has generally been found in previous studies. Therefore, the label Tension is suggested.

The fourteenth measure of this experiment (Rating Time) showed significant correlations with the following scales: Displeasing–Pleasing ($-.54$), Ugly–Beautiful ($-.48$), Uninteresting–Interesting ($-.47$), No Discomfort–Extreme Discomfort ($.47$), and No Pleasure–Extreme Pleasure ($-.52$), all at the .05 level of significance. Correlations with factor scores were not significant. It was also of interest to find that the buildings located farther north were rated to be more commonplace ($.48$), weaker ($-.47$), and less interesting ($-.48$). These correlations were significant at the .05 level.

Comparison with similarity judgments. Experiment 1 provided us with stimulus coordinates on each of four dimensions, derived from similarity judgments. In this experiment, two multidimensional spaces were obtained. First, each architectural style has a mean score on each of the thirteen rating scales. These data can be treated as locations in a thirteen-dimensional space. Second, the estimated mean factor scores represent a location for each stimulus in a three-dimensional space.

It is of interest to determine the degree to which the locations of the stimuli in the INDSCAL similarity space can be predicted from their locations in the thirteen-dimensional scale space. Stewart and Love's (1968) redundancy index was applied to the data, and the significance of the redundancy index was examined with an F test (Miller, 1975). The redundancy from the thirteen-dimensional scale space to the four-dimensional INDSCAL similarity space was .85 ($p < .01$). This represents the proportion of variance in the INDSCAL locations of the stimuli that is accounted for by their locations in the scale space. The redundancy index from the factor space to the INDSCAL similarity space drops to .31 ($p < .01$). The relation between the two spaces can also be tested by the maximum canonical correlation coefficient (R_c). R_c between the thirteen-dimensional scale space and the INDSCAL space came to 0.99, and R_c between the factor space and the INDSCAL space was .93. By means of the χ^2 test (see Bartlett, 1941), these canonical correlations were found to be significant beyond the .01 level.

The above-described relations between the spaces seem to imply that our scales cover or reflect to a substantial extent the attributes that are predominant in perception, as revealed by similarity judgments. We can now ascertain which scales (and factors) are most closely related to the INDSCAL space. Table 16.4 presents the significant multiple correlations (R) of the scales and factors with the INDSCAL space (all four dimensions). Significant product-moment correlations between separate dimensions and individual scales and factors are presented in the table. It can be observed that ten out of the thirteen rating scales show a significant degree of correlation (R) with the INDSCAL space and that each of the thirteen scales is correlated with one of the four dimensions. Dimension 1 is related to only Clear–Indefinite. Dimension 2 has a significant affinity with nine

Table 16.4. *Experiments 1 and 2: significant product-moment and multiple correlations*

Scale	INDSCAL dimensions				
	1	2	3	4	R
1 Unusual–commonplace	—	—	—	.53[a]	—
2 Unbalanced–balanced	—	−.64[b]	—	—	.74[a]
3 Simple–complex	—	—	.64[b]	—	.73[a]
4 Clear–indefinite	.48[a]	—	—	—	.71[a]
5 Disorderly–orderly	—	−.51[a]	—	—	.71[a]
6 Displeasing–pleasing	—	−.80[b]	—	—	.88[c]
7 Weak–powerful	—	−.64[b]	—	—	.72[a]
8 Ugly–beautiful	—	−.71[b]	—	—	.87[c]
9 Uninteresting–interesting	—	−.66[b]	—	—	.79[b]
10 No discomfort–extreme discomfort	—	.54[a]	—	—	—
11 No pleasure–extreme pleasure	—	−.80[b]	—	—	.93[c]
12 Drowsy–alert	—	−.45[a]	—	—	.76[b]
13 Relaxed–tense	—	—	.53[a]	—	—
Factor I (Hedonic tone/arousal)	—	−.62[b]	—	—	.85[c]
Factor II (Order)	—	−.51[a]	—	—	.70[a]
Factor III (Tension)	—	—	—	—	—

[a] $p < .05$
[b] $p < .01$
[c] $p < .001$

out of the thirteen scales, as well as with the Hedonic Tone/Arousal and Order factors. Two scales, Simple–Complex and Relaxed–Tense, showed significant correlations with dimension 3. And finally, Unusual–Commonplace ratings are significantly correlated with dimension 4. Factor III (Tension) did not appear to have much in common with the INDSCAL space.

Experiment 3: Stylistic and technical ratings

In Experiment 3, the buildings were rated on a battery of scales referring to stylistic and technical characteristics. Such batteries were used in previous investigations with paintings (Berlyne, 1975; Berlyne and Ogilvie, 1974) and with music (Hare, 1975). But whereas the scales in the collative and affective battery are appropriate, without change or with only minor changes, for all kinds of stimulus material, the stylistic and technical battery has to vary with the medium under investigation. The scales used in the present battery are listed in Table 16.5. Since ratings on these scales may be difficult or unreliable without some

Table 16.5. *Experiment 3: stylistic and technical rating scales*

Importance of:

1 Aesthetic norms or stylistic conventions characterizing the society or age in which the architect lived (Aesthetic norms)
2 Beliefs or values characterizing the society in which the architect lived (Nonaesthetic norms)
3 The architect's feelings or emotions (Emotions)
4 Expression of the uses or functions of the building (Functional expression)
5 Composition or arrangement of elements (Composition)
6 Vertical dimension
7 Horizontal dimension
8 Ornamentation
9 Surface qualities or textures (Surface)
10 Colors
11 Curves
12 Angles
13 Shapes
14 Straight lines

Note: Words in parentheses are abbreviations by which scales are designated in text.

specialized training and experience, it was decided to use architecture students as subjects.

Subjects

Nine male and seven female subjects were recruited from the Department of Architecture of the University of Toronto. All subjects had completed at least one year of design courses and were taking an elective course on behavioral issues in architecture.

Procedure

Apart from the difference in scales, the procedure was the same as for Experiment 2. However, this time, not every row of the balanced square had a subject allotted to it, and so, in the analysis of variance, the temporal position effect could not be subtracted from the error term.

Results

Analysis of variance and factor analysis. Analysis of variance reveals significant intersubject agreement over buildings for all fourteen scales at .01 level.

Table 16.6. *Experiment 3: stylistic and technical ratings (factor loadings)*

	Factor				
Scale	1	II	III	IV	h^2
1 Aesthetic norms	−.14	−.19	.92	.01	91.0
2 Nonaesthetic norms	−.07	.36	.88	.07	91.1
3 Emotions	.20	.43	−.68	.45	90.3
4 Functional expression	.06	.80	.01	.08	65.2
5 Composition	.05	.27	.05	.85	80.4
6 Vertical dimension	.20	.89	−.08	−.00	84.5
7 Horizontal dimension	−.82	−.23	.05	.10	73.7
8 Ornamentation	.53	−.10	.36	.32	51.9
9 Surface	.66	.39	.05	.40	74.9
10 Colors	.45	−.14	−.02	.67	67.7
11 Curves	.81	−.18	−.30	.41	94.2
12 Angles	−.47	.73	.07	.04	76.9
13 Shapes	.53	.47	−.19	.41	70.4
14 Straight lines	−.88	.11	.30	−.18	90.9
% variance	25.9	20.8	17.5	14.7	

The F value for Rating Time was not significant. Mean ratings on the fourteen scales for each of the twenty stimuli were subjected to a principal-component factor analysis with Varimax rotation. A four-dimensional factor structure emerged, and the factor loadings are shown in Table 16.6. Estimated factor scores are presented in Table 16.2.

The four factors accounted for 78.9% of the variance, and no scale has a high loading on more than one factor. Factor I has a negative loading on Horizontal dimension and Straight lines, and a positive loading on Ornamentation, Surface, and Curves. Flamboyancy seems to be an appropriate label for Factor I. Angular, vertical, and functional expressive features are important for Factor II. The high correlation between the Vertical dimension and the Functional expression ratings may be partly due to the fact that churches outnumbered any other type of buildings in this sample. Inspection of the estimated factor scores for Factor II reveals a cluster of sturdy and massive buildings (Egyptian and Pre-Columbian), as well as two steepled buildings (Art Nouveau and Gothic). The label Monumentality seems to express this variable best.

The negative loadings of Emotions and positive loadings of Aesthetic norms and Nonaesthetic norms on Factor III seem to point to stylistic characteristics or sociocultural importance. The factor might be labeled Conventionality. Factor IV is defined by the same scales, Composition and Colors, as the Decorativeness factor in Berlyne (1975), so this label will be maintained in the present study.

Table 16.7. *Experiments 1 and 3: significant product-moment and multiple correlations*

	INDSCAL dimensions				
Scale	1	2	3	4	R
1 Aesthetic norms	—	—	—	—	—
2 Nonaesthetic norms	—	—	—	—	—
3 Emotions	—	—	—	—	—
4 Functional expression	—	—	—	$-.62^b$	$.63^a$
5 Composition	—	—	—	—	—
6 Vertical dimension	—	—	—	$-.66^b$	$.76^a$
7 Horizontal dimension	—	—	$-.74^b$	—	$.82^b$
8 Ornamentation	—	—	$.59^b$	—	$.71^a$
9 Surface	—	—	—	$-.48^a$	$.63^a$
10 Colors	$-.76^b$	—	—	—	$.79^b$
11 Curves	—	$.46^a$	$.50^a$	—	$.75^a$
12 Angles	—	$-.51^a$	—	—	$.67^a$
13 Shapes	—	—	—	$-.67^b$	$.80^b$
14 Straight lines	—	—	$-.65^b$	—	$.78^b$
Factor I (Flamboyancy)	—	—	$.76^b$	—	$.85^b$
Factor II (Monumentality)	—	—	—	$-.73^b$	$.81^b$
Factor III (Conventionality)	—	$-.45^a$	—	—	—
Factor IV (Decorativeness)	$-.52^a$	—	—	—	—

$^a p < .05$
$^b p < .01$

Product-moment correlations of the scales and factor scores with the measures of approximate latitude, longitude, and year of buildings were significant for only one scale. Nonaesthetic norms were judged to be more important for buildings that are less northerly ($-.44$), more easterly ($.57$), and older ($-.49$), all $p < .05$.

Comparison with similarity ratings. The redundancy from the fourteen-dimensional scale space to the four-dimensional INDSCAL similarity space amounted to .93 ($p < .001$). The redundancy from the four-factor space was also significant ($.52$, $p < .001$). The canonical correlation between the scale space and the INDSCAL space was .99 ($p < .01$) and between factor space and INDSCAL, .94 ($p < .01$).

Table 16.7 presents the significant product-moment and multiple correlations for individual scales and factors. It will be seen that ten out of the fourteen scales and two factors were significantly correlated with the INDSCAL dimensions collectively. Product-moment correlations between separate INDSCAL similarity dimensions and estimated factor scores showed each of the factors to be correlated with a different INDSCAL dimension.

Comparison with collative and affective ratings. The canonical correlation between the two scale spaces came to 1.00 ($p < .01$) and between the factor spaces, .82 ($p < .01$). The redundancy from the stylistic and technical scales to the collative and affective scales was .87, and from the four factors of Experiment 3 to the three factors of Experiment 2, the redundancy was .45. In the opposite direction, the redundancies were .78 and .35 for scale and factor spaces, respectively. All these indices are significant at least at .01 levels.

Product-moment correlations between mean estimated factor scores from Experiments 2 and 3 revealed only two significant correlations. The Hedonic Tone/Arousal factor scores showed a significant ($p < .05$) positive correlation with the Monumentality factor scores (.54). The second significant correlation was found between the Order factor scores and the Flamboyancy factor scores ($-.64$, $p < .01$). The remaining factors, Factor III from Experiment 2 (Tension) and Factors III and IV from Experiment 3 (Conventionality and Decorativeness, respectively), revealed some relations with individual scales, but not with other factor scores.

Discussion

The findings of this study confirm the usefulness of multidimensional scaling used in conjunction with rating scales for research in experimental aesthetics. Moreover, the study points to the value of these procedures for research in environmental and architectural psychology, which, of course, overlaps with applied experimental aesthetics.

In each of the experiments, subjects, from the populations sampled, showed a large degree of consistency in their judgments of the buildings. On the basis of the similarity judgments, the buildings representing architectural styles could be located in a four-dimensional space, accounting for a large percentage of the variance. Considering the complexity of the stimuli, the large number of attributes of the buildings, and the attitudes of the subjects that contributed to the variance, this is a finding that encourages one to go on with the search for the dominant dimensions in the environment.

The mean ratings yielded factors that resembled factors derived from ratings of other kinds of stimuli. However, the major difference was that in the present study, the usual Arousal factor merged with the Hedonic Tone factor. Thus Osgood's (Osgood et al., 1957) Evaluation, Potency, and Activity dimensions – which are related to Hedonic Tone, Arousal, and Uncertainty, respectively – were not as clearly represented as in previous studies (see Berlyne, 1974d).

Canter (1972) reported a similar merger of Osgood's dimensions, referring to several factor analytic studies of verbal responses to architectural stimuli. The factors emerging from the stylistic and technical battery pointed to a four-way categorization of architectural styles.

The canonical correlations and redundancy indices indicated that the attributes reflected by the two scale batteries and the underlying factors play a large part in subjects' similarity judgments and, presumably, their perception of the architectural environment. The product-moment correlations of the dimensions with the scales and factors give us some insight into what attributes and underlying variables may be related to the similarity dimensions.

Dimension 1 correlated with only Clear–Indefinite and Colors. It seems that this dimension is related to Clarity in the design of a building. It is of interest to note that O'Hare and Gordon (1977) also report the emergence of a dimension characterized by Clarity. They employed paintings of landscapes, and it might thus be postulated that this dimension may be especially important in environmental perception – that is, dealing with spatial relations (see also Lynch, 1960).

Dimension 2 ranged from buildings that were rated to be displeasing, weak, and curved to ones that were judged to be more beautiful, balanced, interesting, and angular. The major underlying variable of this dimension seems to be Hedonic Tone/Arousal. However, significant correlations with the Order and Conventionality factors seem to suggest that the degree to which the design of the buildings appeared to adhere to expected norms influenced the similarity judgments, and thus the locations of the stimuli along this dimension. Similarly, in the study by O'Hare (1976; O'Hare and Gordon, 1977) one dimension was found to be related to hedonic, representational, and arousing aspects of the paintings.

The stimulus locations of the buildings along dimension 3 are related to the degree of complexity of and the importance of ornamentation in the building. There is also a relation with the subject's internal feelings – that is, how relaxed or tense he or she feels while looking at the buildings. Thus the frequently found underlying variable of Uncertainty (see Berlyne, 1974d), whose importance was recently disputed by O'Hare and Gordon (1977), appears with the present set of stimuli once again! And, *nota bene,* these stimuli are not "simplified figures" or "non-art forms" (O'Hare and Gordon, 1977, p. 69), which according to these authors are the only stimuli for which the Uncertainty dimension might be of importance.

Dimension 4 was found to be related to attributes determining whether a building was commonplace and belonged to the western part of the world. It might be best interpreted as dealing with the degree of Familiarity that subjects had with an architectural style.

In conclusion, the dimensions and factors revealed in the present study point to independent and dependent variables, which demonstrate great similarity with those found in other studies in experimental aesthetics and architectural psychology.

Acknowledgments

This investigation was supported by research grant S74–1650 to D. E. Berlyne from the Canada Council and scholarship No. 1560 to Anke Oostendorp from the National Research Council. The authors also wish to express gratitude to Dr. P. Anderson, of the Ontario Science Centre, and to the Department of Architecture, University of Toronto, for their helpfulness.

17 Contextual compatibility in architecture: an issue of personal taste?

Linda N. Groat

The problem of how to relate new infill buildings to existing urban settings has in recent years become an important point of debate, both in architectural-design practice and in public policy.

Consider, for example, the proposed entry pavilion for the Louvre in Paris (Hoelterhoff, 1985). Designed by the world-renowned architect I. M. Pei, this glass pyramid structure is to be located at the center of the grand Cour Napoléon, which is embraced by the two major wings of the palace. Although the plans were announced some time ago, now excavation of the site has served to embroil the proposal in a debate that is "shaking up the hearts and minds of all France" (Hoelterhoff, 1985, p. 28).

On the one hand, opponents of the proposal argue that the translucent glass structure constitutes an aesthetic obtrusion in the forecourt of this hallowed and historic institution. On the other hand, proponents of Pei's design argue that the design is not only pleasingly "insubstantial," but also appropriate symbolically. More specifically, the pyramid, it is argued, is an appropriate reference to the Egyptian exploits of Napoleon I, who was responsible for opening part of the Louvre as a museum.

This controversy over the proposed pavilion at the Louvre is significant in at least three respects. First, the opposing arguments exemplify, in many ways, the range of concerns that are frequently expressed in public debate over the appropriateness of new infill schemes. Second, the incident demonstrates the potential impact of this design issue in the public realm. Third, and most important, the controversy raises the question of whether it is indeed possible to identify any normative standard or to establish consensual agreement about what constitutes contextual compatibility in the urban environment. In an important sense, then, the phenomenon of contextual compatibility clearly falls within the area of environmental psychology that is commonly labeled environmental aesthetics.

This paper describes one segment of a larger study on contextual compatibility in architecture. More specifically, this segment of the research was concerned with two issues, which are interwoven throughout this chapter. First, on a sub-

228

stantive level, the intention was to investigate the extent to which certain types of contextual-design strategies may be consistently preferred over others. In other words, the research question being posed was this: Is there any underlying consistency in people's evaluations of contextual relationships among buildings?

The second aspect of the research was methodologically focused; it concerned the manner in which physical attributes of the built environment are specified in empirical research. The argument is made, through the example of this research study, that the concepts and categories that are in architectural discourse (typically, historical analyses and criticism) can be usefully employed as a basis for empirical research on the aesthetic quality of the built environment.

A matter of personal taste or an aspect of basic environmental values?

The question of whether there may be some relatively consistent bases on which contextual relationships are evaluated represents, of course, a specific instance of a more general question: Are environmental preferences a matter of personal taste or representative of some generally consistent environmental values?

This question is commonly raised in discussions of research on environmental preferences. At a general level, S. Kaplan (1979a) has argued that preference judgments are neither arbitrary nor idiosyncratic, but reveal common patterns of aesthetic values. More particularly, with reference to the natural environment, Ulrich (1983) has argued: "There is absolutely nothing in this substantial body of findings to suggest that aesthetic preferences for natural environments are random or idiosyncratic" (p. 107). Similarly, with reference to the built environment, Oostendorp and Berlyne (1978c) argue that "individual differences in taste for architectural styles may not be as large as, especially, art theorists want us to believe" (p. 146).

With respect to the topic of contextual compatibility more specifically, three distinct sets of literature offer different perspectives on the question of preference judgments.

The design literature. Recent architectural discourse on the problem of contextual design offers a wide range of anecdotally based and speculative analyses. Although several distinct points of view can be identified within this literature, most architectural authors tend to emphasize the variability of aesthetic responses. For example, a number of architects and critics have argued that appropriate contextual design is best achieved by leaving the creative architect unconstrained by guidelines or other legislative mandates (e.g., Cavaglieri, 1980). Thus by emphasizing the essentially idiosyncratic nature of each design problem, this argument implicitly suggests that preference judgments of contextual relationships are unlikely to arise from any consistent basis of evaluation.

At the other end of the sprectrum, however, some architectural writers do stress the fundamental importance of certain underlying design principles. Thus, for example, Brolin (1980) argues very persuasively that small-scale details (ornament) usually represent "the critical element" (p. 143) in contextual design. Nevertheless, he undercuts this argument by introducing his analyses as "subjective" (p. 6); and elsewhere, he has taken pains to emphasize that "the whole matter of relating new buildings to old is a matter of taste" (Brolin, 1982).

Design-review guidelines. In the past ten to fifteen years, an increasing number of communities have chosen to establish designated building-conservation districts, typically under the jurisdiction of a design review board, in order to protect and maintain the aesthetic and/or historic quality of these valued urban areas. Moreover, these efforts toward effecting some degree of aesthetic management and control are increasingly being supported by the courts (Bohlman and Dundas, 1980; Brace, 1980; Crumplar, 1974).

By definition, the implicit assumption of such regulatory procedures is that some normative standard for evaluating contextual compatibility can be established and consistently applied to a variety of design proposals (Bowsher, 1978). At the same time, however, many design-guidelines documents have been written in a purposefully open-ended fashion, the implication being that it is not actually possible to identify the specific *types* of contextual-design strategies that are advisable. For example, many guidelines simply require that a certain number or percentage of specified design relationships (e.g., scale, height, volume) be maintained, but it is left up to the designer to chose which of these relationships are the most significant or relevant (Lu, 1980).

In summary, then, two implicit assumptions appear to be inherent in the formulation of many design-guidelines procedures: (1) that it is possible to achieve some consensual agreement about the contextual appropriateness of individual design proposals, but (2) that it may not be possible to specify the *types* of design strategies that are generally most suitable for contextually sensitive situations.

The empirical literature. To date, there has been little or no empirically based research focused specifically on contextual compatibility at the architectural scale. Nevertheless, the existing empirical work on the perceived compatibility between built form and landscape settings does offer some intriguing findings of potential relevance to this investigation.

In combination, two related studies by Wohlwill begin to address the question of the relative consistency in preference judgments of contextual relationships. In the first of these studies (Wohlwill and Harris, 1980), the authors found a high degree of consensus among small groups of respondents ($n = 7$ to 10) in their rank-order judgments of "fittingness." More specifically, nine of the twelve composite rank orders of the five landscape scenes were identical. And in the second study, Wohlwill (1979) investigated the relationship between judgments of compatibility (degrees of appropriateness) and aesthetic judgments of preference (degrees of liking). In this case, the results indicated that there was virtually no difference in the response patterns of the two judgments.

If indeed, as Wohlwill suggests, there is virtually no psychological distinction between judgments of preference and judgments of appropriateness, then it may be reasonable to suggest that the consistency that Wohlwill and Harris (1980) found in rank-order judgments of fittingness might well obtain in a comparable investigation of judgments of preference. Taken together, then, the results of these two studies suggest that preference judgments of contextual relationships may actually be highly consistent among various groups of respondents.

A conceptual framework for the analysis of contextual design

The discussion has revealed that the three sets of literature offer rather different perspectives on the relative consistency of aesthetic judgments. In addition, the discussion has brought to light two distinct aspects of the issue: (1) the consistency among various respondent groups, and (2) the consistency with which particular types of contextual-design strategies are preferred over others. The former is concerned with differences among people; the latter, with differences among physical forms.

It is the latter of these two issues that poses a particularly challenging methodological problem. In order to investigate the extent to which certain types of design strategies are consistently preferred, it is first necessary to establish a systematic procedure for classifying contextual-design strategies.

Typically, however, one of the major stumbling blocks to research on environmental aesthetics has been the lack of precision in specifying the physical attributes of the environment under study. This shortcoming in much of the research to date has been noted by several authors, including Wohlwill (1976), Canter (1977), and Groat (1983a). More specifically, Canter (1977) has argued that specifying the physical constituents of an environment is a much more significant issue than "the research literature would have one believe" (p. 159). He then goes on to add that "there are really remarkably few examples of physical forms having been studied directly for their relationships to psychological and behavioural processes" (p. 159).

In fact, according to Wohlwill (1976), most research on environmental perception has managed conveniently to "sidestep" the issue by selecting sites or views without any attempt to assess them in terms of specified variables. He further argues that the typical combination of semantic differential and factor analytic procedures yields a purely descriptive analysis. The result is that the research then reveals nothing about the role played in the respondents' judgments by any specifiable environmental characteristics.

Why should this be so? Canter (1977) suggests that the reason for such a dearth of studies with a truly environmental focus

appears to be, in part at least, the difficulty of deciding which physical attributes to study. Taken in the abstract, independently of any conceptual framework, there is an infinity of ways of dividing up and measuring physical parameters. . . . So researchers have either selected one which caught their fancy, with disappointing results, or given up because they were spoilt for choice. (p. 159)

Given these very substantial shortcomings in the existing empirical literature and Canter's rather pessimistic comments, it would seem that any adequate analysis of building features relevant to contextual fit would be inordinately difficult to achieve. However, Groat (1983a) has pointed out that one solution is actually implicit in Canter's commentary. Canter suggests that what has been lacking is an appropriate "conceptual framework" for specifying physical features. And yet, as Groat (1983a) argues, "the needed conceptual frameworks are there for any researcher willing to sift through the ever-increasing and bountiful literature on architectural theory . . ." (p. 31).

With respect to the specific issue of contextual compatibility, there is already a very substantial body of architectural criticism that identifies – from an anecdotal and speculative perspective – potentially significant features of contextual design. This literature can thus provide an appropriate basis for generating a conceptual framework for the analysis of contextual-design strategies. The formulation of such a conceptual framework not only serves this practical methodological purpose, but also offers the further advantage that research findings can be described in terms that are common to the design literature, thereby increasing the potential for applicability in architectural-design practice.

For these several reasons, then, a major objective of the initial formulation of this research study was the development of a conceptual framework for the analysis of contextual design. It was derived from a systematic review of the critical literature on contextualism (e.g., Biddle, 1980; Brolin, 1980; Edwards, 1924/1946; Ray et al., 1980; Smith, 1977). Since two of the major sources were edited volumes, this means that the development of the framework was based on a review of more than thirty authors, most of whom are critics, practitioners, or conservation specialists.

In its "final" version, the conceptual framework represents a unique descriptive framework. Although it was derived from existing critical commentary, its actual form bears little or no relation to any one source or even to combinations of sources. Rather, it is the result of an iterative process of "design" that involved several rereadings of the source materials, frequent reevaluations, critical commentary from colleagues, and a search for comprehensive clarity.

An essential feature of this framework (Table 17.1) is that it distinguishes among design attributes according to the degree of control that the architect is typically able to exercise over them. This investigation was, however, concerned with only the design attributes that are under the architect's control – that is, the components of a design strategy. Thus issues such as the insertion of nontypical building uses or extreme contrast in project sizes – although important aspects of contextual compatibility – were excluded from consideration in this study.

Another feature of the framework is that it defines contextual design in terms of both interior and exterior design features. This investigation, however, was concerned with the impact of only exterior design attributes, the most frequently considered aspect of contextual design. As a consequence, each of the urban or campus scenes eventually selected for use in this study was analyzed in terms of only three components of design strategy: site organization, massing, and façade design. All three components are defined in terms of a 7-point scale, from contrast to replication. The constituent physical attributes of the three components are defined as follows.

Site organization. Site organization has to do with the basic spatial pattern that a building imposes on the site. Tactics such as setback distances, landscaping patterns, and circulation pathways contribute to the definition of this spatial pattern.

Massing. The massing of a building is really its volumetric composition, defined in terms of design attributes such as height, shape, and complexity of overall form.

Façade design. Façade design here is used to mean the surface treatment of the planes (i.e., the elevations) that define the shell of the buildings. Manipulation of the façade is rendered not only through such stylistic tactics as Tudor or Georgian motifs, but also through more abstract features such as the proportioning of window openings or the use of color and materials.

Method

Selecting the range of urban scenes

Within the thematic framework of the initial research objective for this study, it would have been theoretically possible for the respondents to view either real-life

Table 17.1. *A conceptual framework for the analysis*
of contextual-design strategies

I. Givens: issues typically beyond the architect's control
   1. Site location: _____
   2. Building type: _____
   3. Size: _____

II. Design parameters: issues partially under the architect's control
   4. Prominence:
      minimum   1—2—3—4—5—6—7   maximum
   5. Definition of context:
      adjacent   1—2—3—4—5—6—7   regional

III. Design strategy: issues typically under the architect's control

A. Space
   6. Exterior site organization:
      contrast   1—2—3—4—5—6—7   replication
      Tactics:
      ____ footprint of the building on the site
      ____ circulation: pathways, etc.
      ____ vehicular access: driveways, parking
      ____ alignment, setback distances and angles
      ____ landscaping: site demarcations
      ____ other
   7. Interior spatial organization:
      contrast   1—2—3—4—5—6—7   replication
      Tactics:
      ____ circulation paths, hallways
      ____ room/area layouts
      ____ level changes
      ____ placements of vertical circulation
      ____ other

B. Massing
   8. Exterior massing
      contrast   1—2—3—4—5—6—7   replication
      Tactics:
      ____ shape, complexity of overall form
      ____ articulation of base, body, top
      ____ roofline, vertical projections
      ____ other
   9. Interior semifixed arrangements
      contrast   1—2—3—4—5—6—7   replication
      Tactics:
      ____ overall configuration of partitions
      ____ arrangements of heavy furniture, etc.
      ____ other

C. Style
   10. Façade design
       contrast   1—2—3—4—5—6—7   replication

Table 17.1 (*cont.*)

Tactics:		
____ overall stylistic attributes		
____ rhythm, proportion of fenestration		
____ color		
____ materials		
____ degree of ornament, detail, relief		
____ other		
11. Interior surface treatment		
contrast	1—2—3—4—5—6—7	replication
Tactics:		
____ overall interior style		
____ shape, proportion of surface details		
____ color		
____ materials		
____ degree of ornament, detail, relief		
____ other		

or simulated environments. However, from a practical and logistical standpoint, it was not possible to identify a sufficiently broad range of exemplary contextual-design strategies in any given geographical locale, particularly within a reasonable radius of the research base. As a consequence, it was necessary to represent the various contextual-design exemplars through the medium of color photographs.

A number of empirical studies have explored the validity of simulation media. Taken together, these studies have demonstrated that responses to color photographs correlate highly with responses to the real environment (Hershberger and Cass, 1974; Howard, Mlynarski, and Sauer, 1972; Seaton and Collins, 1972). More recently, Feimer (1984) investigated the validity of simulation procedures using a respondent sample of over 1,000 people. He concluded that the magnitude of simulation effects "is sufficiently small to be inconsequential for many practical and empirical applications" (Feimer, 1984, p. 77).

The set of color photographs for this study consisted of a range of twenty-five urban scenes, each of which included both a recently designed infill building and several of the immediately adjacent buildings. The primary basis for the selection of these particular buildings was that they would, *in toto,* represent the broadest possible range of contextual-design strategies as identified through the use of conceptual framework.

Overall, the selection of the urban scenes was a multistage process, involving (1) a preliminary search in the professional journals and major books on contextualism; (2) procurement of the color photographs, typically from the architectural firms responsible for the jobs; and (3) a preliminary analysis of the contextual-design strategies.

Once a tentatively "final" set of photographs had been selected, each of the urban scenes was analyzed in terms of the three design-strategy components of the conceptual framework – site organization, massing, and façade design – using the 7-point rating scale. As a result of this analysis, it was then possible to assign a profile score to each of the thirty or so buildings being analyzed. A profile score was defined as the set of three ratings (e.g., 6–6–3 or 5–4–6) on the three design-strategy components.

This analysis of design strategies was conducted by two pairs of expert judges, all of whom had some familiarity with architectural research. Each pair conducted its own analysis and arrived at a consensual profile score. Subsequently, all four judges reviewed the two sets of profile scores and negotiated a final profile score for each urban scene.

The final selection of the contextual scenes was made after this analysis of design strategies. There were two major criteria for the final selection: (1) representation of the broadest possible range of design strategies; and (2) clarity, lack of ambiguity, and consistency of quality in the photographic materials.

The respondents

Two major respondent groups were interviewed for this study: (1) nonexperts, represented by a total of seventy-three individuals at three locations in the upper Midwest; and (2) experts, represented by twenty-four individuals, eight from each of three design review commissions in the metro-Milwaukee area. For reasons pertinent to the research design of the larger study of which this investigation forms a part, the three nonexpert groups were actually users and residents or neighbors at the sites of three of the simulated urban scenes. The three selected case-study sites were the addition to the Farmers and Merchants Union Bank, Columbus, Wisconsin; the Alumni Center at the University of Michigan, Ann Arbor; and the Summit Place town houses, St. Paul, Minnesota.

These three case-study sites represent a variety of site and population conditions, particularly in the following respects: (1) contextual-design strategy, (2) size of town where the project was located, and (3) geographical distribution, all three sites being at least 350 miles apart.

At each of the three sites, a substantial majority of the users and residents ($n =$ 11 or 12) who were available at the time of the site visit were interviewed. The specific proportions at each site were as follows: (1) approximately 66% of the bank employees were interviewed; (2) approximately 60% of the Alumni Center employees were interviewed, this figure representing roughly 80–90% of those available at the time of the site visit; and (3) approximately 80% of the town house residents were interviewed, this figure representing 100% of the residents who were in residence at the time of the site visit. In addition, at each of the two nonresidential sites, every effort was made to interview employees who represented all levels of and job functions in the organization.

With respect to the neighbor groups at each case-study site, however, the respondents represent a very small proportion of the specified neighbor population. Given the inherent limitations of the case-study visits, it was not feasible to implement any sort of meaningful (in a statistical sense) sampling procedure. Therefore, the recruitment procedures for these individuals was based primarily on convenience and efficiency, involving both referrals and blind letters of introduction with follow-up calls.

Finally, with respect to the design review commissions, the respondents represent nearly 100% of the total review-commission membership in the metro-Milwaukee area. Twenty-four of twenty-six commission members agreed to be interviewed. Only one member declined; the other member was hospitalized and could not participate.

In the final analysis, the three nonexpert respondent groups include a broad range of middle-class and upper-middle-class individuals from the upper Midwest. The expert group represented by the substantial majority of review-commission members in the Milwaukee area.

Interview procedures

The entire interview sequence for the larger research study consisted of nine distinct segments and typically required sixty to ninety minutes to complete. Each respondent was interviewed individually in either his or her home or office.

Although several segments of the interview procedure implicitly addressed the research questions discussed earlier in this paper, the analyses derived from one particular interview segment are especially pertinent here. In that segment of the interview, each respondent was asked simply to rank order the twenty-five urban scenes according to his or her preference for the contextual relationship. More specifically, the respondents were asked to establish a rank order based on the extent to which they liked or disliked the relationship between the infill building (which was underlined) and the surrounding context.

In light of the discussion of Wohlwill's (1979; Wohlwill and Harris, 1980) research earlier in this chapter, it is important to emphasize that the respondents were specifically asked to rank order the scenes according to their personal preference judgments. Although it may be that, as Wohlwill suggests, there is no psychological difference between preference and appropriateness ratings, the intention of the research question was to maximize the potential for the respondents to express their idiosyncratic ''tastes.''

Table 17.2. *Ranking scores for contextual relationships:*
case-study respondents

Order	Mean rank	Building number and name
1	3.48	1 East Cambridge Savings Bank addition
2	4.85	7 Frick Collection addition
3	5.78	2 The Alumni Center, U. of Michigan
4	6.94	9 Lincoln Park town houses
5	7.72	25 519 Ashland residence
6	8.43	17 Apartments, Beacon Hill
7	10.44	15 Salem Five Cents Bank addition
8	10.60	23 The Asia Society
9	10.76	10 Law Building, U. of Wisconsin
10	11.59	19 Pacific Heights town houses
11	12.00	12 Citizens Federal Savings addition
12	12.87	14 Residence
13	13.24	24 Frank–Carlsen residence
14	13.39	6 Maryland National Bank
15	15.71	3 Valerio residence
16	16.28	5 Metropolitan Museum of Art addition
17	16.72	20 Portland Public Services Building
18	17.18	18 Dodge Center
19	17.50	16 Allen Memorial Art Museum addition
20	17.61	4 East India Marine Hall addition
21	17.79	13 Summit Place town houses
22	18.15	21 Farmers and Merchants Union Bank addition
23	20.14	8 Enderis Hall, U. of Wisconsin–Milwaukee
24	20.47	11 Beckley residence
25	20.51	22 Library, Mount Mary College

Results

Comparisons of the respondent groups' preference judgments

Tables 17.2 and 17.3 present the composite rank orders for the preferred contextual relationships of the case-study respondents and the review commissioners, respectively. These rank orders were simply computed from aggregating
the rank orders generated by the individual respondents. However, an adjustment
was made to eliminate the influence of the case-study-site respondents' ratings of
their own site.

The two composite rank orders demonstrate that the preference judgments of
the two major respondent groups are very similar. More specifically, eighteen of
the twenty-five rank positions are either identical or only one position reversed,
and only two rankings are reversed by five or more positions. More important,

Table 17.3. *Ranking scores for contextual relationships:*
review commissioners

Order	Mean rank	Building number and name
1	3.02	1 East Cambridge Savings Bank addition
2	3.74	7 Frick Collection addition
3	5.20	2 The Alumni Center, U. of Michigan
4	6.48	25 519 Ashland residence
5	6.85	9 Lincoln Park town houses
6	7.72	17 Apartments, Beacon Hill
7	9.43	23 The Asia Society
8	9.76	15 Salem Five Cents Bank addition
9	10.70	10 Law Building, U. of Wisconsin
10	11.33	19 Pacific Heights town houses
11	12.04	14 Residence
12	13.15	3 Valerio residence
13	13.28	24 Frank–Carlsen residence
14	14.20	6 Maryland National Bank
15	16.04	4 East India Marine Hall addition
16	16.28	18 Dodge Center
17	16.43	12 Citizens Federal Savings addition
18	16.52	20 Portland Public Services Building
19	16.83	5 Metropolitan Museum of Art addition
20	17.52	21 Farmers and Merchants Union Bank addition
21	17.61	16 Allen Memorial Art Museum addition
22	19.26	13 Summit Place town houses
23	20.09	11 Beckley residence
24	20.48	8 Enderis Hall, U. of Wisconsin–Milwaukee
25	21.04	22 Library, Mount Mary College

the case-study respondents' six most preferred relationships are also the review commissioners' most preferred relationships. And similarly, the case-study respondents' three least preferred relationships are also the commissioners' least preferred relationships. Computing a Spearman rank order correlation coefficient, a value of rho $= .961$ was obtained; this is significant at the .02 level.

To compare the overall response patterns among the three case-study respondent groups, rank order correlations were calculated for each pair of case-study locations. The correlation coefficients for each of these three comparisons are as follows: Columbus/Ann Arbor, .900; Ann Arbor/St. Paul, .947; and Columbus/St. Paul, .858. All three correlations are significant at least at the .02 level.

Similarly, to compare the overall response patterns among the three review-commission groups, rank order correlations were calculated for the three commission groups. The correlation coefficients for each of these three comparisons are as follows: Milwaukee/Shorewood, .918; Shorewood/Cedarburg, .902; and Milwaukee/Cedarburg, .914. All three correlations are significant at least at the .02 level.

In summary, then, these several analyses straongly suggest that there is a very high degree of consensus among the various respondent groups and subgroups with respect to their preference judgments of contextual relationships.

Preferred contextual-design strategies

In order to determine the extent to which particular types of design strategies may be consistently preferred over others, it is first necessary to develop a systematic way of analyzing the relationship among all the profile scores generated in the earlier analysis of the contextual-design strategies. This can be achieved through a partial order scalogram analysis (POSA) (Lingoes, 1973).

The POSA of the set of twenty-five selected buildings is presented in Figure 17.1. Each of the twenty-five buildings is represented by a point on the roughly diamond-shaped plot and is identified by its building number. (See Figure 17.2 for the set of twenty-five infill-scenes photographs.) Next to each building number is the profile score of its design strategy. The lines of the plot connect each point (building) to those points with the most similar profile score.

The arrangement among these twenty-five buildings is based on both the quantitative and the qualitative differences in the profile scores. The quantitative dimension by which these buildings are organized is represented by the vertical dimension of the diamond plot. In other words, the buildings with low total scores (i.e., the high contrast strategies) are found at the lower end of the diamond, whereas the buildings with high total scores (i.e., the high replication strategies) are found at the upper end of the diamond.

The qualitative differences among the buildings are represented horizontally. This means that the buildings at the same horizontal level of the diamond are quantitatively similar (i.e., have roughly the same total profile score) but are qualitatively different. So, for example, at the left side of the diamond is building 1 (the East Cambridge Savings Bank addition), with a 5–4–6 profile; at the right edge is building #6 (the Maryland National Bank), with a profile of 6–6–2. Both buildings have similar total scores (15 and 14, respectively), but the design strategy of the East Cambridge bank is manifested in a relatively greater degree of contrast in site organization and massing and a significantly higher degree of replication in façade design.

Figure 17.3 is a reproduction of the original POSA configuration, but in this instance the case-study respondents' rank order of preference has been superimposed on the figure. (The commissioners' rank order has not been reproduced simply because it is so similar as to be redundant.)

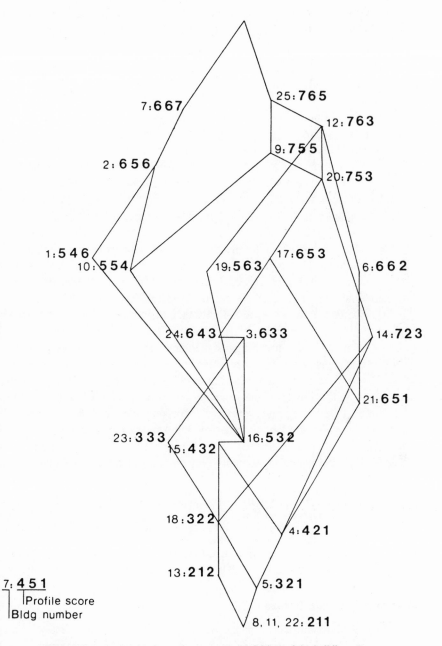

Figure 17.1. A partial order scalogram analysis (POSA) of the building set.

Visual inspection of the rank order designations indicates that the rank order of preference generally moves from the upper-left side of the plot to the lower right. In other words, the most preferred relationships are defined by strategy profiles characterized by a high degree of replication. Conversely, the least liked relationships are typically defined by strategy profiles characterized by a high degree of contrast.

Further inspection of the pattern of rankings on the POSA also reveals some important differences in preferences across the qualitative dimension (represented horizontally on the plot). Even though all the projects on each "rung" of the plot have a similar total profile score, the projects on the left are consistently preferred over those on the right. The difference among these profiles scores is the relative score for each of the three scales. In the case of the preferred projects, the score on the façade-design scale is almost always higher.

Despite the generally consistent pattern of rank orders throughout the POSA plot, there are some relationships that do not entirely conform to that pattern. For example, building 20 (Michael Graves's Portland Public Services Building) is ranked lower than would be expected from its position in the POSA plot. Similarly, both buildings 15 and 23 (the Salem Five Cents Savings Bank addition and the Asia Society) are ranked more highly than their position in the plot would suggest.

These anomalies, of course, suggest that the conceptual framework presented earlier in this paper is simply not sensitive enough to all the design qualities that affect people's preference judgments of contextual relationships. Despite these several anomalies, however, the generally consistent pattern of preferences revealed in the POSA does suggest that the conceptual framework has successfully identified several of the key attributes of contextual-design strategies.

In summary, then, this analysis makes it possible to draw some general conclusions about the nature of generally preferred and disliked contextual-design strategies:

1. The design strategies of the most preferred relationships are characterized by a relatively high degree of replication.
2. The design strategies of the least preferred relationships are characterized by a relatively high degree of contrast.
3. Replication of at least some aspect of façade design is more critical for perceived compatibility than is replication of either site organization or massing.

Discussion and conclusions

Two themes were introduced at the outset of this paper. The primary theme concerns the relative consistency of aesthetic judgments of contextual compatibility. The secondary theme is more methodologically focused and concerns the extent to which the architectural-design literature can be used as a basis for

Building 1
East Cambridge Savings Bank addition
East Cambridge, Massachusetts
Charles G. Hilgenhurst & Associates
(Photo: Patricia Gill)
Profile score: 5–4–6

Building 2
The Alumni Center
University of Michigan
Ann Arbor, Michigan
Hugh Newell Jacobsen
(Photo: John Rahaim)
Profile score: 6–5–6

Building 3
Valerio residence
Milwaukee, Wisconsin
Joseph Valerio
(Photo: Linda Groat)
Profile score: 6–3–3

Figure 17.2 The twenty-five infill scenes.

Building 4
East India Marine Hall addition
Salem, Massachusetts
Philip Bourne
(Photo: Patricia Gill)
Profile score: 4-2-1

Building 5
Metropolitan Museum of Art addition
New York, New York
Roche Dinkeloo Associates
(Photo: Naomi Leiseroff)
Profile score: 3-2-1

Building 6
Maryland National Bank
Annapolis, Maryland
RTKL Associates, Inc.
(Photo: Courtesy of RTKL Associates, Inc.
Profile score: 6-6-2

Figure 17.2 (*cont.*)

Building 7
Frick Collection addition
New York, New York
Harry Van Dyke
(Photo: Brent Brolin)
Profile score: 6–6–7

Building 8
Enderis Hall
University of Wisconsin–Milwaukee
Milwaukee, Wisconsin
Plunkett Keymar Reginato
(Photo: Linda Groat)
Profile score: 2–1–1

Building 9
Lincoln Park town houses
Chicago, Illinois
Bauhs & Dring
(Photo: Linda Groat)
Profile score: 7–5–5

Figure 17.2 *(cont.)*

Building 10
Law Building
University of Wisconsin
Madison, Wisconsin
Roger Kirchhoff
(Photo: Linda Groat)
Profile score: 5–5–4

Building 11
Beckley residence
Milwaukee, Wisconsin
Robert M. Beckley
(Photo: Linda Groat)
Profile score: 2–1–1

Building 12
Citizens Federal Savings addition
San Francisco, California
MLTW/Moore Lyndon Turnbull Whit
(Photo: Robert Coven)
Profile score: 7–6–3

Figure 17.2 (*cont.*)

Building 13
Summit Place town houses
St. Paul, Minnesota
Robert Engstrom Associates
(Photo: Garth Rockcastle)
Profile score: 2–1–2

Building 14
Residence
Boston, Massachusetts
Graham Gund Associates
(Photo: Steve Rosenthal, courtesy of
   Graham Gund Associates)
Profile score: 7–2–3

Building 15
Salem Five Cents Bank addition
Salem, Massachusetts
Oscar Padjen Architects
(Photo: Patricia Gill)
Profile score: 4–3–2

Figure 17.2 (*cont.*)

Building 16
Allen Memorial Art Museum addition
Oberlin College
Oberlin, Ohio
Venturi & Rauch
(Photo: Courtesy of Venturi, Rauch &
    Brown)
Profile score: 5–3–2

Building 17
Apartments, Beacon Hill
Boston, Massachusetts
James McNeely
(Photo: Courtesy of James McNeely)
Profile score: 6–5–3

Building 18
Dodge Center
Washington, D.C.
Hartmann-Cox Architects
(Photo: Courtesy of Hartmann-Cox
    Architects)
Profile score: 3–2–2

Figure 17.2 (*cont.*)

Building 19
Pacific Heights town houses
San Francisco, California
Daniel Solomon and Associates
(Photo: Joshua Freewald, courtesy of Daniel
    Solomon and Associates)
Profile score: 5–6–3

Building 20
Portland Public Services Building
Portland, Oregon
Michael Graves
(Photo: Frances Downing)
Profile score: 7–5–3

Building 21
Farmers' and Merchants' Union Bank
    addition
Columbus, Wisconsin
Gornet and Shearman
(Photo: Valerie Johnson)
Profile score: 6–5–1

Figure 17.2 (*cont.*)

Building 22
Library
Mount Mary College
Wauwatosa, Wisconsin
Pfaller Herbst & Epstein, Inc.
(Photo: Linda Groat)
Profile score: 2–1–1

Building 23
The Asia Society
New York, New York
Edward Larrabee Barnes Associates
(Photo: Nick Wheeler, courtesy of Edwar
    Larrabee Barnes Associates)
Profile score: 3–3–3

Building 24
Frank–Carlsen residence
St. Paul, Minnesota
Sylvia Frank/Peter Carlsen
(Photo: Garth Rockcastle)
Profile score: 6–4–3

Figure 17.2 (cont.)

Building 25
519 Ashland residence
St. Paul, Minnesota
Architect unknown
(Photo: Garth Rockcastle)
Profile score: 7–6–5

Figure 17.2 (*cont.*)

specifying the physical attributes of built form in empirical research on environmental aesthetics.

The second theme has been elaborated in the development of the conceptual framework presented in Table 17.1 and in the subsequent analyses of the contextual-design strategies used in this study. In this regard, the success of the POSA in revealing systematic differences between liked and disliked design strategies strongly suggests that categories and concepts from the architectural literature can be fruitfully drawn on for empirical research. A further advantage is that findings derived from such research can be more easily and directly presented in the architectural and design literatures, thereby increasing the likelihood of application in design practice. In the case ot this research study, for instance, the results have been published in two articles for one of the professional architectural journals (Groat, 1983b, 1984).

The primary theme of this investigation – the consistency of aesthetic judgments of contextual design – has been elaborated, for the most part, through the substantive aspects of the investigation. The several data analyses have demonstrated that there is a much higher level of consistency in preference judgments of contextual compatibility than is customarily suggested in much of the architectural literature. This is true in two distinct respects.

The data analyses have indicated that there is a relatively high degree of consistency in the preference judgments of several diverse respondent groups and subgroups. This finding tends to substantiate the findings of previously cited research in environmental aesthetics (e.g., S. Kaplan, 1979a; Oostendorf and Berlyne, 1978c) and, more particularly, Wohlwill's research on contextualism in landscape settings (Wohlwill, 1979; Wohlwill and Harris, 1980).

The data analyses have also demonstrated that design strategies that embody a relatively high degree of replication, especially in aspects of façade design, are

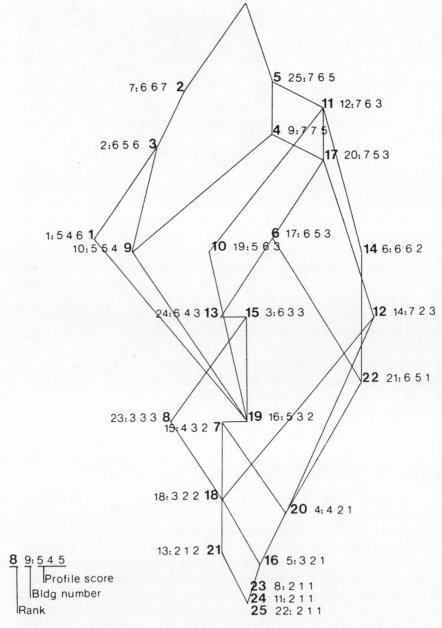

Figure 17.3. The rank order of contextual relationships as superimposed on the partial order scalogram analysis (POSA).

consistently preferred over other types of design strategies. This finding thus lends support to Brolin's argument that small-scale façade details and ornament may be "the critical element" in contextual design (Brolin, 1980, p. 143).

Taken together, these two findings have important implications for architectural-design practice. Within the framework of the current discourse on contextualism in the architectural literature, many architectural practitioners and critics have argued strenuously that buildings can achieve compatibility with their surroundings primarily through replication of site organization and/or massing (e.g., Carlhian, 1980; Johnson, 1984). The results of this research, however, strongly suggest that such design strategies are unlikely to be appreciated by the public (or even by the experts).

And more important, the results of this study also suggest that the most preferred contextual-design strategies incorporate at least some degree of replication of façade design, *in addition to* replication of site organization and massing. But this does not mean that indiscriminate mimicry of façade-design features is necessary for compatibility. Rather, as many of the particular examples in this study (e.g., building 1, East Cambridge Savings Bank addition; 2, University of Michigan Alumni Center) demonstrate, innovative features and imaginative reinterpretations of traditional façade elements can be blended in ways that truly appeal to nondesigners. Thus the essential implication of this research for design practice is that architects must be willing to adopt an evolutionary – rather than a revolutionary – stance toward architecture and urban design.

D. Urban scenes

Editor's introduction

Included in this section are four papers on urban aesthetics, including appraisals of central-business-district scenes (Nasar), residential scenes (Talbot; Nasar), and the commercial strip (Nasar). Because this section does not include a paper on the relevant scales and dimensions for assessing the quality of urban scenes, I have included in this introduction a brief description of my inquiry into this question.

First, a list of seventy-three adjectives referring to environmental affect from Craik's (1966) Landscape Adjective Checklist (see Editor's Introduction to "Natural and Rural Scenes") and from Kasmar's (1970) lexicon were compiled. Fifty lay respondents and twenty graduate students in planning, architecture, and landscape architecture were asked to select from the list the adjectives that they considered to be most relevant to the assessment of urban scenes. For this task, the order of adjectives was varied for each respondent. The most frequently selected scales are shown in Table II.1. Subsequent analysis revealed that ratings of urban scenes loaded on three factors. As was found by Russell, there was a pleasantness (like–dislike, beautiful–ugly, interesting–boring, inviting–repelling, and attractive–unattractive) and an arousal (tense or excited vs. relaxed or calm) dimension. The third factor that emerged, safety, is not unlike Russell's relaxing variable (a mix of pleasantness and low arousal). Thus studies of urban affect might do well to assess pleasantness, arousal, excitement, and safety through scales such as appeal, active, exciting, and safe.

Now consider the papers included in this section. In the first paper, Nasar examines the evaluative responses to central-business-district scenes in Japan and the United States by student respondents from each country. The two groups exhibited similar patterns of response. Preference increased with Order and Diversity, and each group preferred the foreign (more novel) scenes to the native (more commonplace) ones.

Nasar, in the second paper, examines evaluative response to residential scenes. Three perceptual and cognitive dimensions in response to these scenes emerged: Openness, Diversity, and Clarity. The evaluative responses of different

257

Table II.1. *Urban affect adjective list*[a]

Most frequently selected by

Laypersons	Professionals	Combined groups
1 Exciting	1 Exciting	1 Exciting
2 Adventurous	2 Active	2 Alive
Alive	Stimulating	3 Active
Lively	3 Alive	Lively
3 Impressive	Expensive	4 Stimulating
Moving	Powerful	5 Adventurous
4 Active	4 Invigorating	6 Appealing
5 Appealing	Lively	Moving
Stimulating	5 Appealing	7 Fashionable
6 Fashionable	Attractive	Impressive
Friendly	Fashionable	8 Attractive
7 Powerful	6 Adventurous	Invigorating
Brisk	Comfortable	9 Brisk
8 Attractive	Inspiring	Inspiring
Inspiring	Threatening	10 Friendly
Invigorating	7 Good	11 Beautiful
9 Forceful	Moving	Nice
Majestic	8 Brisk	12 Powerful
10 Beautiful	Cheerful	Safe
Pleasant	Dangerous	
Threatening	Impressive	
11 Nice	Lonely	
Rich	Nice	
Safe	9 Beautiful	
12 Warm	Safe	
13 Pretty		
Cold		

[a]Adjectives appear in rank order from most to least frequently selected.

observers responding to different sets of scenes were compared and found to be similar. Scene evaluation improved with increments in clarity, ornateness (a diversity variable), and openness.

In the next paper, Talbot reports three studies that center on urban open space. She interviewed residents in older urban neighborhoods, modern apartment complexes, and elsewhere. Although the respondents spoke favorably of open space (prospect), they did not necessarily prefer large spaces. Instead, they preferred spaces that had trees, bushes, and paths. As was suggested by Nohl, people enjoyed more naturalistic open spaces. The respondents, however, also liked mowed areas, which they associated with smaller spaces.

Nasar centers on the retail signscape in the final paper in this section. Evaluative ratings were obtained in response to photographs of a model of a commercial strip in which the signscape was varied in coherence (by reducing contrast and size) and complexity (number of different sizes, shapes, and colors). As was expected, contrast reduced coherence. Furthermore, as predicted by S. Kaplan's two-process model, evaluation was most favorable for moderate complexity and high coherence.

18   Visual preferences in urban street scenes: a cross-cultural comparison between Japan and the United States

Jack L. Nasar

While some visual preferences in the environment may be influenced by culture, others – universals – may be common to many cultures. Environmental psychologists have examined the relationship between visual aspects of the physical environment and affect (for a review, see Wohlwill, 1976). Acknowledging individual variation in response, some researchers have sought with some success to define general principles that guide preference. However, much of the research has centered on North American and Western European populations. Although these populations are diverse, their common background (as Westerners) may have influenced their preferences. Cross-cultural comparisons with other populations (Canter and Thorne, 1972; Sonnenfeld, 1969) have provided and can provide further insight into the nature of environmental preference. This research represents one such comparison – between Japan and the United States. The aim was to uncover those preferences common to persons in both of these complex and industrialized cultures.

While sharing a status as developed and industrialized nations, Japan and the United States have distinct differences in their cultures and physical environments (Canter and Canter, 1971; Hall, 1966, pp. 49–54; Rapoport, 1977, p. 43). If, despite these differences, common patterns of preference in relation to the visual environment are found among persons from these two nations, then such preferences might be interpreted as representing universals.

In this research, several characteristics of the visual environment were selected as most likely to relate to preferences for urban scenes. These characteristics included novelty, complexity, order, naturalness, openness, upkeep, and prominence of vehicles. The first two characteristics, novelty and complexity, were drawn from the work of Berlyne (1972) and Wohlwill (1976).

An earlier version of this paper was presented at the International Conference on Psychology and the Arts, Cardiff, Wales, September 5–9, 1983. This paper was first published in *Journal of Cross-Cultural Psychology* 15 (1984): 79–93.

Berlyne (1972) described hedonic response as a function, in part, of stimulus properties that generate uncertainty/arousal. He put forth collative variables (those that involve some comparison among stimulus elements) as generating uncertainty/arousal and included as two of these variables, complexity and novelty. According to Berlyne (1972), the relationship between hedonic response and collative properties took the form of an inverted U-shaped function, in which increases in uncertainty/arousal produced increases in pleasure up to a point, after which pleasure decreased.

With regard to novelty, Berlyne (1972) described two kinds: "absolute" novelty, in which the stimulus is unlike anything in the observer's experience; and "relative" novelty, in which the stimulus has elements that to the observer are somewhat familiar but are organized in an unfamiliar way. According to Berlyne (1972), moderate amounts of novelty, such as that engendered in "relative" novelty, would be expected to produce pleasure. This expectation has received some empirical confirmation. Oostendorp and Berlyne (1978a, 1978b) and Gärling (1976a) found the expected relationships between looking time and general evaluation related to rated novelty (usual–unusual and familiar–unfamiliar) of man-made scenes. Canter and Thorne (1972) found that residents of Scotland and Australia preferred unfamiliar to familiar urban scenes. In a footnote, these authors indicate similar findings for residents of Japan.

In contrast, Sonnenfeld (1966) suggested the presence of much individual variation in response. In particular, he found that younger persons preferred exotic natural scenes, while most others preferred more familiar scenes. It seems, however, that his exotic scenes, unlike the man-made novel scenes of Oostendorp and Berlyne (1978a, 1978b) and Gärling (1976a), might have presented extreme levels of novelty rather than moderate or "relative" novelty. A preference for the familiar was also found by Nasar (1980): Two groups 'of persons each preferred scenes from their immediate neighborhoods to scenes from the neighborhood of the other group. However, in this case because the comparison was between two sets of stimuli, both of which were familiar, the results suggest a preference for the extremely familiar over the familiar. The research reported here compared responses to relatively familiar scenes with responses to relatively novel scenes. It was expected that the relatively novel scenes would be preferred. Specifically, Ss from Japan and Ss from the United States, viewing scenes of typical urban streets, some of which were in Japan and some of which were in the United States, were expected to prefer the scenes from the others' country to the scenes from their own country.

Complexity is the other collative variable considered here. Wohlwill's (1976, pp. 45–47) extensions of Berlyne's (1972) work into man-made environments confirmed the optimal level principle as applicable to environmental complexity and preference. It was expected that for respondents from both Japan and the

United States, preference would relate to rated complexity and the relationship would take the form of a curvilinear function.

The remaining environmental properties considered here were selected not so much because of their place in a theoretical framework, but because of the empirical evidence suggesting their contribution to environmental preference. Nevertheless, these variables, which include order, naturalness, upkeep, and vehicle prominence, have an ecological flavor (Gibson, 1979). Each might be associated with possibilities for either avoiding biological dangers or gaining biological benefits. For example, order might afford the observer the capability of making sense of the environment; naturalness might afford the observer an escape from the pollution of urban settings; upkeep might provide the observer with clues about the potential for crime or for structural dangers in the settings; and prominence of vehicles might provide an indication of the chance of casualty. In any case, in various studies, order and related constructs such as organization, legibility, coherence, and fittingness have been found to be related to environmental preference (Ertel, 1973; S. Kaplan, 1975; Lowenthal and Riel, 1972; Wohlwill, 1979); naturalness has been found to be related to environmental preference (Appleyard and Lintel, 1972; Kaplan, Kaplan, and Wendt, 1972; Peterson, 1967); openness, depth, and spaciousness have been found to be preferred attributes to scenes (Flaschbart and Peterson, 1973; Gärling, 1976a; Hesselgren, 1975; Horayangkura, 1978; Wohlwill, 1974); and well-kept and well-maintained environments have been found to be preferred to others (Cooper, 1972; Marans and Rogers, 1973; Peterson, 1967). Finally, Appleyard (1976) demonstrated the influence of traffic on environmental response. In natural settings, of course, some of these variables would be interrelated. It was expected here that for respondents from both Japan and the United States, environmental preference would relate to these attributes taken separately and in those naturally occurring combinations.

Methodology

Because of financial and time constraints, it was possible to examine only a limited selection of cities, street types, and subjects in Japan. For purposes of cross-cultural comparison, similar settings were examined in the United States. While these restrictions limit the generality of the findings, through this study some tentative thoughts about common patterns of environmental preference could be advanced.

Environmental stimuli

This research centered on one type of environmental experience – the view during the daytime of the environment as seen by the driver on a major artery through the center of the city. The decision to examine one kind of experience grew partly from the practical constraints and partly from the investigator's belief that combining different land uses (such as residential and office) and different modes of experience (such as walking and driving) as if they were one might have produced confounded results. Of course, since the study examines only one kind of scene, the findings may have limited generality to other kinds of environmental experience.

Of the various kinds of streets in urban settings, major arteries (those that most residents have experienced) were selected for examination. The impact of their street-scape on the shared aesthetic image of the city (Lynch, 1960) is likely to be more extensive than that of lesser streets. Knowledge about the preferred attributes of such scenes is important, because the informed management, planning, and design of the view from intensively used arteries could contribute more to the improvement of the shared aesthetic image of the city than could similar attention to the view from lesser streets.

The arteries were selected as follows. In each of four cities in Japan and four cities in the United States, three or four residents were asked to identify a major street, one that because of its location and function in the city would be known by most residents. While this was not a scientific survey, the responses in each city were consistent. Residents selected heavily traveled wide arteries through a business center of each city. In the Japanese cities, which unlike those in the United States may have several intense business centers (Canter and Canter, 1971), it was still possible to identify one of the major arteries for study. The streets in each country might be considered comparable in that they all represent highly used and well-known arteries through core areas of the cities.

The eight cities (four in the United States and four in Japan) that were selected for study were:

Tokyo (8,448,000 population)
Osaka (2,682,000 population)
Kyoto (1,468,000 population)
Sendai (628,000 population)
San Francisco, California (3,221,000 population)
Cleveland, Ohio (1,898,000 population)
Cincinnati, Ohio (1,391,000 population)
Columbus, Ohio (1,093,000 population)[1]

The investigator selected from larger cities some that had some diversity in age, structure, economic base, and population size. While the Japanese and American cities are not alike in density, age, or mixture of land uses, these differences reflect socioeconomic and sociophysical differences in the structure of Japanese and American development. The physical characteristics of the street scenes used from each city displayed some variability. The details of these differences are set forth later in this paper.

Mode of presentation

To allow for realistic experiences and meaningful comparisons among environmental preferences of persons from each culture, individuals from each nation should observe the same scene *in situ*. Because of the impracticality of bringing observers from both Japan and the United States to urban scenes in each country, color videotapes and slides of cities in each country were used as the next best alternative. These simulations were brought to the observers in each country.

The videotape was used to give respondents a sense of the movement through the cities. Each videotape presented a three-minute-long drive down one of the streets identified earlier. Efforts were made to keep the driving speed at approximately 20 mph and the car in the middle lane. While such control was possible in the United States, it was less so in some of the Japanese cities, where a dependence on hired drivers produced somewhat faster and more erratic drives. To make each trip appear as a continuous nonstop drive, the investigator stopped the tape whenever the car stopped and restarted the tape when movement resumed. Because the field of vision of the videotape lens was narrower than the actual view experienced from the car, the camera was systematically panned from one side of the street to the middle to the other side of the street. While the systematic movement of the camera may have reduced the reality of the simulation, that movement and loss were considered preferable to the loss in reality from the exclusion of the peripheral elements.

The color slides (35-mm wide-angle views) were shot at roughly ten-second intervals throughout each drive. The slides showed the streets as they might be seen by a driver looking straight ahead down the center of the street. Three slides from each drive were selected at random. If a slide selected had in it the same physical element (such as a building or sign) that was prominent in another that had been selected, the slide was replaced by another random selection. In this way, the slides did not replicate the same scenes. The resulting twenty-four slides were used as stimuli in this research.

Description of the environments

The investigator and a group of three graduate students in city and regional planning assessed physical characteristics of the scenes studied. This group of judges consisted of two American males, one Turkish female, and one Korean female, all with training in architecture.

The judges assessed the environment portrayed in each tape and each slide on a set of 7-point bipolar adjective scales prepared by the investigator. The scales were closed–open, simple–diverse, chaotic–orderly, dilapidated–well-kept, vehicles prominent–vehicles not in sight, and nature (greenery) not in sight–nature

Table 18.1. *Characteristics of the measures of environmental attributes*

	Interobserver reliability (Cronbach α)	r	Variation over slides	
Description			Mean	SD
Nature not in sight–nature prominent	.87	.61	4.35	1.51
Vehicles prominent–vehicles not in sight	.83	.56	3.34	1.25
Chaotic–orderly	.81	.53	4.05	1.02
Dilapidated–well-kept	.79	.48	3.25	.80
Closed–open	.76	.45	4.52	.89
Simple–diverse	.56	.25	3.68	1.06

(greenery) prominent. The investigator presented to the judges a description (verbal and graphic) of the poles of each scale. To reduce order effects, the investigator instructed the judges to vary the order in which they used the scales in their assessments of each slide.

The interobserver reliability scores indicated that the judgments were fairly consistent. The Cronbach α interobserver reliability scores for the four judges ranged from .56 to .87. These are displayed for each descriptor in Table 18.1.

Further inspection of Table 18.1 reveals that there was small but nevertheless noticable variation in the scenes along each descriptive dimension. The standard deviations for the seven descriptor means ranged from a low of .80 for upkeep to a high of 1.51 for naturalness.

Subjects

In Japan, twenty-nine graduate students (eighteen in architectural engineering and eleven in environmental engineering) participated as subjects (Ss) in this study. They were assembled at Osaka University by faculty at both that university and Kyoto University. All but one of the Ss were male.

In the United States, seventeen graduate students in city and regional planning at The Ohio State University participated as Ss in this study. They were assembled by the investigator. Thirteen were male, and five were female.

In both settings, the Ss were approximately twenty to thirty years old.

This research centered on uncovering commonalities in environmental preference. In this regard, the characteristics of the two groups are important. On the one hand, if the groups are select and like subsamples (preprofessional graduate students, primarily male) of the population in each culture, then shared preferences between them could just as well stem from similar preprofessional values as from universals spanning the two cultures. On the other hand, if the groups differ from each other in substantial ways, then shared preferences between them are more likely to represent universals.

While the groups are alike in age, sex, and level of education, they are different in some important ways. First, the Ss in each group were raised in two distinct cultures that have very different languages, environments, religious traditions, and design traditions. Second, the experience of the two groups with their surroundings differed: The Japanese Ss resided in a city almost twice the size and of much greater density than that in which the U.S. Ss resided; and unlike their American counterparts, few of the Japanese Ss owned or operated a car (they relied on mass transit and, in particular, subways and trains). Third, the education of the two groups differed. Not only do the two systems of education differ, with the one emphasizing imitation and the other emphasizing individuality, but also the specific educational experience of the two groups differed: Although both groups were enrolled in professional programs, the Japanese Ss came from technical undergraduate disciplines (such as architecture or engineering), and their graduate course work centered on technical fields (such as urban design and transportation engineering); in contrast, the U.S. Ss came from both hard and soft undergraduate disciplines (including music, geography, psychology, and architecture), and their graduate work consisted of interdisciplinary studies in the social and environmental sciences. While results gleaned from the population subsamples examined here may have limited generality to persons differing in age, sex, or level of education, the differences between the two groups would support the interpretation of common patterns of preference as stemming from broader origins than simply shared values between two like groups.

Procedure

Ss were presented the instructions in written and oral form in their native language. They were told that they would be viewing and responding to slides and videotapes of the view seen while driving on major streets in several cities in Japan and the United States. They were asked to imagine themselves driving through the streets, and they were instructed not to let the kinds of responses requested influence that experience. Finally, they were told that there were no right or wrong answers and that they should just answer with their honest opinions. They were asked to use the 7-point bipolar rating scales pleasant–unpleasant (心地よ・, ー 不快) and interesting–uninteresting (興味がある ー 興味が な・,) to evaluate each tape and slide. These measures were selected to represent the two components ascribed to aesthetic response: attention (interestingness) and pleasure (pleasantness) (Wohlwill, 1976).

It required a two-hour discussion with two Japanese translators to arrive at the best Japanese translation. Nevertheless, it is possible that the translations are not fully equivalent. However, the finding of similar patterns of response in the United States and Japan would be evidence that the translations were equivalent. If different patterns of response were found, the differences could be attributed, in part, to possible confounding influences of the measure.

Ss were asked to hold a viewer (a 3-inch long and ½-inch diameter tube) to one eye and cover the other eye when viewing the scenes. The viewing tube was used to eliminate from view extraneous visual elements in the experimental rooms. Ss were shown the eight tapes followed by the twenty-four slides. In both Japan and the United States, two different orders of presentation were used to mitigate order effects.

Results

Although data were gathered for the twenty-four slides and eight tapes, only the results for the twenty-four slides are reported here. The tapes were used to provide the Ss with a sense of both the surroundings of and the movement through the scenes (shown as slides). The interpretation of the results for all thirty-two stimuli would have been confounded by both the interdependence of the slides with the tapes and the noncomparability of the way in which individuals might respond to single scenes (as slides) and three-minute trips. The small sample size for the tapes (four in each country) prohibited any meaningful analysis of them alone.

Two steps were taken in preparing the affect data for the analyses: First, for the Ss from each country separately, the average pleasantness scores on each slide were calculated, as were the average interestingness scores. Then for each slide, the pleasantness and interestingness scores were combined into a composite.

The use of average scores for each group assumes that the intragroup variability in each sample was small. To test this assumption, Cronbach interobserver reliability scores were calculated separately for each measure (pleasantness and interestingness) and for each group of Ss across the twenty-four slides. In every case (Japanese assessment of pleasantness, Japanese assessment of interestingness, American assessment of pleasantness, and American assessment of interestingness), the results indicated that intragroup variability was small: Every Cronbach α score was above .90. These results suggest that the group means for each sample represent reliable measures.

To test the reliability of combining interestingness and pleasantness into a composite scale, the correlations between these two measures were examined. For both Japanese and U.S. Ss, the two measures were found to be interrelated: The scores on pleasantness and interestingness for the twenty-four slides were

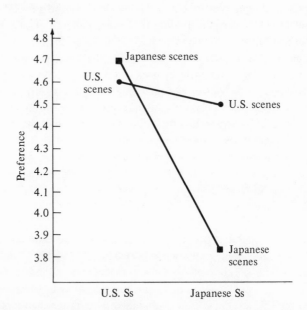

Figure 18.1. The interaction of nationality and country viewed in relation to preference.

highly correlated (U.S. Ss: $r = .71$, 23 df, $p < .001$; Japanese Ss: $r = .85$, 24 df, $p < .001$). These results suggest that the combination of the pleasantness and the interestingness scores would represent a reliable composite. The analyses that follow are based on the composite scores (called composite preferences), which include the scores of U.S. Ss in response to each of twenty-four scenes (twelve from each country) and the scores of Japanese Ss in response to each of twenty-four scenes (twelve from each country).

As was expected, the Ss from Japan and from the United States exhibited similar patterns of response. The twenty-four composite preference scores from each nation showed strong associations with one another ($r = .54$, 23 df, $p < .01$). More important, Ss from each country rated more favorably the scenes from the other country than the scenes from their own country. As can be seen in Figure 18.1, the comparison of composite preferences from each group for native and for foreign scenes indicated an interaction (approaching significance) between nationality of the Ss and country viewed ($F = 3.95$, 1, 44 df, $p = .05$). This difference in preference was more pronounced for the Japanese Ss. With all thirty-two stimuli, this interaction was significant ($F = 4.81$, 1, 60 df, $p < .05$).

Comparing the judgments of the properties of the Japanese and American scenes, the investigator found no differences in the judged openness, diversity, naturalness, and order of each set of scenes. However, the Japanese scenes were described as having significantly more vehicles than the American scenes (vehicles prominent–vehicles not in sight: Japan = 2.87, U.S. = 3.91) (F = 4.90, 1, 23 df, p < .05) and less order (chaotic–organized: Japan = 4.44, U.S. = 3.67) than the American scenes: The latter effect approached significance (F = 3.96, 1, 23 df, p = .06). From these findings (note also the correlations reported below between preference and both prominence of vehicles and order), one might expect the American scenes to receive more favorable evaluations than the Japanese scenes. The analysis revealed no significant difference in the composite preference scores for the scenes from each nation.

For each cultural context, this research examined the first-order correlations between preference for scenes and physical attributes of scenes. As was expected, there were several similarities in the correlates of the expressed preferences of the U.S. and Japanese Ss: Increases in preference for scenes were found to be related to increases in the upkeep, naturalness, absence of vehicles, and scene-order. As can be seen in Table 18.2, for both groups, composite preference had moderate to strong correlations with upkeep (U.S.: r = .53, 23 df, p < .01; Japan: r = .55, 23 df, p < .01), prominence of nature (U.S.: r = .64, 23 df, p < .01; Japan: r = .61, 23 df, p < .01), and prominence of vehicles (U.S.: r = −.40, 23 df, p < .05; Japan: r = .44, 23 df, p < .01). For the Japanese Ss, composite preference displayed a significant correlation with organization (r = .57, 23 df, p < .01), while for the U.S. Ss, this effect only approached significance (r = .35, 23 df, p < .09). One noticeable difference between the two groups was the relationship of contrast to preference: For the Japanese Ss, unlike their counterparts in the United States, composite preference was associated strongly with contrast (r = .52, 23 df, p < .01).

Stepwise multiple regression procedures (Helwig and Council, 1979, p. 391) were employed to examine the relationship between composite preferences and several environmental attributes in combination. Three models were developed, one for the Japanese Ss, one for the U.S. Ss, and one for the two sets of Ss. Variables were added to a model only if the F value for their addition was significant (p < .05). The composite preference scores of each group of Ss were treated as criterion variables in the models. Factor-based scores for the attribute measures were treated as predictor variables: To arrive at these predictor variables, the investigator condensed the assessments of physical properties of the slides through principal-axis factor analysis and then calculated factor-based scores. The derivation of these factor-based scores is reported first.

The factor analysis produced a three-factor solution, which accounted for 76% of the trace. Each factor had an eigenvalue greater than 1.0. The varimax rotated solution was selected for use because the interpretation of it was clearer than that of the unrotated factor solution. Those variables with substantial loadings (i.e., greater than .50) on the first factor, which accounted for 33% of the trace,

Table 18.2. Correlations between scores on scales describing and evaluating the environment

Assessment	Nature	Order	Upkeep	Open	Diverse	Vehicles	Contrast	U.S.Ss
Nature not in sight–nature prominent								
Chaotic–orderly	.36							
Dilapidated–well-kept	.53[b]	.71[b]						
Closed–open	.25	-.07	-.14					
Simple–diverse	.02	-.14	.07	.27				
Vehicles not in sight–vehicles prominent	-.07	-.37	-.21	-.18	-.22			
Low contrast–high contrast	.24	-.10	.20	.40[a]	.73[b]	-.28		
Composite preference (negative–positive)								
American Ss	.64[b]	.35	.53[b]	-.15	.26	-.40[a]	-.32	
Japanese Ss	.44[b]	.57[b]	.55[b]	.25	.52	-.61[b]	-.52	.54[b]

[a] $p < .05$
[b] $p < .01$

included chaotic–orderly (.90), dilapidated–well-kept (.88), and nature not in sight–nature prominent (.57). Those variables with substantial loadings on the second factor, which accounted for 27% of the trace, included simple–diverse (.87), low contrast–high contrast (.85), and vehicles prominent–vehicles not in sight (−.55). Finally, those variables with substantial loadings on the third factor, which accounted for 16% of the trace, included closed–open (.77) and nature not in sight–nature prominent (.71). These factors might be designated as representing Order, Diversity, and Openness. Each factor had significant first-order correlations with Japanese composite preference ($r = .53$, 23 *df, p* < .001; $r = .59$, 23 *df, p* < .01; $r = .46$, 23 *df, p* < .05, respectively) and with American composite preference ($r = .61$, 23 *df, p* < .01; $r = .40$, 23 *df, p* < .05; $r = .41$, 23 *df, p* < .05, respectively). For the first two factors, the factor-based scores were calculated by multiplying the scores on each important measure in a factor by that measure's factor loading, adding those weighted scores together, and dividing that sum by the sum of the loadings considered. As the third factor included a variable (nature not in sight–nature prominent) with substantial loading on the first factor, that variable was removed, and the scores on the remaining scale, open–closed, were used unaltered.

The three sets of scores were then regressed onto the composite preference scores. For each group of Ss, a similar pattern of preference emerged: Preference was associated with increases in Order and increases in Diversity. For the U.S. group only, preference was associated with increases in Openness as well. As the predictor factors were orthogonal, potential problems with multicolinearity were considered as negligible. The resulting equations are reported below.

U.S. student preference = −.13 + .55 Order + .22 Diverse + .35 Openness
$$\text{Cumulative } R = .66, 4, 23 \ df, p < .001$$
Japanese student preference = 1.51 + .43 Order + .19 Diverse
$$\text{Cumulative } R = .58, 4, 23 \ df, p < .001$$
Pooled Preference = 1.96 + .49 Order + .17 Diverse
$$\text{Cumulative } R = .47, p < .001$$

The direction of the relationships between factor scores and preference in each model was the same. A Chow test was undertaken to establish whether the two separate regression models for Japan and the United States could be pooled. For this test, a model for American preference was calculated with only two independent variables, as in the other equations: Preference = 1.69 + .56 Order + .15 Diverse. The results ($F = 4.47$, 3, 42 *df, p* < .05) indicate that the separate equations could not be pooled. A comparison of the preference scores of each group shed some light on this finding. The U.S. Ss responded with higher preference ($\bar{x} = 4.7$) than did the Japanese Ss ($\bar{x} = 4.1$) ($F = 7.03$, 1, 23 *df, p* < .05). That is, the difference in preference was one of intensity, not direction.

Finally, it was expected that the relationship between diversity and preference would take the form of a nonlinear function. A visual inspection of the scatter plots provided no evidence that a curvilinear relation was present. Furthermore, the test of the fit of quadratic functions indicated that no quadratic function adequately fit the data.

Discussion

The results provide evidence that young educated adults from two distinct cultures share preferences in relation to certain visual aspects of street scenes along major arteries. Further investigations would be required to demonstrate whether the same patterns found here would emerge for different kinds of scenes or for different groups (such as persons from less developed countries, or persons of different age, sex, or education from those here).

Nevertheless, the results may be interpreted as supporting the universality of certain environmental preferences. The groups differed in culture, language, religious traditions, environmental experience, social experience, and educational emphasis. If the commonalities in preference between the two groups result from the similar composition of the Ss (adult college-educated males) in the groups, then the effect of selection and education must be considered as powerful enough to overrule the effect of differences in culture, language, religion, environmental experience, and educational emphasis. It is more plausible that the commonalities found here are of broader origins: Possibly they represent universals.

The results provide support for Berlyne's (1972) position that persons prefer relative novelty to familiarity. Ss were presented relatively commonplace scenes in cities. The finding that Japanese Ss preferred the American scenes to those in Japan and that the reverse was true for U.S. Ss is best explained as resulting from the difference in the perceived novelty of the same scenes by each group. Although novelty was not measured, there does not seem to be a plausible alternative explanation to the results found.

Perhaps these results would not apply to an older population. Sonnenfeld (1966) contends that younger persons prefer novelty, whereas others prefer familiarity. However, his findings are based on responses to some scenes that were much more exotic than those examined here. We might conclude, then, that relative novelty (as was examined here) is preferred and that extreme novelty is disliked by most persons, with the exception of those who, like the young, have a higher tolerance for such extremes. However, an empirical examination of the responses of older persons to relatively familiar and relatively novel scenes (such as those examined here) would be needed to verify this expectation. It is possible also that extremely familiar scenes would be preferred to others. Consider Nasar's (1980) finding that for two groups of persons, each preferred scenes from its immediate neighborhood to scenes from the other group's neighborhood. To better understand environmental preferences in regard to familiarity or novelty, it might be necessary to examine the responses of persons to scenes that cover the full continuum, including scenes seen daily, scenes seen less regularly, commonplace scenes never seen by the observer, novel scenes never seen by the observer, and extremely novel scenes never seen by the observer.

With regard to the other collative variable considered – diversity – the results here are ambiguous. The expected nonlinear relationship between assessed diversity and preference did not appear. Instead, the factor Diversity was found to have a linear relationship with preference. These unexpected results may stem from measurement problems. In the scenes selected, increases in diversity were correlated with decreases in the prominence of vehicles. Furthermore, the interobserver reliability in the assessment of diversity was relatively weak (.57). These contaminations of the diversity variable confound the interpretation of its relation to preference.

The expected preference for ordered scenes, natural scenes, well-kept scenes, and open scenes was supported by this cross-cultural comparison. For both sets of Ss, preference was associated with changes in organization, naturalness, upkeep, and openness. While correlations do not indicate cause, there is no reason to suspect some other (unmeasured) attribute as producing the associations found. However, there is a problem in deriving cause here, because of the interrelationships found among the independent variables (naturalness, order, and upkeep with one another and openness with naturalness). This interdependence makes possible a number of alternative explanations of the role of these variables in affect, and raises some researchable questions. Perhaps order produces preference, and upkeep and nature contribute to perceived order; but would another explanation of the workings of these three variables be better? Perhaps openness is preferred, and preference accounts, in part, for preferences for natural scenes; or does a preference for naturalness account for the preference for openness in the urban scenes examined? Further investigation might be needed to arrive at a correct interpretation of the role of each of these variables in environmental preference.

Finally, preference was found associated with traffic. Appleyard (1976) found that residential streets with heavier traffic were judged as less preferred, less safe, noisier, and higher in air pollution than other streets. In this research, unlike Appleyard's, Ss only saw the vehicles: They neither heard them, saw them moving, nor smelled their exhaust fumes. If the correlation between vehicle prominence and preference was causal, it is likely that the effect would be more pronounced for the observer experiencing the vehicle *in situ,* where sound, motion, and odor would be present.

In conclusion, this research has identified several common patterns of preference among select Ss from two distinct cultures. If these findings are found stable for other groups and if causality is established, then urban-design solutions that incorporate moderate novelty, increased diversity, increased contrast among buildings, good maintenance, order, more vegetation, and reduced prominence of vehicles might produce favorable reactions among the public, who are the consumers of urban-design projects. Because the variables cited represent perceived rather than actual qualities of environments, designers may be able to alter perceptions of these attributes without altering the attributes directly. For example, a public space might be made to appear more open through designs that accentuate the horizontal rather than the vertical dimension. Thus relatively inexpensive alterations in the semifixed features of urban spaces may achieve considerable effects.

Acknowledgments

This research was supported by a grant from the United States–Japan Consortium for Environmental Design and Planning Education. The generous assistance of Kunihiro Narumi and Yukio Itsukushima and many others in Japan, and the contributions of Ernest Arias in gathering some of the data and of Mal-Soon Min in coding the data are acknowledged.

Note

1. Japanese population figures are from *Europa Yearbook: A World Survey* (1980), and United States figures are from the U.S. Bureau of the Census (1980).

19 Perception and evaluation of residential street scenes

Jack L. Nasar

In recent years, public interest and policy (Carp, Zawadaski, and Shokrin, 1978; National Environmental Policy Act, 1969) have created a demand for objective information on perceived quality of the environment. Such information may be of particular importance for housing, because people typically spend more time in their homes than in any other place in the city (Chapin and Hightower, 1965) and because housing typically occupies more area than does any other land use in a city. Thus the home has been described as a primary territory that is "central to the day-to-day lives of occupants . . . [and] psychologically important" (Altman and Chemers, 1980, p. 129). Presumably, then, the visual quality of housing and the neighborhood is of some importance.

Nevertheless, some may question whether empirically based guidelines for visual quality are needed. For private housing, it might be argued that the market will produce a desirable outcome: People will select homes that they like. However, free choice at the individual level may result in visual disorder at the neighborhood level: Public regulations (such as zoning, design review, and building codes) are needed to control development. For public housing, the potential resident has little influence or choice with regard to the visual milieu of the housing, and the evidence (see Groat, 1982; Hershberger, 1969; Newman, 1972) suggests that the educated intuition of the design professional may not result in a milieu that fits the visual needs of the residents, especially if they differ in life style and social class from the professional. This paper reports several studies aimed at uncovering visual-quality needs in housing scenes as these scenes are experienced by the public on a daily basis. These studies examined public perceptions of and preference for residential street scenes in relation to judged attributes of the scenes.

Several attributes may have particular relevance to aesthetic response. They include Naturalness, Openness, Diversity, and Organization. Research suggests that individuals tend to prefer natural to built elements, openness to enclosure, order to disorder, and some diversity (Appleton, 1984; S. Kaplan, 1979a; Ulrich, 1983; Wohlwill, 1976). For design application, it is important to determine

275

whether these attributes are prominent in the public's perception and evaluation of housing scenes as experienced on a daily basis.

Methodological concerns

One approach to examining dimensions of perception has involved the factor analysis of verbal descriptions of environments (Canter, 1969; Collins, 1969; Hershberger, 1969; Küller, 1975; Lowenthal and Riel, 1972). This approach has been criticized because the results may be an artifact of the investigator's selection of scales. In response to this problem, similarity ratings, multidimensional scaling, and cluster analysis have been suggested as alternatives (Groat, 1981; Horayangkura, 1978; Oostendorp and Berlyne, 1978a). Through examining scene attributes in relation to the scene locations in the derived space, one can derive labels for the dimensions and clusters. Thus this research used similarity ratings and multidimensional scaling to derive dimensions of perception. Labels for the dimensions were obtained by examining scene attributes in relation to the derived multidimensional space.

Other methodological choices were also made with an eye toward applicability. Thus the research sought a realistic mode of environmental experience, a diverse and representative sample of respondents, a relevant set of environmental attributes, and a relevant set of scales to measure the emotional quality of the environment.

Wide-angle color photographs and slides of roadside residential scenes were used as stimuli. A surrogate for on-site experience was necessary because of the cost and impracticality of obtaining responses to a variety of housing scenes on site. Color photographs and slides were chosen because of the evidence that responses to such simulations accurately reflect responses on site (Hershberger and Cass, 1974; Shafer and Richards, 1974; Zube, Pitt, and Anderson, 1974). Nevertheless, to test the validity of the simulation, responses were also obtained on site at selected sites.

A diverse sample was obtained. Research indicates potential differences in environmental preferences in relation to sociodemographic characteristics of the observer (Sonnenfeld, 1966; Zube, Pitt, and Evans, 1983). Such differences point to two research strategies: one aimed at identifying group differences in preference, and the other aimed at finding commonalities in preference across the groups. In either case, a diverse sample of respondents is called for. This research examined commonalities in response because of the relevance of such information to the design of public spaces.

A comprehensive set of attributes (including those cited earlier as likely influences on aesthetic response) was assessed. The full list is detailed later in this paper. One can measure attributes through direct physical measurement or through ratings by judges. Ratings were used here, because it was thought that if the physical and perceived characters differ, the perceived character may have more impact on evaluative response. Because the attribute scores are the result of ratings, they are perhaps more accurately labeled as perceptual and cognitive attributes of the scenes.

Four aspects of evaluative response were assessed: pleasantness, interest, safety (from crime), and desirability as a place to live. This includes two dimensions – hedonic value (pleasantness) and involvement/arousal (interest) – that are considered to be important aspects of aesthetic response (Berlyne, 1974c; Wohlwill, 1976). This also includes the three components of evaluative response found by Russell (this collection). Recall that he found pleasantness as one of two main factors. Mixtures of pleasantness and the nonevaluative arousal produced two subfactors of evaluation: excitement (interest) and distress (safety). Finally, two of the variables – judged safety from crime and judged desirability as a place to live – were included for their likely relevance to behavior in urban settings: Fear of crime may influence urban spatial behavior (Newman, 1972), and judged desirability of housing as a place to live may give a better indication of actual housing choices than would judgments of affective quality.

This research has three phases. In the first phase, dimensions of perception are derived; in the second phase, perceptual and cognitive correlates of those dimensions are examined; and in the third phase, evaluative response is examined in relation to the perceptual and cognitive aspects of the scenes. All three phases use the same housing scenes as stimuli, so that sample is described below.

The housing scenes

The sample of stimuli included wide-angle (28-mm) color photographs and slides of two sets of thirty residential scenes in the city of Pittsburgh, Pennsylvania. To obtain this sample, I selected at random twelve observation points and viewing directions within each of the eight residential-land-use categories in the city. Each scene was photographed as it would be seen by passers-by. To mitigate potential biases related to photographic technique (Hochberg, 1966), each scene was photographed in controlled conditions: clear autumn days, eye-level views down the road lining up but excluding cars on the left side of the road, and sunlight behind the camera. Of the eighty-four scenes photographed, twenty-four (three from each land use) were eliminated because of problems with focus, lighting conditions, distortion, or movement.

Because previous success with color slides and photographs (Hershberger and Cass, 1974; Shafer and Richards, 1974; Zube et al., 1974) does not guarantee success for the photo simulations here, the responses to the photographs were compared with on-site responses at a randomly selected set of the scenes. The

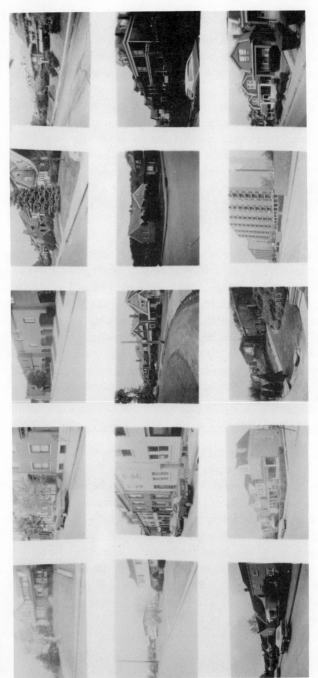

Most preferred

278

279

Least preferred

Figure 19.1. Slide set 1 (from most preferred to least preferred).

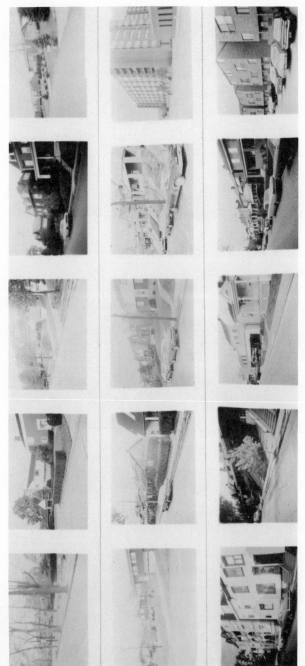

Most preferred

280

Least preferred

Figure 19.2. Slide set 2 (from most preferred to least preferred).

on-site responses were obtained from groups of individuals who lived within a block of each scene examined and who indicated familiarity with the scene. In contrast, the responses to the photographs were obtained from individuals who, because they resided in a city over 100 miles away from the sites, were likely to be unfamiliar with the scenes. Thus evaluations *on site* by observers *familiar* with each scene (each observer only responding to one scene) were compared with repeated-measure evaluations of *photographs* of the same scenes by observers *unfamiliar* with the scenes. The results revealed significant correlations between the responses on site and the responses to the photos. Furthermore, no significant differences between the groups were found. It seems likely that the responses to the simulation accurately gauge day-to-day experience.

Phase 1: Dimensions of perception

Respondents

Twenty-three individuals (nine males and fourteen females), members of various neighborhood groups in Harrisburg, Pennsylvania, took part in this phase of the research. All of them were high-school graduates. Otherwise, the sample was relatively diverse: It included individuals with ages ranging from twenty-one to fifty and with reported incomes ranging from $3,000 to $27,000 per year. Seventeen percent were nonwhite. Seventy-four percent were married, and of those, 35% had children. Because the sample is selective and small, the results must be interpreted with care. Note, however, that comparison of the findings in Phase 1 with those in Phase 3, where responses by a separate and larger sample of respondents were obtained, can shed light on the generality of the Phase 1 results.

Procedure

Interviews were done on an individual basis. The sixty photographs were shuffled and placed on a surface so that all were in view. Respondents were instructed to attend to physical features of the housing, and to sort the scenes into as many groups as necessary so that scenes in any group were more similar to one another than to scenes in any other group. Socioeconomic information was obtained after the task was completed.

The investigator tabulated the frequency with which each pair of scenes was grouped by the respondents. The resulting frequency (from 0 to 23) represented an ordinal measure of the perceived similarity of each possible pair of scenes. Torsca 9 nonmetric multidimensional scaling (Young, 1968) was applied to these data. This procedure maximizes the correspondence between the order of the observed similarities among stimuli (the scenes) and the order of distances derived in a conceptual space. Torsca 9 was used because of evidence that it has the best chance of arriving at an absolute minimum stress value (Subkoviak, 1975).

Of course, individuals may group scenes for different reasons, such that the accumulated similarity scores may be confounded by individual differences. Presumably, this confounding would enlarge the error term, so that a common multidimensional solution would be less likely to adequately fit the data and would be less likely to relate to perceptual and cognitive attributes of the scenes. If, despite individual differences, a multidimensional scaling solution is found with systematic relationships to judged scene features, the resulting dimensions may have greater generality.

Results

A three-dimensional solution was found to adequately fit the data. The mapping of the Kruskal (1964) stress values from one to six dimensions (0.40, 0.16, 0.12, 0.10, 0.08, and 0.07, respectively) revealed a pronounced elbow at two dimensions. Nevertheless, a third dimension was added because this additional dimension beyond the elbow sometimes holds useful information. The Kruskal stress value of 0.12 for this three-dimensional solution indicates at least a "Fair" fit to the data (Kruskal, 1964) and perhaps a good fit, given the large ($n = 60$) sample size.

Phase 2: Labeling the dimensions

What is the meaning of the dimensions? This phase of the research examined the fit of scores for the perceptual and cognitive scene attributes to scores of each scene on each axis of the multidimensional space.

Measurement of scene features

Seven-point bipolar scales were used to obtain ratings of the scene features. Definitions of the poles of each scale were provided to enhance the reliability of the judgments. The attributes that were measured were selected for their potential relevance to aesthetic response. Some scales assessed collative attributes (cited by Wohlwill, 1976): diversity (diverse–simple, ornate–plain, and colorful–dull) and novelty (commonplace–unusual). Others assessed S. Kaplan's (1975) mystery (the promise [much–little] of new information from more involvement in the scene) and coherence/identity (the organization [organized–disorganized] of the elements in the scene, fittingness [fitting–unfitting] of buildings to one another, and clarity [clear–ambiguous] of the building's use and parts). Others assessed enclosure (S. Kaplan, 1975; Küller, 1975) and naturalness (Kaplan and Kaplan, 1982; Peterson, 1967): closed–open, amount of vegetation (prominent–not in sight). Others assessed the prominence of various building features – shape, surface texture, verticals, roofing tile, and bricks – as did Oostendorp and Berlyne (1978a). The prominence of buildings and of sky was also assessed. Finally, several nuisance or social-status factors (Küller, 1972; Peterson, 1967) were assessed, including dilapidation (well kept–dilapidated); prominence of poles, wires, and signs; and prominence of automobiles.

Judges

In order to find labels for the dimensions that would have relevance to design professionals (i.e., the individuals who specify such dimensions in design), I had design professionals make the judgments. Eighty-one professional planners, architects, and upper-level students in these fields were contacted to describe the scenes. Although the sample is neither random nor representative, there is little reason to expect their descriptions to differ systematically from those of other professionals in similar-sized cities.

Procedure

To reduce the task for each judge, the scales were divided into four forms with between five and six scales per form, and the sixty slides were divided at random into two groups of thirty. Each judge was assigned at random to rate one of the two sets of thirty slides on one of the four forms. Because of uncertainties in scheduling interviews, the number of ratings per scale varied from between four and fifteen. The professionals were shown one slide for use in testing and refining their understanding of each scale. The thirty slides were shown rapidly to help anchor judgments. Then each slide was presented for about ten seconds to be rated.

For each scale, the interobserver reliabilities were calculated. The scores indicate relatively consistent ratings among the judges. The scores exceeded .80 for vegetation prominence, diversity, building prominence, brick prominence, vertical prominence, sky prominence, upkeep, and color; .70 for ornateness, openness, roof-tile prominence, and surface-texture prominence. The remaining scales – novelty; shape prominence; fittingness; prominence of poles, wires, and signs; organization; clarity; prominence of autos; and mystery – had scores of .64, .63, .53, .53, .44, .42, .38, .38, respectively.

To fit the judged attributes to the axes of the multidimensional scaling space, PROFIT (Caroll and Chang, 1966) – a regression, property-fitting procedure – was used. This procedure uses as input the x, y, and z coordinates of each scene in the multidimensional scaling space and the mean ratings for each scene on each descriptor scale. The analysis produces product-moment correlations between each descriptor and its fitted vector, directional angles of these fitted vectors to normalized space, and the angles between these fitted vectors.

Examination of the correlations between descriptors and their vectors revealed that each scale had a good fit to its vector. The best fits (highest correlations) were found for vegetation prominence, upkeep, diversity, color, ornateness, building prominence, and openness. For vegetation prominence and upkeep, the correlations exceeded .70; for diversity, color, ornateness, building prominence, and openness, the correlations exceeded .60; for organization and mystery, the correlation exceeded .50; for fittingness, vertical prominence, and shape prominence, the correlations exceeded .40; and for sky, novelty, and clarity, the correlations exceeded .28.

Examination of the angles between each descriptor vector and the multidimensional scaling axes suggests which vectors best fit each dimension. The descriptor vectors with the smallest angles to the first dimension include diversity (-17 degrees), mystery (23 degrees), color (32 degrees), and ornateness (34 degrees). This dimension was labeled Diversity. The descriptor vectors with the smallest angles to the second dimension include sky (-17 degrees), buildings (20 degrees), vegetation (-23 degrees), openness (24 degrees), and verticals (29 degrees). This dimension was labeled Open/natural versus Closed/man-made. The descriptor vectors with the smallest angles to the third dimension include clarity (18 degrees), fittingness (29 degrees), brick (-34 degrees), and roofing tile (-34 degrees). This dimension was labeled Clarity/fittingness.

The analysis of the angles between the descriptor vectors indicates that the vectors for each of the three dimensions – Diversity, Open/natural, and Clarity/fittingness – are orthogonal. For example, the vectors for diversity, building prominence, and clarity have angles of 85 degrees, 88 degrees, and 80 degrees between one another. In sum, the results support the expectation that diversity, organization, and openness represent shared perceptual and cognitive attributes of the housing scenes. The next question is whether these attributes influence scene evaluation.

Phase 3: Affective appraisals of scenes

This phase of the research assessed the relationship between evaluative ratings of the scenes by the public and the perceptual and cognitive attributes of the scenes.

Instrument

The instrument included four 7-point bipolar scales for assessing the scenes: Unpleasant–Pleasant, Interesting–Boring, Desirable–Undesirable as a place to live, and High–Low rate of robbery, burglary, and assault. The positive and negative poles were varied to reduce response set. Respondents were also requested to indicate their sex, race, education level, marital status, family size, and income. For each sample, the four scales were highly interrelated ($rs > .67$, $p < .01$). Thus the mean score on the four scales for each slide was calculated for use as a composite evaluative score.

Respondents

One hundred and four individuals from six neighborhood groups in Harrisburg, Pennsylvania, volunteered to participate in this study. They were assigned at random to one of the two sets of thirty slides. The data from eight respondents were dropped because of obvious miscodings or omissions. This resulted in usable data from forty-seven individuals for one set of slides and from forty-nine people for the second set. The samples included respondents of diverse socioeconomic characteristics. There were more males (53% and 62%) than females; many more whites (about 85%) than nonwhites; many more with some college education (80%) than with no college education; more singles, divorced, and widowed (about 60%) than married; and more without children (55% and 59%) than with children. Incomes varied, with some respondents (21% and 30%) earning less than $10,000 per year, some (about 20%) earning more than $20,000 per year, and the plurality (48% and 57%) earning between $10,000 and $19,999 per year.

Procedure

Respondents were interviewed in groups at scheduled meetings of their organizations. They were told that they would be viewing thirty scenes and that they should imagine themselves in each scene. They were asked to rate each scene on each of four scales and to provide the sociodemographic information. They were shown the full set of slides to help anchor their judgments and were given one additional slide for training. Slides were presented in one of two orders to mitigate order effects. Human-subject requirements for confidentiality, anonymity, and informed consent were followed.

Results

As expected, for the two sets of scenes and respondents, common patterns of response emerged. In both cases, composite evaluation improved with increases in ornateness ($rs = .75, .44, ps < .01$), upkeep ($rs = .54, .62, ps < .01$), and vegetation prominence ($r = .40, p < .05; r = .66, p < .01$). When the two samples were combined ($n = 60$), composite evaluation maintained its significant correlations with these variables (rs = .59, .56, and .52, ps < .01), and it also related to decreases in the presence of poles, wires, and signs (r = −.44, p < .01) and in the prominence of automobiles (r = −.39, p < .01), and to increases in color (r = .37, p < .01), openness (r = .37), and diversity (r = .26, p < .05).

A maximum R^2 stepwise regression procedure (Helwig and Council, 1979, p. 391) was used to find the relationship between combinations of the perceptual and cognitive attributes and evaluative response. The mean scores of selected descriptors were treated as predictor variables, and the mean scores on composite evaluation were treated as the criterion variable. Only predictor variables with significant (p < .05) values were included in the model. After the two models were developed, Pearson product-moment correlations were computed for the obtained scores from each slide set (the mean evaluation scores) and the predicted scores (derived from the multiple regression equations). In both cases, the predicted and obtained scores were similar. The correlation between the scores predicted by the first equation and the scores obtained for the second sample was r = .72, p < .001, and the correlation between the scores predicted by the second equation and the scores obtained for the first sample was r = .68, p < .001. Furthermore, a Chow test yielded no significant difference (F = 1.8, *df* 4, 52, p < .05) between the two equations. Thus a stepwise regression analysis was applied to scores for all sixty scenes. The resulting equation is displayed below, with the variables listed in the order in which they were added:

Favorable evaluation = 2.8 + .66 Ornateness + .35 Upkeep − .38 Ambiguity − .22 Closed

$$R^2 = .64, p < .0001$$

A sensitivity analysis reveals that changes in Ornateness produce the greatest change in evaluative response. This is followed by changes in Ambiguity (clarity preferred), and then Upkeep (well-kept preferred) and Closed (openness preferred).

Discussion

The results of Phases 1 and 2 of the research indicate three perceptual and cognitive dimensions for the housing scenes: Diversity, Openness/naturalness, and Clarity/fittingness. Although the generality of these dimensions may be limited by the small sample of respondents, the findings in Phase 3 provide separate confirmation of the relevance of the same dimensions: Marker scales for each of the three dimensions (Ornateness for Diversity, Ambiguity for Clarity/fittingness, and Enclosure for Openness/naturalness) were found associated separately and additively with evaluative response. That these findings emerged for two sets of somewhat varied respondents responding to two different sets of thirty scenes strengthens the interpretation that they are important visual properties of housing scenes. Furthermore, these features have emerged as salient in other studies: Diversity in studies by Oostendorp and Berlyne (1978a) and Küller (1972); organization (unity, clarity) in studies by Horayangkura (1978) and Oostendorp and Berlyne (1978a); and openness (urbanness, enclosure) in studies by Horayangkura (1978), Küller (1975), Gärling (1976a), and Hershberger and Cass (1974). Notably, Horayangkura did not have a measure of diversity, and Oostendorp and Berlyne did not have a measure of openness. Although some studies (Gärling, 1976a; Hershberger and Cass, 1974) have found complexity and organization merged into one dimension, this may result from measurement artifact where environmental complexity is confounded with disorder (Kaplan and Kaplan, 1982). The use of ornateness (a measure of visual richness) in this study may have overcome this problem.

With regard to scene evaluation, the results suggest that several processes relating to form and meaning may influence response. On the one hand, variables such as diversity (which likely increase uncertainty) enhance preference; on the other hand, variables such as clarity (which reduce uncertainty) enhance preference. In addition, openness/naturalness and nuisance may influence evaluation through meanings associated with them. These variables may provide cues to the observer about such things as social class and safety. Whether accurate or not, such interpretations would likely influence evaluation. Furthermore, these variables may also affect judgments of formal features. For example, the presence of built features and nuisances may reduce coherence. This, in turn, may affect preference through these effects on form rather than on meaning. Thus preference for housing scenes may relate to variables that carry both formal and associational characteristics that are desirable. Formal aspects of the features may engage attention (ornateness) and aid in "making sense" (clarity, openness, reduced nuisances). Associational aspects may suggest something about the character of the residents (such as their social class) and the perceived activities that go on in such areas. Further study of both formal and associational processes may well produce useful findings for design.

The findings for scene evaluation here are also consonant with other research on environmental aesthetics. Diversity has been repeatedly found to be an important influence on aesthetic response (Berlyne, 1974d; Wohlwill, 1976). The exact nature of the relationship – linear or nonlinear – has varied. Although the findings here suggest a linear relationship, it is possible that the scenes did not include a sufficient range of complexity to capture a downturn in preference. Spaciousness or urbanness has emerged in a variety of studies as being related to preference (Hershberger and Cass, 1974; S. Kaplan, 1975; Peterson, 1967; Ward and Russell, 1981b). Clarity or coherence has also been found to be an important influence on preference (Kaplan and Kaplan, 1982; Oostendorp and Berlyne, 1978a). Finally, upkeep or related social-class variables have been found to be related to environmental preference and response (Küller, 1975; Marans, 1976; Peterson, 1967). Because people typically experience their urban surroundings intermittently over a long period of time, further study of preference in relation to such experience might be worthwhile. Nevertheless, the similarity found here between the responses on site by observers familiar with the scenes and the responses to photos of the same scenes by observers unfamiliar with the scenes suggests that the patterns of results are likely to be applicable to typical day-to-day experience.

In conclusion, the perception and evaluation of housing scenes was related to four perceptual and cognitive attributes of the scenes. Increases in visual richness, use clarity, and openness and naturalness and decreases in the prominence of nuisances such as poles, wires, and signs related to improvements in rated visual quality. Venturi (1966) and some postmodern architects argue that ambiguity and complexity are desirable elements in architectural design. The results here contradict their view on ambiguity: Respondents evaluated scenes more favorably when the use and the elements were unambiguous. The findings confirm the desirability of complexity. Perhaps Venturi's call for "complexity and contradiction" in architecture should be replaced by a call for complexity and clarity.

20 Planning concerns relating to urban nature settings: the role of size and other physical features

Janet F. Talbot

Introduction

In most aesthetic endeavors, physical size is only a minor consideration. Works that are considered aesthetically pleasing can exist at both ends of the scale of size: both large, grand, and powerful, and small and exquisite, with subtle nuances. If anything, there is some suggestion that size opposes aesthetic ends, that creating something large implies a lack of adequate control and design. Large things are gross and crude, "full of sound and fury, signifying nothing" (Shakespeare, *Macbeth*). In many artistic domains, the small form is considered the most demanding and the closest to perfection: in poetry, the haiku or sonnet; in music, the string quartet, for example.

The relationship between aesthetics and the applied area of environmental planning has always been somewhat problematic (Porteous, 1982). Physical planners may wish to meet aesthetic ends, to create spaces that are artistically pleasing, yet in physical planning, the emphasis is on function, on what is necessary in order that an environment meet a specific set of human needs. Aesthetic considerations, although often a concern of environmental planners, have been assumed to have a lower priority.

This paper considers size and its meaning to environmental planners both aesthetically and functionally. As far as size is concerned, the assumption may have been that size simply has to be a priority for physical planners, despite its relative unimportance in other forms of aesthetic endeavor. Unlike artists, who can begin projects with size as just one of many "free," or flexible, components of the final design, physical planners often work in situations where size is "fixed," or predetermined, from the outset.

Since the most basic function of planners is to divide land into smaller sections that serve different purposes, it is not surprising that their first question is often how much land, rather than what type of land, will meet a specific need. Examples of the use of size as a primary criterion in planning include the practice of establishing minimum lot sizes and minimum front setbacks for houses of different sizes. Public parks are often considered to have minimum sizes, as well. In both building and open-space planning, the general assumption has been that bigger is better: Larger residential lots on which larger houses can be built are legally protected from incursion by smaller houses and lots, which are defined as having nuisance value simply because of their relatively small size (Toll, 1969). And larger public parks are often favored by professionals over smaller open areas, partially because of the administrative efficiency gained by locating greater numbers and varieties of sports fields in one place (Wilkinson, 1983).

Planning with an operational preference for large spaces has long been the established practice of urban planners and park directors. Zoning regulations and open-space standards institutionalized this policy, which was rarely questioned as the general assumption on which more specific actions were based (Buechner, 1971; Cranz, 1982; Eckbo, 1964).

Yet, as with many other expansive and relatively expensive notions about appropriate ways to achieve the "common good" that were more prevalent a decade or two ago than they are now, questions have recently been raised concerning the viability as well as the desirability of the "bigger is better" method of planning (Hester, 1984; R. Kaplan, 1983). This assumption, as powerful as it has been in shaping urban environments, has never been examined in a research context before. With both financial and physical constraints on planners increasing, there is an increasing acknowledgment of the need to learn how sizes of physical spaces actually are interpreted by people.

This paper presents the results of a series of research studies that focus on questions of size. These studies examine the importance that individuals place on size as well as on other physical qualities of real places in the city. Both residential neighborhoods and urban open areas are examined in these studies, and various questions about the design and arrangement of these spaces are considered. After reviewing these findings, insights from the data are used to develop a more theoretical discussion of the importance of size as one component of urban spaces. In the concluding section of the paper, a preliminary set of guidelines is presented that may be helpful to managers and planners, in considering size in combination with other environmental and human elements in more effectively managing open spaces in the urban environment.

Framework for the studies

Although the three studies differ from one another in some respects, they have many methodological similarities. In each of the studies, people were asked to rate their preferences for black-and-white photographs of real places. Most of the specific sites shown in the photographs were located in the nearby urban environment, although these places were not necessarily familiar to the participants. For the first two studies, the photographs were printed as portions of longer question-naires that were left at individual residences for people to complete and return; for the third study, the photographs were used in an interview format. In two of the studies, additional data about the specific surroundings in which the partici-pants lived were known and were included in analyzing the impacts of neigh-borhood environments.

Study One: Older urban neighborhoods

The first study focused on questions relating to the physical make-up of urban residential areas (Frey, 1981). The 369 participants lived in a variety of neigh-borhoods, including fairly homogeneous single-family and multiple-family areas, as well as mixed areas incorporating various "nonconforming" uses and land parcels that predated the areas' current zoning designations. By examining the preferences and reactions of people living in such varied environments, the study was able to test the standard zoning assumption that uniformity in struc-tures and lot sizes is generally preferred and is, in fact, functionally superior to living in more varied physical surroundings.

The study also examined the impact of the natural environment on people's feelings about their neighborhoods. While including many questions about the importance of nearby bits of nature, the design of the study also enabled the data to illuminate a question of central interest to planners: What are the impacts of living close to large public open spaces? Half of the study participants lived next to public parks, while the other half were more distant. Comparisons of the responses of these residents offered insights concerning the functional meaning of parks to the people who live near them.

The results of this study suggest that living near large open spaces and living in areas with land parcels of similar sizes are neither necessarily preferred nor appreciated by people. On average, the participants said they preferred to live in areas with moderate degrees of variety, both in the size of yards and in the distance of houses from the street (mean preferences were 3.2 and 3.0, respec-tively, on a 5-point scale, with 5 = "very much" variety).

Neighborhood satisfactions were assessed through rating scales that were com-bined into five clusters through the use of dimensional analysis procedures. These clusters reflected the participants' satisfaction with neighborhood Services (access to and variety of services), Outdoor Recreation (availability of parks and recreation areas), Trees (number of trees and shrubs), General Activity (variety and friendliness of people, variety of buildings, activity level, availability of gardening opportunities), and Lack of Problems (feelings of security and safety, recent changes, amount of traffic, noise and litter).

In general, people with parks nearby were not more satisfied with their sur-roundings than people without such close access to public open spaces. Of the satisfaction measures described earlier, only the Outdoor Recreation cluster reflected higher satisfaction among the participants who lived close to parks.[1] Other impacts of park access depended on how frequently the residents visited these facilities. The General Activity, Lack of Problems, and Trees satisfaction clusters were rated lower by frequent park visitors who did not live close to a park, and were rated higher by frequent park visitors who lived close to a park. For the other half of the sample, who reported visiting parks less often, neighborhood satisfaction levels were quite similar, regardless of how close they were to parks.

In contrast to these results, other measures of nature availability reflected stronger impacts on residents' feelings of satisfaction. These more significant components of access to nature included living in areas that had many trees (which resulted in higher satisfactions on the Services, General Activity, and Trees clusters) and having nearby outdoor spots in which to sit and watch wildlife (which affected satisfactions with Outdoor Recreation, General Activity, and Trees). Having surroundings close to home that included a variety of self-defined "special" nature spots (such as a favorite tree, a wildlife spot, or a well-landscaped area that was "especially important or enjoyable") was also quite valuable. People who had such spots closer to home rated their neighborhoods higher than did the rest of the sample on all five clusters of satisfaction items.

Additional evidence relating to the value of small nature elements was seen in the results of an open-ended question about what the participants would miss the most about their neighborhood if they moved away. The 128 participants who mentioned something about their natural surroundings reflected higher satisfactions than did the rest of the sample on the Outdoor Recreation, Trees, and General Activity clusters. Again, the comments that the participants made most often described specific natural elements such as trees, a garden spot, or a place to watch the birds, rather than mentioning any particular enjoyment of parks or other large open areas.

Study Two: Modern apartment complexes

The second study was somewhat more focused, dealing with treatments of open spaces within large apartment complexes (Kaplan, 1983). One of the agencies supporting this study was the local Planning Department, which was interested in being able to give developers some guidance beyond the common requirement that a certain amount of open space be left within a multiple-family complex. The 268 participants completed questionnaires that were distributed at 9 apartment complexes in Ann Arbor. The specific physical characteristics of the outdoor areas at each complex were known and were included in analyses, as well as the particular view of the outdoors from each respondent's dwelling unit.

The results of this study provide a strong contrast to planners' assumptions that large spaces are generally preferred. Designs for open areas around multiple-family structures often include large expanses of mowed land, and this was, in fact, a common view for the survey respondents. However, the view of large mowed areas was the least preferred of all the natural-area views, rated as less preferred than a view of woods or a few trees; of a landscaped area, gardens, or a park; or of unmowed fields (mean preferences, on a 5-point scale, ranged from 4.0 to 4.8 for these natural areas, while the mean preference for large mowed areas was 3.8). At the other extreme, all natural-area views were strongly preferred to outdoor views that included constructed components, such as power lines, play equipment, and streets (the mean preference for views of play areas was 3.1; preferences for views of elements such as parking lots, streets, and power lines ranged from 1.4 through 1.7). The photograph ratings complemented this pattern of results. Photographs of open mowed areas received lower preference ratings than did photographs of other open-space treatments (preferences for the photograph clusters showing open mowed areas adjacent to apartment buildings averaged 2.0, while more landscaped areas averaged 3.1 and natural areas in which nearby buildings were only minimally visible received ratings averaging 4.1). Thus the results from both verbal and photograph ratings show that large open spaces in general were not particularly preferred by the residents of multiple-family complexes, as contrasted with other, more landscaped settings.

Rather than size itself, the significant ingredient in natural spaces that received high preference ratings and that made positive impacts on the satisfaction levels of people living near them was the inclusion of trees, in particular, as well as bushes and other landscaping elements. These ingredients screen the view of less preferred elements, such as buildings, fences, and other people, and they buffer noise. In addition, they define an area as an outdoor place to sit in and enjoy, a place where one might see wildlife and where seasonal changes in the details of one's natural surroundings are reflected.

Study Three: Urban open spaces

The third study focused directly on size as the major question of interest. This study explored whether people can readily determine the physical sizes of outdoor areas, and whether their preferences for different places are related to how large these places are (Talbot and Kaplan, in press).

In this study, fifty-six participants were asked to determine the relative sizes of fifteen urban outdoor areas and to rate the areas for preference. Each area was represented by a set of four photographs. The areas ranged from 0.1 acre to 17 acres in actual size, and were quite varied in appearance, including well-manicured and parklike areas as well as more overgrown, natural settings.

Table 20.1. *Comparison between physical factors affecting size and preference ratings*

Physical features noted (with sample comments)	Percentage mention		
	Affecting size	Lowering preference	Raising preference
Features perceived as making places look smaller			
Buildings, concrete (sidewalks, buildings in the background, "in the city")	27	25	—
Fences (fences at the borders)	18	14	—
Houses (residential areas, proximity to houses)	14	9	—
Enclosure (no openness, closed in, feeling of confinement)	10	12	—
Mowed areas (manicured, developed)	12	—	16
Features perceived as making places look larger			
Spaciousness, wide open (clear, lots of space, spread out, sweeping area, can see far)	48	—	16
Trees (big trees, woods, lots of trees in the distance)	29	—	38
Trails (pathways, good for hikes)	12	—	12

The findings of this study revealed that people were fairly accurate in ranking places according to their relative size (the correlation between the actual sizes and the average size ratings was .59, $p \leq$ .05). More important, however, there was not a significant correlation across the fifteen settings between the participants' size judgments and their preference ratings.

Additional data from this study suggest a possible source of confusion concerning the relationship between physical size and environmental preference. After completing each of the photograph-sorting tasks, the participants were asked to explain what characteristics of the areas had affected their ratings. Despite the lack of correspondence between the actual ratings of size and preference, there was a high degree of similarity among the participants' answers concerning what factors made places look large or small, and what things made places more or less preferred.

Table 20.1 lists all the characteristics that were mentioned by at least five of the fifty-six participants as having affected their size judgments. The percentage of the sample who noted each characteristic is listed, along with the percentage who mentioned each of these physical attributes as having affected their preference ratings (either as contributing to or as detracting from their preference for the settings). Both an enclosed or a confined appearance and evidence of constructed components, such as buildings or fences, were said to make places look smaller and to be less preferred. At the other extreme, places with trees and pathways and places that appeared to be "spacious" were said to look large as well as being highly preferred. Only one feature was mentioned as affecting size and preference ratings in opposite directions. This was the appearance of open, mowed areas: Seeing mowed, grassy areas was mentioned as suggesting that a place was small, but was also mentioned as a characteristic of highly preferred places.[2]

As the results shown in Table 20.1 indicate, seven of the eight physical characteristics that were mentioned as contributing information about the sizes of places were also thought to affect preference judgments in similar directions. Perhaps ironically, the findings of this study suggest that although larger natural settings in reality are not preferred over smaller areas, people are similar to professional planners and park managers in believing that large places and well-liked places share many characteristics.

Discussion

It has been noted before that people are good at quickly recognizing what they like, but not as good at verbalizing the reasons underlying their preferences (R. Kaplan, 1979b). The issue of size seems to be a good illustration of this point. As the results of the third study indicate, people say that they like large places, and they point out similar characteristics when describing places that they like and places that are physically large. Although this suggests that people think of the places they like as being large regardless of their actual size, the results of the rating tasks in this study indicate that physical sizes are not related to people's preferences for different places.

The findings of the first two studies complement these results. Access to the kinds of large, open spaces that commonly exist in single-family neighborhoods and in multiple-family apartment complexes is not generally preferred. In addition, having access to these places has little impact on people's feelings of satisfaction with their surroundings. These two studies do show, however, that people respond strongly to smaller and more natural areas. The close availability of trees, of "special" nature spots, of gardens, and of other individually scaled natural settings have strong and widespread impact on neighborhood satisfaction levels.

Despite these findings about the value of discrete natural elements, it is important to recognize that people do prefer natural areas that they perceive as being "spacious" and that these preferred areas share certain physical attributes. These attributes, while not necessarily related to physical size, could be described as reflecting the coherence of a place: the sense of substance or meaning that a setting reflects; the sense that the individual elements within the space are complementary to one another and essential to the place itself; and the sense that the place as a whole has an identity that makes it distinct and valued (Kaplan and Kaplan, 1982; S. Kaplan and Talbot, 1983). Rather than being particularly large, an area that is perceived as being spacious is big enough to feel comfortable within and allows the individual a variety of potential involvements.

This sense of spatial coherence implies both inward and outward linkages. The physical qualities of a spacious area suggest some continuity between that space and the larger environment surrounding it. While distinct as an area itself, the space also echoes the meaningful elements in the environmental patterns seen elsewhere. Similarly, a spacious area has distinct subdivisions, different regions that also have unique qualities and that are clearly linked to one another. The complementary arrangement of these interior regions illustrates the way in which the entire space is related to its larger surroundings.

Additional qualities of settings that are perceived as spacious are related to issues of structure and scale. An individual's involvement with a place will be facilitated if the structure of the setting is clearly evident. To have a visible structure, a setting must have different parts that are easily identifiable, linkages among the different parts that are clearly visible, and functional relationships among those parts that are readily apparent. In addition, a human scale is needed in the smaller regions, communicating a sense that entrance into the setting is easily accomplished and that many choices are available to the individual.

Supportive evidence

Although not previously examined directly for their insights about size, the findings from a number of other research studies correspond to the data and discussions presented here. A study of the value of extended wilderness experiences suggested that the size or remoteness of a wilderness area was not critically important in determining the extent of individual benefits from such experiences (S. Kaplan and Talbot, 1983). Rather than any specific environmental quality, the purposes with which individuals entered the setting made more differential impacts in shaping the quality of environmental interactions and in determining the benefits experienced.

Other studies suggest that, contrary to planners' assumptions, people prefer smaller rather than larger spaces in at least some circumstances. In a study of potential designs for a very small urban plaza, spatial arrangements showing one undivided area were less preferred than arrangements in which the space was parceled into five or six small sitting areas (R. Kaplan, 1978a). And two separate studies of Detroit inner-city residents suggested that they prefer areas that are small in size. In one photograph-based study, residents gave the highest preference ratings to small parks in which developed property was clearly visible beyond the open spaces (Talbot and Kaplan, 1984). In the other study, trees on residential streets were preferred over larger wooded areas (Getz, Karow, and Kielbaso, 1982). Finally, in a very different context, Newman's (1972) research has illustrated the problems that result from designing low-income housing projects surrounded by large undifferentiated open spaces.

Applications

If size is not to be the primary basis for planners' decisions, how, then, should they deal with questions about the appropriate sizes of spaces for different uses? How can the creation and maintenance of coherent settings, of settings that feel spacious and that express a richness of personal affordances no matter what size they are, be encouraged? Rather than relying on specific minimum sizes, the data presented here suggest that planners and designers can be more flexible, using guidelines such as the following:

- Encourage a moderate degree of variety in the treatments of adjacent land parcels, encompassing complementary sizes and shapes of lots as well as interesting mixtures of building placements and styles.
- Since research indicates that natural elements strongly influence the perception of large regions, manage the landform, trees, and other plantings in such a way that a strong image of the urban natural landscape is promoted. Patterns of foliage soften harsh edges of buildings and contribute visual continuity to large-scale urban settings, which are often sorely lacking in coherence.
- Create and maintain the unique quality of each setting. Protect the physical elements that are most unique and valued at each site – an old, distinctive tree, a sharp incline, or a winding pathway, perhaps – and add elements that enhance the character of the existing space.
- Within an individual setting, create distinct, well-defined regions with clearly differentiated characteristics and functions. Carefully lay out the pathways and other connections among the different interior regions, since they will suggest similar linkages between the entire setting and the larger surroundings.
- Capture a sense of mystery within the setting, maximizing the potential for partially screened views that are quickly resolved as an individual moves through the area. This gives multiple definitions to a setting and adds to its perceived size, since the individual experiences different facets of the space while continuing to move within it.
- Provide multiple entrances into the setting. Again, this enriches – rather than limits – the connections between the immediate setting and the larger environment. It also enhances the perceived substance or meaning of the setting, since it increases the opportunities available to the individual.
- Imagine the individual within the setting and create a number of partial enclosures within each space. Each setting should offer a number of small, com-

fortable, somewhat sheltered spots. Each should be a distinct area within which the individual can appreciate a meaningful bit of nature, such as a group of plantings or a low tree.

Conclusion

Despite the recent popularity of the "small is beautiful" philosophy (Schumacher, 1973) in many planning disciplines, planners of urban open spaces have clung to notions that it is always preferable to have more land. The data reviewed here suggest that, while feelings of spaciousness are valued by people, natural settings that are physically large are not necessarily preferred. The desired perception of spaciousness can be achieved in settings that are relatively small, as the result of manipulating physical elements within a setting to increase its coherence and to maximize the perceived opportunities for individual involvement.

Acknowledgments

Work on this paper was supported in part through cooperative agreements with the U.S. Forest Service North Central Forest Experimental Station, Urban Forestry Project. Rachel Kaplan conducted the study of the modern apartment complexes. Both Rachel Kaplan and Stephen Kaplan have collaborated in the development of the theoretical concepts as well as in the research studies discussed in this paper.

Notes

1. Differences in satisfaction levels were assessed through Analysis of Variance and Student's t tests, $p \leq .02$.
2. The condition of grassy areas is apparently a significant issue that elicits strongly divergent responses: The presence of unmowed, open grassy areas, while not frequently noted in relation to size judgments, was mentioned almost as frequently as the presence of mowed areas as being a feature of preferred settings.

21   The effect of sign complexity and coherence on the perceived quality of retail scenes

Jack L. Nasar

Because commercial strips occur along major arteries and are seen regularly by the public, they have a substantial impact on the visual image of the city. Unfortunately, this impact is often negative. Herzog, Kaplan, and Kaplan (1976), for example, found that for five categories of urban scene (cultural, contemporary, commercial, entertainment, and campus), people most disliked scenes of commercial strips. Similarly, residents and visitors most frequently cited roadside commercial strips as visually blighted areas in a city (Nasar, 1979).

Clearly, urban commercial strips produce visual overload (Rapoport and Hawkes, 1970), and a major factor contributing to this overload is the signscape (i.e., the multiplicity of signs that the viewer can comprehend in a single view). In a study of public responses to commercial scenes from which various features had been removed, Winkel, Malek, and Theil (1970) found that signs were highly noticeable;[1] and when Nasar (1979) asked the public to describe the physical elements that most reduced visual quality, people most frequently cited signs and billboards.

The sign problem is similar to the tragedy of the commons (Hardin, 1968). What seems beneficial to each individual alone is detrimental to all the individuals together – the community. With signs, each merchant attempting to call attention to his or her establishment seeks a distinctive sign that presents a desirable image and stands out from the surroundings. When seen alone, each sign may present a favorable image and attract attention; but when many such signs are placed side by side, the result is often chaos. Thus planners and legislators must consider the combined visual effect of signs en masse – the signscape.

Cities and commercial areas attempt to control the signscape through such mechanisms as sign ordinances and design review. To help frame guidelines for

An earlier version of this paper was presented at the Ninety-third Annual Convention of the American Psychological Association, Los Angeles, 1985.

sign ordinances and review procedures, empirical research has centered on rec-ognition, recall, and traffic safety (see Carr, 1971; Ewald and Mandelker, 1977; Tunnard and Pushkarev, 1981, pp. 277–326). Although design professionals (Lozano, 1974; Rapoport and Hawkes, 1970) theorize about urban visual quali-ty, empirical analysis of perceived visual quality in the signscape lags behind.

This paper provides empirical information on the effect of signscape features on perceived visual quality. This research centered on the commonalities in perceived visual quality rather than on individual differences. Individuals, of course, may differ in their judgments of perceived visual quality; but since many people experience the signscape, the shared judgments of visual quality are more important.

The definition of perceived visual quality

Perceived visual quality is a psychological construct: It involves subjective as-sessments. Such assessments have primary reference to either the *environment* (as would ratings on variety or colorfulness of a scene) or to people's *feelings* about the environment (as would ratings of the pleasantness or excitement of a scene). The former are called *perceptual/cognitive* judgments, and the latter are called *emotional* judgments (Ward and Russell, 1981b). Although perceived visual quality may depend, in part, on *perceptual/cognitive* factors, it is, by definition, an *emotional* judgment involving evaluation and feelings. As a result, perceived visual quality is measured here through evaluative judgments of the signscape.

To be relevant, such judgments must center on the dimensions of evaluation that people actually use in evaluating their surroundings. What are these salient dimensions of environmental evaluation? Ward and Russell (1981a) examined this question.[2] Using a variety of research strategies, they consistently found three aspects of environmental evaluation as salient: pleasantness, excitement, and calmness. One aspect – pleasantness – is a purely evaluative dimension. The other two aspects – excitement (exciting–dull) and calmness (calming–distress-ing) – involve both arousal and evaluation. For example, an exciting place is more pleasant and arousing than a dull one; and a calming place is more pleasant and less arousing than a distressing one. Based on Ward and Russell's findings, the present research assessed the pleasantness, excitement, and calmness of various signscapes.

Presumably, evaluation influences behavior such that people are more likely to visit and linger in a place perceived favorably and avoid one perceived negatively (Mehrabian and Russell, 1974; Osgood, 1971; Ulrich, 1973). Although eval-uative responses, alone, may not predict actual behavior, the combined appraisal of evaluative responses and expected behavior gives a good indication of actual behavior (Osgood, 1971). Thus in this research, respondents were asked to indicate the degree to which they would want to visit, shop in, or spend time in various signscapes.

Environmental components of perceived visual quality

Two signscape features – complexity and coherence – were chosen for study, because of their relevance to sign control and because of their likely influence on perceived visual quality. Complexity is defined as the amount of variation in the scene, and coherence is defined as the degree to which the scene hangs together.

What is the relevance of these features to sign control? Sign ordinances may specify the height, shapes, and colors of signs, and the style, number, and placement of letters. In doing this, they control complexity. Sign ordinances also restrict the placement of signs (flat on the building, perpendicular to the building, or free standing) and the size and brightness of and movement in signs and letters. Although such standards do not directly control coherence, they control a related variable – contrast (the degree to which signs stand out from or contrast with their surroundings).[3] Reductions in the contrast (in color, texture, size, and shape) of focal man-made elements increase the perception that those elements are compatible with their natural surroundings and, as a result, enhance the perceived coherence of the scene (Wohlwill, 1979, 1982; Wohlwill and Harris, 1980). Such findings for natural surroundings may not apply to the signscape, but Sorte (1971) has found that the judged coherence of signs (temporary features) is less than that of buildings (permanent features). Thus increases in sign contrast can exaggerate the reduction in coherence. In one of three studies reported here, the effect of sign contrast on coherence was tested.

What is the relevance of complexity and coherence to perceived visual quality? Perceived visual quality has been described as the product of two fundamental human needs: the need to be *involved,* and the need to have the scene *make sense* (Kaplan and Kaplan, 1982). The environment must be *involving* to attract human attention, and it must *make sense* for humans to operate in it. Complexity and coherence play major roles in satisfying these human needs.

By definition, complexity creates uncertainty, which, in turn, elicits involvement to reduce the uncertainty (Berlyne, 1972; Lozano, 1974; Rapoport and Hawkes, 1970; Wohlwill, 1976). The relationship between complexity and involvement (arousal from uncertainty), thus, should be direct and monotonic. This form of relationship has been consistently supported by empirical findings for measures of involvement such as looking time and interest (Berlyne, 1972; Wohlwill, 1976). Conversely, hedonic tone (pleasantness or beauty) has been posited as having an inverted U-shaped (non-monotonic) relationship to complexity. With increases in complexity, hedonic tone should increase to a point –

an optimal level of arousal – after which hedonic tone decreases. Too little complexity is monotonous and boring; too much is chaotic and stressful (Berlyne, 1972; Ulrich, 1983; Wohlwill, 1976). A middle level of complexity is the most pleasant. The empirical findings for complexity have been inconsistent, perhaps because of methodological artifact (Kaplan and Kaplan, 1982; Wohlwill, 1976). For example, some studies have failed to control natural covariates of complexity, such as dilapidation, poles and wires, and vegetation; others have not used a sufficient range of complexity for the downturn in pleasantness to emerge; still others have overlooked the possibility of nonlinear relationships. When these problems have been eliminated, the expected effects of complexity have generally emerged (see Wohlwill, 1976).

Now consider coherence. For a scene to make sense, it needs unity, patterning, organization, or something that helps it hang together. This "something" has been called coherence (Kaplan and Kaplan, 1982). By aiding comprehension, coherence should reduce uncertainty and increase hedonic tone (Kaplan and Kaplan, 1982; Wohlwill, 1976). These relationships have been consistently confirmed in empirical research (Groat, 1983; S. Kaplan, 1975; Küller, 1972; Nasar, 1983, 1984).

In sum, the literature indicates that complexity and coherence (achieved through reductions in contrast with the surroundings) should influence pleasantness and arousal in predictable ways. Complexity should increase arousal (i.e., increase excitement and decrease calmness), and coherence should decrease arousal. Pleasantness should be highest for moderate complexity and high coherence. As a result, this research examines the following research questions:

1. Does signscape evaluation have commonalities across various groups (shoppers and museum visitors, males and females, young adults and older adults, and shoppers and merchants)?
2. Does excitement increase with complexity and decrease with coherence, and does calmness decrease with complexity and increase with coherence?
3. Is pleasantness highest for a middle level of complexity and for a high level of coherence?
4. What are the combined effects of complexity and coherence (i.e., do they combine to accentuate the effect found for each separately)?

These questions were addressed in three studies, two of which also provided information for use in developing guidelines for the signscape of a ten-block commercial area (henceforth called the study area) adjacent to The Ohio State University. The studies were instigated by dissatisfaction with the visual image by the public and merchants in the area (Gordon and Nasar, 1984), and by a related request from an area task force for assistance in revising the sign controls for the area. One study assessed signscape evaluations by shoppers and potential shoppers in the area, and the other assessed signscape evaluations by merchants in the area.

Both studies assumed that coherence could be manipulated by reductions in signscape contrast. I tested this assumption in a separate study, in which I also tested the effect of signscape complexity on coherence. Kaplan and Kaplan (1982) suggest that complexity in the built environment is frequently disordering. This effect may be more pronounced for signscapes because complexity may interfere with legibility. Thus I expected that complexity would also reduce coherence. Because all three studies obtained responses to the same set of color photographs of simulated roadside signscapes, this simulation is described first.

Signscape simulation

For this research, a scale model (constructed earlier) of a street-side strip was employed. I photographed the model in color with nine different signscapes (Figure 21.1): combinations of three levels of complexity (least complex, moderately complex, and most complex) and three levels of contrast (least contrast, moderate contrast, and most contrast). Each sign carried only a label of a service or shop (such as hotel, photo, drugs) and a logo. The model had simulated buildings, paving, sidewalks, street furniture, automobiles, vegetation, people, and background skyline. These and other characteristics of the scene and labels were held constant across the conditions, so that each scene differed only in signscape complexity and contrast.[4]

Complexity was manipulated by altering the amount of variations among the signs: their location (height and centering along façade), shape, color, direction (vertical, horizontal, or diagonal), and lettering style. The "least complex" signscape consisted of signs of the same shape, color, and lettering style mounted at one location and direction. The "most complex" signscape had signs that differed on all these attributes. The "moderately complex" signscape had moderate variation in shape, color, lettering style, location and direction, but less variation than did the "most complex" signscape.

Contrast was manipulated by altering the size and lettering style of the signs, and the contrast of the color and material of the signs and the lettering in relation to the background. The "least contrast" signscape had the smallest signs, the smallest lettering, and natural colors. The "most contrast" signscape had the largest signs, largest lettering, brightest colors, and highest color contrast between letters and their background. The "moderate contrasting" signscape had middle-sized signs and lettering, and less intense coloring.

Several issues about the generality of the simulation should be noted. First, do color photographs accurately depict on-the-street experience? The research suggests that they do: Responses to color photographs and slides have consistently been found to accurately gauge responses obtained at the places portrayed in the photographs (Evans and Wood, 1981; Shafer and Richards, 1974). Second,

Least complex; least contrast (most coherent)

Moderately complex; least contrast (most coherent)

Most complex; least contrast (most coherent)

Figure 21.1. The tested signscapes.

Least complex; moderate contrast (moderately coherent)

Moderately complex; moderate contrast (moderately coherent)

Most complex; moderate contrast (moderately coherent)

Fig. 21.1 (cont.)

Least complex; most contrast (least coherent)

Moderately complex; most contrast (least coherent)

Most complex; most contrast (least coherent)

Fig. 21.1 (*cont.*)

might different simulated environments affect the results? Signscape evaluation may vary with the context such that individuals may be more willing to accept highly complex and contrasting signscapes along intensely developed major arteries than along minor roads, with low-intensity development and historic buildings. The simulation here was not a replication of the study area, but it was similar to the study area: A four-lane commercial strip with a mixture of older (but not necessarily historically significant) buildings. Furthermore, as was suggested by Kaplan and Kaplan (1982), respondents were not asked to evaluate the *model* per se, but to evaluate what it would be like to be in *signscapes* similar to those in the simulation. Thus responses to the model should accurately reflect responses to the study area. Finally, will results from the survey apply to other study areas? There is little reason to suspect that results would differ for areas similar to the one examined here. However, for areas with different street sizes, intensities of development, building styles, or historical significance, the results may differ. Comparison studies are needed to determine the generality to other study areas.

Study 1: The effect of signscape features on coherence

In this study, the effect of signscape contrast and complexity on judged coherence was examined.

The sample

Twenty-one adults (ten males and eleven females from twenty to forty-five years old) in the ten-block commercial area participated in the study. Interview sites were selected at random from a map of the area. Then at different times of the day and week, an interviewer went to one of the sites and stopped the third passer-by for an interview. If rejected, the interviewer stopped the next passer-by of the same sex as the previous person until an interview was obtained. On completion of an interview, the interviewer proceeded to the next preselected site and stopped the third passer-by of the opposite sex from the previous respondent. Over 95% of those approached agreed to participate.

Procedure

Before each interview, the photographs were shuffled (to vary their order) and handed to the respondent. Respondents were told that their answers were confidential and anonymous and that there were no right or wrong answers. They were asked to rank order the nine scenes for coherence (the degree to which the scene hangs together) from most coherent (hangs together very well) to least coherent (hangs together very poorly).

Table 21.1. *Median scores and summary statistics*[a] *for judged coherence (1 = least coherent, 9 = most coherent)*

Level of contrast	Level of complexity			
	Least	Moderately	Most	Combined
Least	7	6	4	6
Moderate	6	5	3	5
Most	4	4	5	4
Combined	6	5	4	

[a]Contrast: $\chi^2 = 5.86$, $2df$, $p < .05$
Complexity: $\chi^2 = 5.94$, $2df$, $p < .05$
Contrast × complexity: $\chi^2 = 16.37$, $8df$, $p < .05$

Results and discussion

Did contrast and complexity in the signscape reduce coherence? Friedman's two-way analysis of variance was used to compare the rank scores for coherence for differences across the signscape conditions (see Seigel [1956] for a description of this test). The analyses were blocked (or stratified) for subjects and for the alternative treatment variable (i.e., the analysis compared the rankings within the separate blocks). Thus, for example, tests for effects of contrast considered the rankings of contrast in each level of complexity and for each subject as blocks. A nonparametric analysis was chosen because the data were ordinal.

The scores and summary statistics for each condition are displayed in Table 21.1. The results confirmed the expected main effects of contrast and complexity. For contrast, judged coherence was lowest for the most contrasting signscape, higher for the moderately contrasting signscape, and highest for the least contrasting signscape. For complexity, judged coherence was highest for the least complex signscape, lower for the moderately complex signscape, and lowest for the most complex signscape. Unexpectedly, a significant interaction between contrast and complexity revealed that the effects of each independent variable were reduced at high levels of the other independent variable. Thus Table 21.1 shows that although judged coherence varied inversely with complexity, it was relatively stable with variations in complexity in the most contrasting signscape. Similarly, although rated coherence varied inversely with contrast, decreases in contrast did not yield increases in coherence in the most complex signscape.

The findings suggest that in general, perceived coherence can be affected through manipulations of signscape contrast. Thus in the following studies, I refer to the contrast conditions in terms of coherence. Keep in mind, however, that at high complexity, the manipulations in contrast had little effect on coherence. High coherence should change the perception of high *physical* complexity to the aesthetically valued *moderate* complexity. But because coherence did not vary with contrast in the most complex signscapes, the *perceived* complexity of those signscapes was not moderated by reductions in contrast. Thus I expected hedonic value in the low-contrast conditions to be highest for *moderate* (not high) complexity.

The findings also indicate that signscape complexity reduced coherence. Does this mean that reduced complexity is another surrogate for coherence? No. By definition, complexity produces uncertainty, which may reduce coherence. Complexity, however, should differ from coherence in its effect on evaluative response. Coherence, by helping the viewer make sense of a scene, should improve perceived visual quality, whereas complexity, by involving the viewer, should enhance scenic quality only up to a point, after which added complexity reduces scenic quality.

Study 2: Signscape evaluations by the public

The sample

Ninety-two people, representing present and potential shoppers and passers-by in the commercial area, were interviewed. The sample of present users consisted of seventy people stopped in the ten-block area. The sampling procedure was the same as that in Study 1. The sample of potential users consisted of twenty-two visitors to a local museum. Museum visitors were selected because a Center for the Visual Arts, which will be built in the area, is likely to attract museumgoers to the area. The sampling procedure was the same as that in Study 1, except that all interviews took place near the museum entrance.

For each sample, over 90% of those approached agreed to participate. The combined sample had forty-nine males and forty-three females, seventy-six of whom were between the ages of eighteen and twenty-four, twelve between the ages of thirty-five and fifty-five, and four who were either younger or older.

Procedure

Before each interview, the photographs were shuffled. Respondents were told that there were no right or wrong answers and that their answers would be kept confidential and anonymous. They were asked to look at each of the nine scenes, imagine how they would feel to be in places that looked like each scene, and respond to the scene, not to the quality of the photograph.

They were asked to select the scene they would most like to visit, shop in, and linger in. They were also asked to rank order the scenes separately in terms of their pleasantness, excitement, and calmness (unpleasantness, boredom, and distress for half of the respondents). Because the order of the scales may influence response, twelve interview forms were used, each having the four scales arranged in a different order. To mitigate order effects, these forms were rotated through respondents. No more than ten people responded to any one order of scales.

Results

Commonalities in response. The responses of various groups within the sample were first compared for similarities. For each respondent, there were three sets of nine rank order scores (i.e., the comparative ratings of each scene on each scale). Spearman rank order correlations among the sets of nine median scores of the three response scales for each group were obtained. In almost every case, the rankings of the scenes across the groups were similar. The scores of the museum sample and the on-the-street sample were highly correlated ($r = .97$ for pleasantness; $r = .99$ for excitement; $r = .95$ for calmness: $p < .01$); the scores of males and females were highly correlated ($r = .86$ for pleasantness; $r = .98$ for excitement; $r = .99$ for calmness: $p < .01$); and the scores of those aged eighteen to thirty-four and those aged thirty-five to fifty-five were highly correlated on two of the three scales ($r = .95$ for excitement; $r = .82$ for calmness: $p < .01$).

Although highly correlated, one set of scores can be systematically higher or lower than another set. Thus the responses of each group were also tested for differences through the Mann-Whitney U test, a nonparametric alternative to the *t* test for differences in the distribution (see Seigel [1956] for a description of this test). Eighty-one comparisons were made (one for each of the three adjective ratings of each of the nine scenes by each of the three pairs of groups: museumgoers versus those interviewed on the street; males versus females; and younger versus older). Only six significant differences emerged, and they were all differences in intensity rather than in direction of response. In the six differences, the older group responded more favorably than did the younger group.

Because the results indicate commonalities among the groups in their evaluative responses to the signscapes, the analyses that follow use composite scores from the full sample. Note, however, that separate analyses were conducted for each group and produced essentially the same results as those reported here for the composite data.

Effects of complexity and coherence. What influence do complexity and coherence have on scene evaluation? Recall that each respondent rank ordered the nine scenes in terms of pleasantness, excitement, and calmness. The median-rank scores for each scale were tested separately for differences across the

Table 21.2. *Treatment medians and summary statistics[a] for evaluations by the public*

Treatments	Adjective scales		
	Unpleasant (1)–pleasant (9)	Boring (1)–exciting (9)	Distressing (1)–calming (9)
Signscape coherence			
Least coherent	3	6	3
Moderately coherent	5	5	5
Most coherent	6	4	6
Signscape complexity			
Least complex	3	2	6
Moderately complex	6	6	6
Most complex	5	7	4

[a]Coherence: Pleasant $\chi^2 = 47.21$, $2df$, $p < .01$
　　　　　Exciting $\chi^2 = 93.39$, $2df$, $p < .01$
　　　　　Calming $\chi^2 = 129.98$, $2df$, $p < .01$
Complexity: Pleasant $\chi^2 = 46.10$, $2df$, $p < .01$
　　　　　Exciting $\chi^2 = 295.47$, $2df$, $p < .01$
　　　　　Calming $\chi^2 = 77.77$, $2df$, $p < .01$

signscape conditions through Friedman's two-way analysis of variance with blocking for respondent and for the alternative treatment (complexity or coherence) variable.

Table 21.2 displays the medians and summary statistics for each scale. In general, the findings reveal the expected effects of coherence and complexity on the evaluative responses. First, look at the results for coherence in Table 21.2. Pleasantness is highest for the most coherent signscape, lower for the moderately coherent signscape, and lowest for the least coherent signscape. Excitement is lowest for the most coherent signscape, higher for the moderately coherent signscape, and highest for the least coherent signscape. Calmness is highest for the most coherent signscape, lower for the moderately coherent signscape, and lowest for the least coherent signscape.

Now look at the results for complexity in Table 21.2. Pleasantness is significantly higher for the moderately complex signscape than for either the least complex or the most complex signscape. Excitement is highest for the most complex signscape, lower for the moderately complex signscape, and lowest for the least complex signscape. Unexpectedly, calmness is equally high for moderate and low complexity.

The interactive effects of complexity and coherence and the summary statistics are displayed in Figures 21.2 to 21.4. For pleasantness and calmness (Figures 21.2 and 21.4), the inverted U-shaped function relative to complexity emerges;

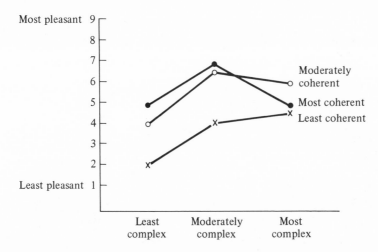

Figure 21.2. Median pleasantness scores of the public for complexity and coherence. (Coherence × complexity $\chi^2 = 108.60$, *8df, p* < .01)

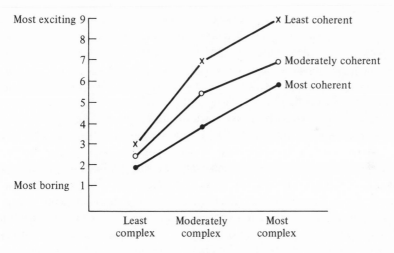

Figure 21.3. Median excitement scores of the public for complexity and coherence. (Coherence × complexity: $\chi^2 = 345.24$, *8df, p* < .01)

coherence accentuates the favorable effect of moderate complexity. Moderate complexity is rated as most pleasant and calming, and these favorable ratings are more pronounced in the moderately coherent and the most coherent signscapes than in the least coherent signscape for pleasantness and for calmness. In addition, pleasantness and calmness of the three levels of coherence show less variation in the most complex signscape. Recall, however, that for the most complex signscapes, the manipulations of coherence were relatively unsuccessful.

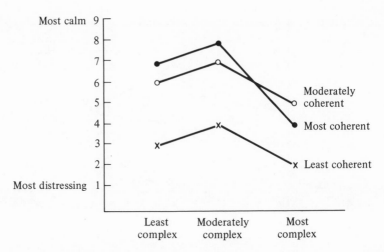

Figure 21.4. Median calmness scores of the public for complexity and coherence. (Coherence $\times$ complexity: $\chi^2 = 138.44$, $8df$, $p < .01$)

For excitement (Figure 21.3), the expected increases in excitement from complexity were reduced by coherence. The difference in excitement for complexity was largest for the least coherent signscape, smaller for the moderately coherent signscape, and smallest for the most coherent signscape.

Finally, consider those scenes that respondents would most like to visit, shop in, and spend time in. The percentage of the sample selecting each signscape as the most liked for these purposes is shown in Table 21.3. Significantly more respondents selected the moderately complex signscape than either the least complex or the most complex one. More respondents also selected the most coherent signscape than either the moderately coherent or the least coherent one, but this difference was not statistically significant. A significant interaction between coherence and complexity, however, revealed that more people selected the moderately complex and most coherent signscape than any other. Unexpectedly, a relatively large number of people selected the least coherent and most complex signscape as the most desirable place to visit, shop in, and spend time in. This unexpected result is indicative of some individual differences in response.

Discussion

The results support the two-process model of environmental evaluation (Kaplan and Kaplan, 1982; Ulrich, 1983; Wohlwill, 1976). In one process, coherence decreases arousal, and high coherence is the most pleasant. In the other process,

Table 21.3. *Percentage of public selecting scenes as most liked to visit, shop, and spend time[a] (n = 92)*

	Complexity			
	Least (%)	Moderate (%)	Most (%)	Total
Coherence				
Least	3	8	22	33
Moderate	1	14	13	28
Most	3	25	11	39
Total	7	47	46	100

[a]Complexity: $\chi^2 = 19.31$, $2df$, $p < .01$
Complexity $\times$ coherence: $\chi^2 = 10.53$, $8df$, $p < .05$

complexity increases arousal, and moderate complexity is the most pleasant. Coherence decreased arousal (excitement and distress), and pleasantness was highest for high coherence. Complexity boosted arousal (excitement), and pleasantness was highest for moderate complexity. Unexpectedly, calmness was equally high for moderate and low complexity. This finding simply suggests that calmness was influenced by both the evaluative and the arousal components. In sum, the moderately complex and most coherent signscape was judged as moderately exciting, most calming, and most pleasant. In addition, the finding that this signscape was the most liked as a place to visit, shop in, and spend time in suggests that it may attract users.

A question that remains is whether merchants would respond in similar fashion. On aesthetic grounds, they might favor moderate complexity and high coherence for their signscape. If they place a greater emphasis on attracting attention, however, they might favor high complexity and low coherence for their signscape. A third study used a method similar to that in the other two studies to examine the merchants' responses to the signscapes.

Study 3: Signscape evaluations by merchants

The sample and procedure

Thirty-three merchants in the ten-block commercial strip adjacent to The Ohio State University were interviewed. The interviewer stopped at every other store and requested an interview with the store owner or manager. Ninety-five percent of those approached took part in the study.

Table 21.4. *Median rank scores and summary statistics[a] for merchant preferences for their area (1 = dislike, 9 = like)*

	Complexity			
	Least	Moderate	Most	Combined
Coherence				
Least	2	3	4	3
Moderate	3	5	4	4
Most	5	5	4	5
Combined	3	5	4	

[a]Complexity: $\chi^2 = 10.36$, $2df$, $p < .01$
Coherence: $\chi^2 = 6.92$, $2df$, $p < .05$
Complexity $\times$ coherence: $\chi^2 = 23.09$, $8df$, $p < .01$

The merchants were asked only to place the scenes in order from the one they would most like their strip to resemble to the one they would least like their strip to resemble. This survey was shortened, because a pretest had revealed that the merchants would not complete the longer survey.

Results

Friedman's two-way analysis of variance blocking for subjects and for the alternative treatment was employed to compare the rank scores across the signscape conditions. The results indicate that the merchants, like the potential shoppers, responded most favorably to moderate complexity and high coherence for their strip. The median scores and summary statistics in relation to coherence and complexity are displayed in Table 21.4. A significant main effect of coherence revealed that the most coherent signscape was rated as most desirable for the area, the moderately coherent signscape as less desirable, and the least coherent signscape as least desirable. A significant main effect of complexity revealed that the moderately complex signscape was rated as more desirable for the area than either the most complex or the least complex signscape. A significant interaction between complexity and coherence revealed that moderate complexity and high coherence were most desirable, except in the most complex signscape, where variations in coherence had little effect on desirability. This last effect may result from the reduced variation in coherence found in Study 1 for the most complex signscape.

Discussion

The results provide further evidence of the desirability of moderate complexity and high coherence for signscapes. The signscape that the merchants liked most for the area was the same one that the public had rated as most pleasant and as the one they would most like to visit, shop in, and spend time in (Tables 21.2 and 21.3).

The effects attributed to coherence may result directly from the manipulations of signscape contrast rather than the associated changes in coherence, but this is not supported by the data. At high complexity, where manipulation of contrast had limited effect on coherence, the manipulations of contrast had little effect on pleasantness. Although additional study is needed to clarify this process, it is safe to say that reductions in signscape contrast are likely to increase coherence and improve the perceived visual quality of a scene.

Relevance

The study area is a four-lane-artery commercial district that allows large, colorful, and varied signs (as can be seen in Figure 21.5). Yet both the merchants and the public liked less complexity and more coherence for the area. Thus both results confirm the earlier finding of dissatisfaction with the visual quality of the study area (Gordon and Nasar, 1984) and suggest the strategy to improve that image: Reduce complexity, and increase coherence.

The data on public and merchant evaluations of the signscapes also proved useful in garnering support for a revised code. The findings of the studies, presented to the merchants and the community, showed that the then-applicable sign regulations and individual decisions resulted in a signscape that the merchants and their customers disliked. At their request, I used the findings to devise revisions to the sign-control ordinance for the area. Subsequently, the merchants and the community task force approved changes (for the most part based on the findings) in sign controls for the area, and they forwarded these recommendations to the city council for adoption as a special overlay ordinance applicable to only the ten-block study area. Furthermore, although the code has not yet been adopted by the city council, several merchants have requested guidance for the redesign of signs and have tried to adhere to the proposed ordinance. The results of these changes are shown in Figure 21.6. The recommended revisions for the sign ordinance covered a number of elements.

First, they dealt with process. I recommended that a design review board be established to serve in an advisory capacity, that a schedule be developed for

Figure 21.5. Signscape before this research.

amortization of existing signs within five years, and that the replacement signs be required to meet new sign guidelines. The task force adopted these recommendations, but gave the design review board more than advisory power; new signs are required to pass design review.

Beyond process, I dealt with standards. The old ordinance allowed relatively low coherence in the signscape. It allowed free-standing signs to be mounted up to 35 feet high and up to 8 times the square root of the area (height × width) of the building façade. Graphics that project from buildings could be as large as 6 times the square root of the building façade area and could project up to 40% of the distance between the curb and the right of way (but only up to 6 feet into the right of way). Wall graphics (those mounted flat on buildings) could be as large as 3 to 8 times the square root of the building façade area. No additional limitations were placed on lettering size, color contrast, or lighting.

To increase coherence, I recommended that free-standing signs be reduced to no more than 50% their old allowable size, that the allowable size of projecting graphics be reduced by at least 50%, and that the allowable projection be reduced by at least 30%. I also recommended that lettering height be limited to 14 inches (still readable from passing cars), that internal and neon illumination be prohibited, and that a set of acceptable low-contrast colors for letters and signs be

Figure 21.6. Signscape after this research.

developed. The revised ordinance actually goes further and prohibits free-standing and projecting graphics altogether. The guidelines for design review indicate that colors, materials, and lighting in signs should be "restrained and harmonious" with the building and site, and that signs should have "good scale" and "not compete for attention."

The old ordinance allowed variety in the signscape by having limitations only on the height and size of signs. Shapes, colors, locations, number of letters, and logos were up to each merchant. To moderate complexity, I recommended that only a limited set of rectangular proportions be allowed, that letters mounted directly on the building be prohibited, that colors be allowed to vary within limits specified in a color chart to be developed, and that the number of syllables and symbols per store be reduced. In the actual revised ordinance, the code for design review specifies that every sign have "good proportion" to its surroundings, that the number of graphic elements be held to a minimum to convey the sign's message, and that logos conform to criteria for all other signs.

The findings and recommendations may have limited applicability elsewhere. The merchants in the study desired a pleasant milieu created by a signscape that was moderately complex and highly coherent. In different situations, other goals may be appropriate. The findings of this research, however, suggest ways to satisfy other goals. For example, if high excitement was desired – say, for a major entertainment area – the findings for excitement might guide the decisions in creating the signscape. Those findings suggest that to create an exciting milieu, high complexity and low coherence might be encouraged. For high excitement and some pleasantness, the findings suggest that either low coherence and moderate complexity or moderate coherence and high complexity should be encouraged. Different signscape configurations can be selected for different purposes. Of course, the actual effects of the new signscapes should be monitored to determine if the desired goals are achieved.

In conclusion, this research demonstrates principles that govern evaluative response to signscapes. These principles are consistent with the view that the aesthetic value depends on unity in variety. Planners might do well to develop empirical data through studies like the ones discussed in this paper to guide local decisions. Such data can provide scientific bases for decisions, garner public support for decisions, and strengthen the planners' case in the face of challenges to codes and regulations. Ultimately, such an approach should contribute to improvements in the image of the city by enhancing the perceived visual quality of the signscape.

Acknowledgments

Parts of this research were supported by The Ohio State University Office of Research and Graduate Studies. I am grateful to Steven I. Gordon for his comments throughout the research, to Abir Mulick for assembling the model, to Hugo Valencia for assembling and photographing the various sign configurations, and to Al Rezoski and Adam Prager for gathering and coding the data.

Notes

1. Winkel et al. (1970) report that poles and wires may be more of a problem than signs, but the use of black-and-white photographs in that study artificially reduced the impact of signs.
2. Although other studies have examined dimensions of meaning, the Ward and Russell (1981) research is the most relevant to environmental assessment. Other research has either overlooked the physical environment (Osgood, Suci, and Tannenbaum, 1957) or confounds perceptual/cognitive and emotional responses (Canter, 1969; Hershberger, 1969; Küller, 1972). Ward and Russell (1981), in contrast, focused on the dimensions of *emotional* meaning in relation to the *physical environment*.
3. The same features may be controlled for both contrast and complexity, but the manner in which they are controlled differs. For contrast, the size and contrast of the feature is controlled, whereas for complexity, the variability is controlled. Thus when an ordinance reduces the size of letters, it controls contrast; but when it specifies one size of lettering, it limits complexity.
4. The use of the scale model was selected as one way to obtain the control desired. For a different and perhaps more realistic simulation, one could airbrush actual photos of a roadside strip of interest, or employ new computer graphics systems that capture live images and alter them relatively easily.

E. Natural and rural scenes

Editor's introduction

The studies in this section examine the perception of and preference for natural and rural scenes. As in the previous section, this section does not include a paper aimed at deriving dimensions of affect for such scenes, but I briefly describe Craik's (1966) Landscape Adjective checklist (a comprehensive list of adjectives for assessing the landscape).

Craik first had 35 students describe each of 50 landscapes. From the 1,196 descriptors used, he culled those adjectives that were used most frequently (at least 6 times). The resulting list is shown in Table II.2. I have divided the list into adjectives that refer primarily to affective state and adjectives that refer primarily to scene features. Of course, such a division is not always clear. For example, adjectives such as *active, lonely, lifeless, cool,* and *cold* could refer to a subjective state or to a scene property.

The list can be used to obtain a comprehensive assessment of a landscape or to develop a shorter list of descriptors for a particular purpose. For example, one study in this section (Nasar, Julian, Buchman, Humphreys, and Mrohaly) used this list to develop a shorter list of descriptors. As with urban scenes, three factors – liking, excitement, and safety – emerged.

Three of the papers (Fenton; Herzog; Nasar et al.) in this section have relevance to Appleton's (1975a) theory on prospect and refuge. The results suggest that although prospect and refuge may be important features in environmental response, they may not always enhance perceived scene quality. In the first two studies, Fenton (for natural scenes) and Herzog (for field-and-forest scenes) employ similarity ratings to derive salient features of scenes. Both find dimensions relating to prospect and/or refuge. Among Fenton's five dimensions (open grassland–enclosed forest, barren–verdant, land–water, natural–influenced, and walking path) is a prospect dimension (open grassland–enclosed forest) and a refuge dimension (barren–verdant, where barren scenes lack places for refuge). Preference varied along the five dimensions. Preference was most consistent for natural over man-influenced scenes and for scenes with pathways (mystery). Otherwise, preference for open or closed, barren or verdant, and land or water

323

Table II.2. *The landscape adjective checklist*

Affective scales

Active	Exciting	Lifeless	Sad
Adventurous	Expansive	Living	Secure
Alive	Forceful	Lonely	Serene
Awesome	Free	Lovely	Strange
Beautiful	Fresh	Majestic	Threatening
Boring	Friendly	Moving	Tranquil
Calm	Gentle	Nice	Ugly
Challenging	Gloomy	Peaceful	Unfriendly
Comfortable	Good	Placid	Uninspiring
Content	Happy	Pleasant	Uninteresting
Cool	Harsh	Powerful	Uninviting
Dangerous	Imposing	Pretty	Violent
Dead	Inspiring	Refreshing	Warm
Depressing	Invigorating	Relaxing	
Dull	Inviting	Remote	
Eternal	Lazy	Restful	

Perceptual/cognitive scales

Arid	Extensive	Mountainous	Sloping
Autumnal	Falling	Muddy	Slow
Bare	Farmed	Mysterious	Smooth
Barren	Fast	Narrow	Snowy
Big	Flat	Natural	Soft
Black	Flowery	Nocturnal	Spacious
Bleak	Flowing	Noisy	Sparse
Blue	Foamy	Old	Springlike
Bright	Forested	Open	Stark
Broad	Glacial	Orange	Steep
Brown	Golden	Pastoral	Still
Burned	Grassy	Plain	Stony
Bushy	Gray	Pointed	Stormy
Changing	Green	Pure	Straight
Clean	Hard	Quiet	Summery
Clear	Hazy	Ragged	Sunny
Cloudy	High	Rainy	Swampy
Cold	Hilly	Rapid	Tall
Colorful	Hot	Reaching	Thick
Colorless	Humid	Reflecting	Towering
Contrasting	Icy	Rich	Tree-studded
Crashing	Isolated	Rocky	Unusual
Crisp	Jagged	Rolling	Vast
Cut	Large	Rough	Vegetated
Dark	Leafy	Round	Watery
Deep	Light	Running	Weedy
Dense	Low	Rushing	Wet
Deserted	Lumpy	Rusty	White

Table II.2 (*cont.*)

Perceptual/cognitive scales			
Desolate	Lush	Sandy	Wide
Destroyed	Marshy	Secluded	Wild
Dirty	Massive	Shadowy	Windy
Distant	Meadowy	Shady	Wintry
Drab	Misty	Shallow	Wooded
Dry	Moist	Sharp	Yellow
Empty	Monotonous	Sliding	
Eroded	Motionless	Slippery	

scenes varied somewhat with the respondents. Herzog found three salient aspects of his scenes: Unconcealed Vantage Point (no refuge), Concealed Vantage Point (refuge), and Large Trees. Notably, the first two classes of scenes had open vistas. Refuge (concealment of the vantage point) did not differentially affect preference. Large Trees was the most preferred category. Notably, Large Trees was characterized by, among other things, coherence. Examination of Large Tree scenes revealed preference related to identifiability, mystery, and, in contrast to Appleton's theory, enclosure.

In the third study, Nasar et al. manipulated and tested responses to prospect and refuge. The results revealed that rated pleasantness and arousal did not differ across the conditions. Rated safety was more favorable for the open view than for the closed one; and an interaction between sex and observation point revealed that males and females responded differently to refuge.

The remaining studies in this section (Orland; Kaplan and Herbert) consider the role of naturalness, novelty, and familiarity in environmental preference. Orland compared preferences of rural residents (familiar) and urban residents (less familiar) to rural scenes. As in other studies, both groups showed similarities in response, preferring natural to human-influenced scenes; but the rural group responded more favorably to the scenes than did the urban group, presumably because the rural scenes were more familiar to the rural group. Kaplan and Herbert compared responses by Australians and Americans (from Michigan) to natural and rural scenes from Michigan. The preferences of the two groups were highly correlated and similar. Furthermore, four categories of scenes were identified: Residential, Vegetation, Pastoral, and Manicured Landscapes. Both groups responded less favorably to rural houses in the natural context, and responded favorably to Vegetation and Manicured Landscapes. Nevertheless, an initial look at the data suggested a preference for the familiar: The Michigan group responded more favorably than did the Australian group (and recall that the scenes were from Michigan). Further inspection, however, revealed that this difference was accounted for by unfavorable Australian responses to one kind of scene: Pastoral/Open Rural scenes. Notably, this kind of scene should be the most recognizable (familiar) to the Australian sample. On the one hand, the familiar is liked; on the other hand, the familiar is disliked. Thus it seems that familiarity and novelty are complex phenomena and that their effects may vary with the kind of scene, the degree and level of familiarity and novelty, and the kind of respondent. Further study sensitive to these complexities is needed.

22 Dimensions of meaning in the perception of natural settings and their relationship to aesthetic response

D. Mark Fenton

The physical environment has typically been defined either in terms that are independent of the perceiving individual (Barker, 1965; Brunswik, 1956; Wohlwill, 1973) or in terms of individual perception and constructs of the environment (Boulding, 1956; Ittelson, 1976; Klausner, 1971; Wapner, Kaplan, and Cohen, 1973). Research in environmental aesthetics, for example, which most often treats the physical setting as the principal determinant of preference, defines setting variables through the use of objective measures (Shafer, Hamilton, and Schmidt, 1969) or normative judgments (S. Kaplan, 1975; Wohlwill, 1968, 1976). In contrast to Moore's (1977) assertion that "there is an underemphasis on the role of the environment and of the physical environment in particular" (p. 16) in environment-behavior studies, and Wohlwill's (1973) statement that there is a "widespread tendency to define the environment in subjective personal terms, rather than objective, physical ones" (p. 166), it is argued that research in environmental aesthetics has overemphasized the influence of objective setting variables. This is not to say the role of such variables is insignificant, but that these variables have been studied in isolation from the actively perceiving and cognizing individual. This research, unlike previous research in environmental aesthetics, attempts to identify the underlying dimensions through which individuals construe the natural environment and the relationship of these dimensions to aesthetic response.

The objective specification of stimulus features in studies of landscape evaluation and preference has been viewed by researchers developing these techniques as particularly advantageous, in that the scenic beauty of an area can be quantified and therefore easily compared with other specifiable resources of a natural area or with other regions. Similarly, if physical attributes of a landscape that either enhance or detract from the scenic quality of an area can be identified in objective and quantifiable terms, then such attributes can be directly considered

This paper was first published in *Australian Journal of Psychology* 37 (1985): 325–39.

by landscape architects and planners in the creation and preservation of scenic resources. The landscape-evaluation studies of Anderson, Zube, and MacConnell (1976), Pitt (1976), and Zube, Pitt, and Anderson (1974) are characteristic of those approaches adopting an objective and quantitative approach to the assessment of landscape quality. For example, the twenty-three landscape dimensions used as predictors of scenic-resource value in the Zube et al. (1974) study consisted of variables such as absolute relative relief, mean slope distribution, percentage of tree cover, and topographic texture. Similarly, Pitt (1976), in evaluating the scenic-resource quality of stream sites, used such objective physical indices as depth, width, flow velocity, and turbidity, where turbidity in this case was measured as the weight of "the residue remaining on a 45 micron millipore filter after a 300 ml sample from each stream had been passed through the filter" (p. 148).

The landscape-preference techniques developed by Shafer (1969), Shafer et al. (1969), and Thayer, Hodgson, Gustke, Atwood, and Holmes (1976) also make use of objectively quantifiable landscape attributes in the assessment of landscape preference. These studies are characteristic of the use of a theoretical and methodological technique that assumes "that people's behaviour patterns may be related to what they perceive in the environment" and that one "way to measure perception would be to test for a correlation between physical properties of an environment and some index of behaviour within that environment" (Shafer, 1969, p. 74). In this approach, the objective measures of the physical setting are obtained by overlaying photographic representations of the environment with grid squares and summing the number of grid squares overlaying areas of distant, intermediate, and immediate zones of vegetation, nonvegetation, sky, and water. Perimeter measures are also taken for each of these zones by summing the number of boundary edges of each grid square that is adjacent to a different zone. These measures are then used to predict individuals' aesthetic response to photographs of natural landscapes.

In some cases, "the investigator may need to resort to human judges to assess dimensions of the environment that have an objective referent in principle, but nevertheless are not susceptible to direct measurement via physical indices" (Wohlwill, 1976, p. 63). The use of judges to rate attributes of the landscape selected by the investigator has become widely used in a number of landscape-preference studies, particularly those of Kaplan, Kaplan, and Wendt (1972), R. Kaplan (1973), Ulrich (1977), and Wohlwill (1968, 1976). A similar methodological technique is used in studies of landscape evaluation, where most often the existence of landscape attributes is dependent on the judgment of the researcher who is developing the technique or of a small group of experts (Gilg, 1974; Linton, 1968; Wallace, 1974). In most instances, these studies are still attempting to measure the objective attributes of the landscape, but instead of quantifying these variables using physical indices, quantification has been attempted through a normative-judgment approach – that is, through consensual judgments on rating scales (Walker, 1980).

The dichotomy between objective and perceptual definitions of environmental variables, or structural and content determinants of preference, has a historical parallel in much of the research in experimental aesthetics, where since Fechner's (1876) early research, "stimulus," rather than "environmental" or "setting," variables have been defined in both objective and perceived terms. Experimental research into the properties of the golden section used objective measures of the lengths of lines as predictors of preference (Angier, 1903; Eysenck and Tunstall, 1968). Similarly, the notions of order and complexity in Birkhoff's (1933) aesthetic measure and the collative property of complexity employed in much of the research associated with Berlyne's (1971) approach to experimental aesthetics are defined in objective terms – that is, as the number and lengths of lines, dots, and angles within a stimulus array. While there was continuing development of both theory and method into the objective properties of stimuli and their relation to preference, there was a simultaneous and rather different development of approaches that considered judgments of preference and aesthetic response to be dependent on the perception and constructs of the stimulus by the individual. The theories of empathy, in particular those of Lipps (1900) and Kainz (1962), rely on the outcome of an interactive relationship between individual and stimulus, as Pratt (1961) has described the properties "felt into" the stimulus as being "subjective in origin, but by a kind of simultaneous association [interaction] by inference built up through countless repetitions, they are seen and heard as aspects of objects located outside the body" (p. 75).

In contrast to the research in art and experimental aesthetics, landscape-preference studies nearly always define the determinants of landscape quality through the use of objective quantification (Shafer et al., 1969) or through the use of normative judgments (S. Kaplan, 1975; Wohlwill, 1968, 1976). Research in landscape preference needs to examine the role of the perceived or construed environment as a determinant of aesthetic response. Kreimer (1977) has emphasized in a critical analysis of environmental-preference methodologies that current research approaches have often assumed that an individual's perception and constructs of the environment have a "direct" correlate with the "real" external environment. Similarly, Russell and Ward (1982), in the most recent review of the field of environmental psychology, note the "continuing debate between the environmental psychologists taking a cognitive approach and those emphasising the study of the objective physical environment" (p. 664).

The objective of the present research was to identify the important underlying dimensions of meaning or content that individuals were using to discriminate among natural settings and, in addition, to examine the relationship between these dimensions and judgments of aesthetic quality. The definition of content properties, as opposed to the structural – that is, informational – (S. Kaplan, 1975) and objective properties, is based on the phenomenological meaning of the environment to the individual and is not determined a priori by the investigator.

From a psychological perspective, an important theoretical approach to environmental aesthetics is the functional-evolutionary approach proposed by S. Kaplan (1975), which has generally sought to determine whether specific informational properties of the physical setting were related to aesthetic response. However, Kaplan and Wendt (1972) and S. Kaplan (1975) have also suggested additional determinants of landscape preference, which rather than reflecting a concern with the processing of environmental information, emphasize the content of the information being processed (Kaplan and Kaplan, 1982). These properties have been referred to as "primary landscape qualities" (Kaplan and Wendt, 1972) or as "content" properties (Kaplan and Kaplan, 1982), and include elements of the setting having important evolutionary significance, such as the presence of water, trees, green foliage, and nature. While the role of the content properties in landscape preference is of undeniable importance, the research has tended to underemphasize the importance of these factors.

Pairwise similarity ratings and Kelly's (1955) repertory grid procedure have been used to identify the important dimensions by which individuals perceive the environment. Although both methods require that a comparison be made among different settings, the similarity task is predominantly nonverbal, whereas the repertory grid methodology requires the use of verbal labels in the identification of similarities and differences among environmental settings.

The nonverbal approach to eliciting dimensions of content is usually achieved through the method of pairwise similarity judgments, which involves subjects being presented with pairs of stimuli and required to evaluate the similarity or dissimilarity between pairs. This method, in addition to a nonverbal emphasis, allows for a phenomenological classification of stimuli that is not biased by the researcher's a priori conceptual definitions of the perceptual dimensions. The method has been used to elicit dimensions of content for architectural stimuli (Oostendorp and Berlyne, 1978a), urban environments (Gärling, 1976a; Horayangkura, 1978), and the molar physical environment (Ward, 1977; Ward and Russell, 1981a, 1981b). There is, however, a lack of evidence regarding the underlying dimensions of content for scenes of the natural environment. Obviously, any comparisons among studies that use nonverbal similarity judgments must take into account the type of stimulus material being used. For example, one would not expect an urbanism dimension, as found by Horayangkura (1978), to occur with stimuli made up of scenes of natural settings. But an examination of the content dimensions derived by Ward (1977) and by Ward and Russell

(1981b) for stimuli composed of urban, natural, and "mixed" environmental settings indicates that similar dimensions may be derived in the present analysis. For instance, the five dimensions derived by Ward and Russell (1981a) were variations in "natural water scenes versus man-made land scenes; natural versus man-made; open versus enclosed; natural, open scenes versus man-made, enclosed scenes; and natural water scenes versus natural land scenes" (p. 129). Similar dimensional solutions were also found by Ward (1977), which included variations in naturalness, scale, open versus enclosed, and land versus water.

A number of studies have also used Kelly's (1955) repertory grid methodology to elicit an individual's phenomenological descriptions of the environment. This method, as is the case with the nonverbal pairwise similarity method, has been principally used to elicit subjective descriptions of urban and built environments (Harrison and Sarre, 1975; Honikman, 1976), and its use has been underemphasized in the context of natural-environmental-perception research. From the perspective of an interactional approach, Nysdet (1982) has emphasized the usefulness of the method in obtaining information on the content of individuals' perceptual representations of the environment.

Pairwise similarity judgments and the repertory grid method were both used in the present study to examine characteristics of the perceived environment. Both techniques were used because it was felt that the different methods might provide different content dimensions, given that the repertory grid methodology is a much more verbal method than the multidimensional scaling similarity task. In addition, judgments of liking were also obtained for natural settings in an attempt to determine whether the dimensions of content that individuals were using to discriminate among natural settings were related to aesthetic response.

Method

Subjects

The sample was composed of fifty-two volunteer subjects from an introductory behavioral-science class, which included fifteen males and thirty-seven females. All subjects received one and a half percentage points toward their final grade for participation in the experiment. The mean age of the sample was twenty-four.

Selection of photographic stimuli

A sampling methodology proposed by Fenton (1981) was used to select photographic stimuli from a coastal mangrove forest located within an environmental park. A portable video recorder, with a standard zoom set at 50 mm, and breta mount were used to record landscape scenes from a walk through the natural setting. An assistant was used to operate the video recorder, while the experi-menter walked along the trail recording through a continuous scanning action views of the landscape. Using this procedure, a one-hour, twenty-three-minute recording of the mangrove environment was made. The videotape was then measured in seconds, and random intervals were selected to locate twelve scenes within the mangrove environment. The tape was then audio-dubbed, and a re-cording kept of the location of each scene. The video recording was then played back through the camera while in the field, and through the use of an earphone, each randomly chosen scene was located and color slides were taken, using a 35-mm still camera mounted on a tripod (Canon AE-1; f1.8, 48 mm). This pro-cedure partly overcomes bias in the selection of scenes by the experimenter and allows the findings to be generalized to the natural environment of interest.

Experimental apparatus

Color slides were shown to subjects in an experimental laboratory using two Kodak S-AV slide projectors, both of which were fitted with 35-mm wide-angle lenses. The projectors were placed on stands, at a height of 1.2 meters, and placed 2.70 meters directly opposite two whitewashed walls. This enabled two slide images to be projected onto the walls, each image measuring 3.00 meters by 2.50 meters. The effect of increasing the size of the image was to enhance apparent depth (Gibson, 1947). Between the two projectors was placed a table, with microcomputer (Apple IIe), video monitor, disk drive, and Itoh parallel printer. The microcomputer was connected to both S-AV projectors, which allowed computer control over slide presentation.

Procedure

Subjects were required to complete three independent tasks, which included judgments of similarity, construct elicitation and rating, and judgments of liking. Instructions to subjects for each task and the order of presentation of tasks, which were randomized for each subject, were controlled by internal programming of the microcomputer.

Pairwise similarity judgment. There were sixty-six pairwise similarity judg-ments that could be made among the twelve stimuli. In order to reduce the time required to complete the task, an "incomplete data design" (MacCallum, 1978) was used to obtain judgments of similarity among the sixty-six stimulus pairs. For example, each subject completed thirty-three pairwise similarity judgments, which represented 50% of the total number of possible pairwise judgments. Each set of thirty-three pairs was made up of alternating cells of a lower triangular similarity matrix, resulting in one subject completing 50% of the matrix, and the next subject, the remaining 50%. The presentation of each set of thirty-three pairs of stimuli was alternated from one subject to the next, and the order of presentation of stimulus pairs was randomized among subjects. Subjects were required to rate every pair of stimuli on a 7-point bipolar similar–dissimilar rating scale.

Construct elicitation. Twenty independent and randomly selected pairs of stimuli were used to elicit constructs from each subject. Prior to the elicitation of constructs, subjects were given a series of instructions, which were displayed on the VDU screen:

You will shortly be asked to describe the way in which two scenes of a natural environment differ. In describing the way in which the two scenes differ, imagine yourself as being actually in the setting. Feel free when describing differences between settings, to use any descriptions you can think of, whether they are differences in the types of things you can do in each setting, or differences in objects and things found within each setting, or differences in how the settings look. Remember not to use the same descriptions twice.

When the subject was fully acquainted with the procedure, two slides were projected simultaneously onto the walls in front of the subject, and the video monitor was prompted for one pole of a construct by stating:

(TYPE NO if you can't describe a difference)
In what way do the two scenes differ?
One scene is (or has) . . .

The opposing pole was prompted for by the statement:

While the other scene is (or has) . . .

Each bipolar construct was elicited using the same instructions; however, the phrase "is (or has)" was randomly alternated by internal programming. For example, instead of "or has" being enclosed in parentheses on one occasion, the next occasion may have resulted in "or is" being enclosed in parentheses. This reduced to some degree the effects of instructional set on the elicitation of constructs. A maximum of twenty constructs could be elicited for each subject using this procedure. After the elicitation of constructs, each of the twelve stimuli, which were presented in a random order for each subject, was rated on a 7-point scale, of which the anchors were the positive and negative poles of each construct.

Results

Multidimensional scaling analysis: ALSCAL

Fifty-two matrices, composed of dissimilarity measures among twelve pairs of stimuli, for each subject were analyzed by ALSCAL's (Young and Lewyckyj,

Table 22.1. *Stress and R^2 values for two through six dimensions: ALSCAL*

Dimensions	Stress	R^2
2	.667	.821
3	.113	.870
4	.092	.878
5	.060	.924
6	.058	.905

1981) weighted multidimensional scaling option (WMDS). The data were treated as ordinal, continuous, and matrix conditional, and two through six dimensions were derived. The stress and R^2 values for the dimensional solutions are given in Table 22.1.

An examination of the stress and R^2 values for each dimensional solution given in Table 22.1 indicates no substantial improvement in fit after a five-dimensional solution. A five-dimensional solution was therefore selected to account for the variation in the perceived similarity space.

Principal components analysis: PREFAN

Each individual's matrix of ratings of stimuli on constructs were combined into a "super-grid" (Harrison and Sarre, 1975), composing a rectangular matrix of 12 stimuli by 824 constructs. This matrix was analyzed through the use of PREFAN (*Grid Analysis Package*, 1981), which is capable of analyzing a group of grids aligned by element or stimuli, but not by construct. The Bartlett test (1950), used as the criterion for selecting the number of significant components, indicated that eight components accounted for a significant proportion of the variance ($\chi^2 = 31.89$; $df = 9$, $p < .05$). The roots and percentage of variance accounted for by each component are given in Table 22.2.

A comparison of the principal components and WMDS analyses

A comparison of the two solutions was performed by correlating the dimensional coordinates from the WMDS solution with the loadings from the principal components analysis. This procedure was feasible because WMDS, like the principal components analysis, rotates the dimensions so as to account for the maximum amount of variation among stimuli. The dimensional coordinates were therefore meaningful, unlike some multidimensional scaling procedures in which the coordinates are used simply to define the location of points in a multidimensional

Table 22.2. *The percentage of variance accounted for by the first eight significant components*

Component	Root	As percent
1	2347.15	25.86
2	1516.50	16.71
3	1003.41	11.06
4	869.66	9.58
5	621.68	6.85
6	605.41	6.67
7	494.45	5.45
8	484.26	5.34

space. Table 22.3 gives the intercorrelations between the ALSCAL and the PREFAN solutions.

Aesthetic response and dimensions of content

Through the use of a property-fitting program, PREFMAP (Carroll and Chang, 1977), the five-dimensional content space, derived from the WMDS procedure, was used to predict each individual's liking ratings. The PREFMAP analysis has four phases, in which properties – in this case, judgments of liking – may be located in a multidimensional space as vectors (phase four), circular ideal points (phase three), elliptical ideal points (phase two), or elliptical ideal points based on a rotation of the initial stimulus coordinates (phase one). If the analysis is started at phase two, then the properties are located in the unrotated stimulus space; however, if the analysis commences at phase one, then the properties are

Table 22.3. *Intercorrelations among the ALSCAL dimensions and the PREFAN components*

ALSCAL dimensions	PREFAN components							
	1	2	3	4	5	6	7	8
1	−.71	.21	−.10	.56	−.24	−.19	.09	.03
2	−.43	−.71	−.06	−.05	.40	.09	−.02	−.28
3	.28	.10	−.79	−.06	.11	−.24	−.10	.17
4	−.43	−.11	.10	−.63	−.50	−.13	−.17	.05
5	−.47	.58	.01	−.41	.44	.06	.11	.12

Note: Correlations are Pearson product-moment correlations.

Table 22.4. *PREFMAP summary table: location of fifty-two subjects' liking ratings into the five-dimensional content space (in direction cosines)*

	Dimensions					
Subject	1	2	3	4	5	R
Subject 1	0.36	0.36	−0.14	0.81	0.24	0.64
Subject 2	−0.16	0.77	−0.17	0.59	−0.02	0.83
Subject 3	−0.21	0.06	0.71	0.66	0.11	0.87[a]
Subject 4	0.40	0.64	−0.63	0.02	−0.19	0.90[b]
Subject 5	−0.15	0.57	−0.20	0.58	0.53	0.86[a]
Subject 6	0.63	0.38	−0.61	−0.14	−0.26	0.86[a]
Subject 7	0.81	−0.21	0.24	−0.46	0.15	0.61
Subject 8	−0.40	0.45	0.60	0.20	−0.49	0.51
Subject 9	0.24	−0.16	−0.47	−0.15	0.82	0.69
Subject 10	0.25	−0.34	−0.73	0.47	−0.26	0.76
Subject 11	−0.67	−0.38	0.17	0.30	−0.54	0.74
Subject 12	0.60	−0.02	−0.68	−0.36	0.19	0.89[b]
Subject 13	0.97	−0.23	0.02	0.11	0.02	0.77
Subject 14	−0.43	−0.32	0.01	−0.07	−0.84	0.86[a]
Subject 15	−0.39	0.25	−0.46	0.52	−0.55	0.75
Subject 16	0.18	−0.19	−0.20	0.90	−0.29	0.50
Subject 17	0.74	−0.46	−0.44	−0.11	−0.18	0.70
Subject 18	0.21	0.00	0.53	0.49	−0.66	0.78
Subject 19	0.26	0.18	−0.13	−0.54	−0.77	0.75
Subject 20	−0.57	0.43	0.22	0.08	−0.66	0.83
Subject 21	0.25	0.58	−0.28	0.59	−0.41	0.91[b]
Subject 22	−0.21	0.55	−0.31	−0.15	−0.73	0.86[a]
Subject 23	0.58	−0.36	0.31	0.07	0.66	0.88[b]
Subject 24	0.52	0.07	−0.18	0.75	−0.35	0.93[b]
Subject 25	−0.52	0.52	−0.45	0.46	0.20	0.94[b]
Subject 26	0.53	0.42	−0.51	0.52	−0.10	0.93[b]
Subject 27	−0.54	−0.46	0.57	0.19	−0.36	0.77
Subject 28	0.43	−0.50	−0.27	−0.60	−0.37	0.76
Subject 29	−0.39	−0.28	0.34	−0.18	−0.78	0.88[a]
Subject 30	0.36	−0.26	0.34	−0.30	−0.77	0.85
Subject 31	0.05	0.76	0.13	0.25	−0.58	0.96[b]
Subject 32	0.67	0.46	−0.21	0.45	−0.30	0.76
Subject 33	0.30	0.11	−0.48	−0.51	−0.64	0.91[b]
Subject 34	0.07	−0.90	0.04	−0.33	0.23	0.73
Subject 35	0.04	−0.26	−0.92	0.27	0.06	0.87[a]
Subject 36	−0.71	−0.30	−0.48	0.01	−0.42	0.68
Subject 37	0.07	0.69	0.56	0.45	0.05	0.74
Subject 38	0.66	0.42	−0.56	−0.03	−0.27	0.96[b]
Subject 39	−0.16	−0.32	0.70	−0.01	−0.61	0.82
Subject 40	0.89	0.24	−0.30	0.18	0.17	0.97[b]
Subject 41	0.56	0.55	0.19	−0.19	0.56	0.60
Subject 42	−0.79	−0.25	−0.29	−0.41	0.25	0.51

Table 22.4 (*cont.*)

Subject	Dimensions					R
	1	2	3	4	5	
Subject 43	0.52	−0.08	−0.63	0.19	−0.54	0.85
Subject 44	0.60	−0.17	0.69	−0.26	0.26	0.91[b]
Subject 45	0.42	0.34	−0.35	0.76	0.08	0.92[b]
Subject 46	0.54	0.76	−0.23	−0.08	−0.27	0.96[b]
Subject 47	0.81	0.46	−0.03	0.35	0.04	0.87[a]
Subject 48	0.56	0.30	−0.07	0.46	−0.61	0.83
Subject 49	0.28	−0.34	0.22	−0.86	−0.07	0.98[b]
Subject 50	0.05	0.70	−0.49	0.39	0.33	0.85[a]
Subject 51	0.86	0.30	−0.34	0.06	0.23	0.81
Subject 52	0.35	0.56	−0.74	−0.04	−0.12	0.94[b]

[a] $p < .10$
[b] $p < .05$

located in the rotated stimulus space. The PREFMAP analysis was initiated at phase two in order to avoid rotation of the space, which had been accomplished through the WMDS analysis. A comparison of the correlation values among phases indicated that for the majority of subjects, phase four – that is, the vector model – was the most suitable to represent each subject's measure of aesthetic response. Table 22.4 gives the direction of each subject's liking vector in the five-dimensional space. The correlation values, (R), represent the degree of fit between each subject's regression line and the stimuli, and although significance levels are indicated, they should be interpreted with some caution because of nonrandom sampling.

Discussion

Interpretation of the dimensional space

The results obtained in the analyses indicate that similar dimensions of content are obtained when using either the nonverbal pairwise similarity task or the verbal repertory grid task. This is clearly evident when inspecting Table 22.3, which shows that high correlations were obtained between the first four AL-SCAL dimensions and the first four principal components. This finding indicates a stable dimensional space, whether derived nonverbally, through similarity judgments, or verbally, through the elicitation of constructs. In addition, and with respect to the ALSCAL analysis, all but six individuals had R^2 values exceeding .90, indicating that the group space accounted for a significant proportion of the variance in the individuals' personal space. This indicates that individuals tend to use the same underlying dimensions of meaning when construing the natural environment. However, it is suggested that the consensus among individuals in the use of these dimensions may be due in large part to a relatively homogeneous subject sample. With a very different subject population, it is possible that somewhat different dimensions of meaning would be used to discriminate among natural settings. Landscape experts – including designers, architects, and park officials, for instance – might use very different dimensions from those used by individuals with no specific knowledge of natural settings. A clear direction for future research is to examine differences in the phenomenological meaning of the setting among different subject populations. It must also be remembered that the dimensions of meaning are used by individuals only in relation to this particular natural setting. Such dimensions may not be appropriate when construing very different natural settings.

The first four content dimensions were identified through an examination of the distribution of scenes on each dimension and component. In the case of the WMDS space, the spatial configuration was also examined, and in the principal components analysis, the constructs that defined the end points of the component were also examined.

Through this procedure, the first content dimension to be identified in both the ALSCAL and the PREFAN solutions were scenes that varied from open grassland through enclosed forest. This dimension is similar to the dimension of open versus enclosed found by Ward and Russell (1981a), with respect to scenes comprising a mixture of urban and natural settings. The clear distinction between forest and grassland may reflect, as Tuan (1974) has suggested, an innate genotypic predisposition to discriminate between the savanna and the forest environments. From an evolutionary-theoretical position, this would appear reasonable, considering the importance of the savanna and forest environments in early human evolution (Orians, 1980) and that movement from the forest to the savanna environment produced not only adaptive biological changes, but also changes in preference for particular landscapes (Kaplan and Kaplan, 1982).

The second dimension, identified through an inspection of the distribution of scenes and constructs in the spatial configuration, was identified as a barren versus verdant dimension. Scenes identified as barren were construed, for example, as being dead, stark, hot, gray, and less fertile. In contrast, those scenes identified as verdant were construed, through an examination of the construct loadings, as being alive, green, cool, and fertile. The occurrence of this dimension was unexpected because previous studies had not identified this dimension in the perceptual scaling of different environmental settings. It is suggested that this dimension may be important in discriminating among only natural settings.

A factor analysis of semantic differential ratings of natural settings by Calvin, Dearinger, and Curtin (1972) identified a natural–starkness factor, which, as in the present analysis, comprised scenes that were warm and fertile versus scenes that were cold and barren. From an evolutionary-theoretical perspective, such a dimension of content would be expected, since the identification of natural settings having the potential to support life would have considerable adaptive and functional significance.

The third dimension that was identified through an examination of the AL-SCAL and PREFAN solutions was a land versus water dimension. The dimension could also be seen as comprising variation in vegetation from trees to bushes; however, it is suggested that this is directly related, and superordinate, to the land versus water dimension. This dimension appears to be equivalent to Ward and Russell's (1981a) natural water versus natural land dimension, suggesting that this may be an important cognitive dimension by which individuals actively construe the natural environment. The importance of water in the environment as a determinant of preference has also been recognized by Kaplan and Kaplan (1982), who refer to it as a "primary landscape feature." Again, the importance of such a dimension underscores the appropriateness of an evolutionary approach to landscape perception and valuation, which recognizes that the potential to identify water in the environment would obviously have tremendous survival value for the species (Orians, 1980).

The fourth content dimension to be identified was a natural versus man-influenced dimension. The dimension reflects not only the presence of humans in the environment, but also environments that are potentially useful and suitable settings for exploitation and action. It appears that the suitability of the environment for humans is dependent on what the environment affords (Gibson, 1982). For example, constructs loading on this component indicate that the existence of water, shady trees, open ground, and cleanliness appear to afford camping, picnicking, adventure, and "sitability" (Gibson, 1982), respectively. It is suggested that the naturalness versus man-made or man-influenced dimension that has been identified in a number of previous studies (Kaplan et al., 1972; Wohlwill, 1976; Zube et al., 1974) may be more complex than is currently thought, inasmuch as it may indicate not only concrete evidence of humans, but also the potential usefulness of the environment to people. The ability to discriminate scenes in the natural environment in terms of their usefulness and previous interaction with people would undoubtedly have been important in the context of human evolution. Such an ability would have been a successful adaptive advantage to an often-hostile environment.

The fifth dimension, when inspecting the PREFAN solution, was clearly identified, on the basis of the loadings of scenes and constructs, as a walking-path dimension. It included a number of scenes that suggested the possibility of movement into the scene and included, for example, the constructs inviting accessibility, walking, walking into, walking path, and path. It appears to be similar to S. Kaplan's (1975) mystery dimension, which through the existence of pathways allows for further inspection and exploration of a scene. Although there is a moderately strong correlation between this component and the fifth dimension extracted in the ALSCAL solution, the fifth dimension of the AL-SCAL solution also appears to include elements of the barren versus verdant component, suggesting that fertile and verdant environments elicit greater approach behavior than do more barren environments.

The remaining three content dimensions are specific to the PREFAN analysis, and while accounting for a significant proportion of the remaining variance, as defined by the Bartlett test (Bartlett, 1950), the relative contribution of these components to the variance explained is comparatively small when compared with the previous components.

Aesthetic response and the dimensions of content

The relationship between judgments of aesthetic response and the dimensions of content are indicated in Table 22.4. When interpreting Table 22.4, a positive cosine on dimension 1 indicates a preference for open-grassland settings, while a negative cosine indicates a preference for enclosed forests. With respect to dimension 2, a positive cosine indicates a preference for verdant settings, while a negative cosine indicates a preference for barren settings. A preference for water scenes is indicated by a negative cosine on dimension 3, while a positive cosine indicates a preference for land scenes. On dimension 5, a negative cosine indicates a preference for scenes with walking paths, while a positive cosine indicates a preference for scenes with no visible pathways.

Most important, the findings indicate considerable individual differences in aesthetic response. For example, an examination of Table 22.4, giving the location of each subject's liking ratings in the five-dimensional content space, indicates that each subject differentially weights each of the content dimensions in terms of its underlying salience to liking. As an example, subject 1 appears to like natural as opposed to man-influenced settings; subject 2 likes verdant settings, while subject 3 likes natural land as opposed to natural water settings. Although there is considerable between-subject variation in preference for natural settings, a number of overall trends are apparent. In particular, the majority of individuals tend to prefer scenes characterized by open grasslands (dimension 1), which are verdant (dimension 2), have water (dimension 3), are natural (dimension 4), and have pathways (dimension 5). Given the proposed underlying evolutionary significance of the content dimensions, it is not surprising that individuals tend to prefer settings with such content properties.

Although the functional and adaptive significance of the content properties may be important to environmental preference, and they may indeed be "indirect" perceptions of environmental affordances (Gibson, 1982), they may also, in the context of traditional experimental aesthetics, represent the ecological properties of the setting – that is, those "associations with biologically noxious or beneficial conditions" (Berlyne, 1971, p. 69). Given the very functional significance of these dimensions, it would not be unreasonable to assume that these properties have the potential to modify arousal levels and therefore behavioral response to the natural setting. It may well be that the collative properties of experimentally derived "geometric-technical" stimuli (Werner, 1963) are important in modifying arousal levels and consequent aesthetic response, but it is also suggested that when investigating "environments" and not "stimuli," the ecological or content properties may be important determinants of aesthetic response through their effect on an individual's arousal level.

In addition, the very fact of subjects being able to impose some meaning on and identify characteristics of these natural settings would, it is suggested, reduce uncertainty and enable the individual to "make sense" (S. Kaplan, 1975) of the available information. As S. Kaplan (1975) has indicated, considerable frustration is often evident on the part of individuals who are unable to identify or categorize visual forms. This suggests, as indicated earlier, that the meaning, content, or ecological properties (Berlyne, 1971) of natural settings may be important determinants of aesthetic response, through their differential effects on an individual's arousal level. It is suggested that settings that are difficult to identify or categorize, or that are meaningless, would increase arousal levels above optimal levels. However, being able to impose meaning on the natural setting, and in the process make sense of the available information, would reduce uncertainty and, consequently, arousal levels to an optimal level. In addition, if there is a reasonable fit between physical setting and perceived meaning, then this would allow for greater predictability and the possibility of interaction and perceptual involvement in the setting.

A clear direction for future research in the perception and valuation of natural settings is the identification of the underlying dimensions through which individuals construe natural settings and the salience of these dimensions to judgments of aesthetic quality. Although this research examined one specific natural setting, future research should attempt to ascertain whether similar dimensions of meaning occur with respect to other very different natural settings and subject populations. In addition, the finding of considerable variation among subjects in judgments of aesthetic response needs to be investigated further. Variation in aesthetic response does suggest the possibility that each individual may be basing his or her judgments on very different properties of the physical setting. It may well be that some individuals' judgments of aesthetic quality may be determined

by the informational (S. Kaplan, 1975) or objective (Shafer et al., 1969) properties, while other individuals may be basing their judgments on the content or
perceived meaning of the setting. This may well be the individual-difference
dimension of most importance to environmental aesthetic response, but it has not
been addressed as such because investigators to date have assumed that the
determinants of aesthetic response were relatively invariant across subjects.
Hopefully, future research that addresses these problems will bring us a little
closer to an adequate and meaningful assessment of the aesthetic quality of
natural landscapes and environments.

Acknowledgment

I gratefully acknowledge the assistance of Joseph Reser for his helpful comments
on a draft of this article.

23   A cognitive analysis of preference for field-and-forest environments

Thomas R. Herzog

Natural environments fascinate human beings (R. Kaplan, 1983; S. Kaplan, 1977; S. Kaplan and Talbot, 1983). They like natural environments better than urban environments (Kaplan, Kaplan, and Wendt, 1972). Not surprisingly, many attempts by planners to improve urban environments have involved introducing elements of the natural environment into the urban setting. Herzog, Kaplan, and Kaplan (1982) have shown quite clearly that nature within an urban setting forms a distinct category in the minds of observers, a category rated higher in preference than all the other urban categories they investigated. Thus it is known that nature in general is highly valued and that it can be used to improve urban environments.

In a sense, however, we have just scratched the surface. Consider how much better a job urban planners could do in using nature effectively if they knew more about it. The more detailed information needed is of two kinds. First, it seems unlikely that nature is one homogeneous category. Surely, a broad sampling of scenes will reveal several categories of nature that are differentially preferred. Second, there must be some specific identifiable variables that account at least in part for reactions to natural environments. If such variables can be isolated and described, then they can be manipulated by planners.

The research reported in this paper was guided by a theoretical framework that specifically addresses the concerns just discussed. It has been called the informational approach by Levin (1976). It is a blend of current thinking in cognitive psychology and reasonable evolutionary speculation. Probably the foremost spokespersons for this approach have been the Kaplans (S. Kaplan, 1979a; Kaplan and Kaplan, 1978, 1982). The key idea is that humans evolved in environments wherein spatial information was crucial to survival. Hence, the study of environmental preference should concentrate on the types of environments and the kinds of cognitive processes that would be especially important to such an organism.

This paper was first published in *Landscape Research* 9 (1984): 10–16.

One aim of this study, therefore, was to identify kinds of environments, or content categories, that underlie preferences for natural environments. Methods for discovering such categories have been described by R. Kaplan (1972). This initial effort was restricted to field-and-forest settings typical of those found in the midwestern United States. Using very similar materials, S. Kaplan (1979b) had made a beginning in identifying natural-environment categories. Some are based on quite specific contents, such as buildings in a natural setting. Other categories deal with more general spatial properties of natural scenes: open spaces that are undefined in terms of depth cues; spacious scenes that are well structured in depth; enclosed spaces; blocked views. Given our evolutionary history as processors of spatial information, it should not be surprising that such spatial properties might be important in defining natural-environment categories. However, there is no guarantee that the spatially defined categories of one study or series of studies will generalize to different populations or environments. Hence, continued research is needed to validate and expand our knowledge about categories of natural environments.

A second aim of this study was to investigate predictor variables that might help account for natural-environment preferences. The predictor variables used in this study were also derived from the informational approach outlined earlier. That approach stresses two general cognitive processes that would have been important to evolving humans, making sense and involvement. Making sense refers to the process of structuring the environment so that way finding is possible and one can predict what is likely to happen in a given setting. Involvement refers to the process of engaging and maintaining one's interest in an environment. Since both processes were presumably crucial to survival, environments that permit both to function successfully should be highly preferred.

Six predictor variables relevant to the cognitive processes just discussed were investigated in this study. Two of them involve environmental features that should aid the making-sense process. They are the degree of order or structure present in the immediate environment (coherence) and the extent to which the larger setting is well structured in depth (spaciousness). Two other predictors deal with features that should aid the involvement process. They are the amount of information or the number of elements present in the immediate environment (complexity) and the extent to which the environment seems to promise further information if the observer could walk deeper into it (mystery). Physical features such as curved pathways and partial concealment by foliage should enhance mystery. Finally, two predictors that cut across both the making-sense and the involvement processes were investigated. They are the smoothness of the ground surface (texture), which should both facilitate the acquisition of relevant structural information and encourage exploration, and the sense of familiarity evoked by a scene (identifiability), which should aid making sense by allowing the observer to place the scene into a known category, but may also lead to premature exhaustion of interest. With urban environments, past research suggests that the making-sense component of identifiability tends to dominate (Herzog, Kaplan, and Kaplan, 1976, 1982).

Much more is known about some of these predictor variables than others. Mystery, for example, has been found to be reliably associated with high preference in natural environments (R. Kaplan, 1975). Spaciousness also apparently plays a powerful role. S. Kaplan (1979b) reported that spatially well-structured scenes received high preference ratings. Scenes that are open but not well defined spatially received low preference ratings, as did blocked views. Kaplan also suggested that texture may contribute to the spatial structuring that seems to affect preference. In particular, blocked views often involve a very coarse-grained texture (tall grass, rough foliage), while well-structured scenes tend to have a more fine-grained ground surface. Complexity has an uneven track record as a predictor of preference for natural environments (R. Kaplan, 1975). In those few cases where it has been assessed, coherence has shown little empirical relationship to preference for natural scenes (R. Kaplan, 1975), although it may contribute positively to preference for rivers and riverside environments (R. Kaplan, 1977a). Identifiability, defined as a sense of familiarity (as opposed to actual familiarity), has not been empirically related to preference for natural environments. S. Kaplan (1975) noted that graphics of outdoor scenes that observers were unable to "identify or categorize" evoked "anger and hostility." In summary, past research suggests that several of these predictor variables may be related to preference for natural environments.

The major objectives of this study, then, were to investigate the category structure of field-and-forest environments and the role of predictor variables in accounting for preference. A further variable of interest was viewing time. Often, people get only a passing glance at a given environment, but occasionally they may have time to stop and look more closely. Does it make a difference in their reactions? To find out, both brief viewing exposures and relatively longer ones were used in this study.

Method

Participants

The sample consisted of 247 introductory psychology students of both sexes at Grand Valley State College. Participation was voluntary and earned students extra credit toward their course grade. Sixteen groups of from five to twenty-seven participants were run. The first twelve groups viewed each scene for 15 seconds, while the last four groups, run two years later, viewed each scene for either 20 or 200 milliseconds.

Stimuli

The settings consisted of 100 color slides of field-and-forest environments from various locations in Michigan's lower peninsula. The scenes typically contained relatively flat terrain and generous amounts of either green foliage or ground cover. Spatially, they ranged on a continuum from large open areas through smaller clearings to densely vegetated settings within a forest. No scenes contained either water or people. The only obvious human influence in any of the scenes was the presence of pathways in some of them.

Procedure

All participants rated each of the 100 scenes for one of seven variables. All ratings utilized a 5-point scale ranging from 1 = "not at all" to 5 = "a great deal." The six predictor variables were identifiability, coherence, spaciousness, complexity, mystery, and texture. *Identifiability* was defined as "how much of a *sense of familiarity* (rather than actual familiarity) you have for this scene. How easy would it be for you to get to know this scene?" *Coherence* was "how well the scene 'hangs together.' How easy is it to predict from one portion of the scene to another?" *Spaciousness* dealt with "the *feeling of spaciousness* that the scene conveys. Ask yourself how much room there is to wander around in." A *complex* scene was one that "contains a lot of elements so that it promises further information if only you could have more time to look at it from your present vantage point." By contrast, *mystery* was present when a scene "promises further information if you could walk deeper into it." *Texture* referred to "how fine-grained the ground surface is," or "if the ground surface is obscured by objects in the foreground, then rate the scene on how fine-grained the surface of the obstruction is." These variables are discussed in more detail by S. Kaplan (1975). The criterion variable, *preference,* was defined as "how much you *like* the scene, for whatever reason."

For the first twelve groups of participants, each scene was presented for fifteen seconds. Five practice slides preceded the 100 scenes, and a brief intermission occurred halfway through each session. Four of the groups rated the scenes for preference; four groups rated for identifiability, coherence, and spaciousness; and four groups rated for complexity, mystery, and texture. In the latter eight groups, each participant rated for only one predictor variable, with approximately one-third of each group rating each variable. Four different orders of scene presentation were used for the four groups that rated for preference and for each combination of predictor variables. Final sample sizes were twenty-one for texture, twenty-two for each of the other five predictor variables, and seventy-five for preference.

The last four groups rated the scenes for preference, but saw them for only very brief durations. The viewing time was 20 milliseconds for two of these groups and 200 milliseconds for the other two. The brief viewing times were achieved by using an electronic shutter mounted in front of the projector lens. Two different orders of scene presentation were used for each of the two brief-viewing-time conditions. Aside from the change in viewing times, the format for each session was the same as for the first twelve groups. Final sample sizes were twenty for the 20-millisecond duration and twenty-one for the 200-millisecond duration.

Results and discussion

Categorization of scenes

To discover the categories that influenced the participants' responses, the preference ratings of the 100 scenes in the 15-second condition were subjected to a nonmetric factor analysis, the Guttman-Lingoes Smallest Space Analysis III (Lingoes, 1972). The analysis yielded three dimensions, or clusters of scenes. Dimensional composition was determined by including all scenes with a factor loading greater than .40 on a given dimension and no loading greater than .35 on any other dimension.

The dimension with the greatest number of scenes (thirty-six) included primarily open spaces. All of S. Kaplan's (1979b) types (open, undefined; spacious, well-structured; enclosed; blocked views) were represented. The one feature common to almost all the scenes was that the vantage point of the observer (i.e., the camera) is out in the open. Hence, this was named the *Unconcealed Vantage Point* dimension.

By contrast, the second dimension, consisting of thirteen scenes, consistently had the vantage point of the observer located inside the forest. In many (but not all) of these scenes, the observer is looking into a clearing or an open area from inside the forest. One gets a strong sense of watching from concealment. This was named the *Concealed Vantage Point* dimension.

The last dimension consisted of ten scenes and also involved a vantage point inside the forest. Unlike those in the second dimension, however, these scenes are dominated by large old trees. The scenes contain either an abundance of such trees or fewer such trees with trunks of unusual shape. Nine of the scenes have this feature. Pathways are also common, fairly clear examples appearing in six of the scenes. The best examples of this dimension involve pathways bordered by

Figure 23.1. Scene from the Unconcealed Vantage Point category.

tall trees. Given the dominance of large trees, this dimension was called simply *Large Trees.*

Without further analysis, these categories immediately suggest two interesting observations. First, as anticipated, spatial structure is important in defining the categories. However, it is not the amount of space or even how the elements within the space are organized that matters. Rather, the key factor seems to be how the space is organized with respect to the observer. The space either does or does not afford a concealed vantage point to the observer. Appleton (1975a) has written eloquently about the concept of *refuge,* which involves those features of an environment that provide an opportunity for concealment. Recently, Woodcock (1982) has distinguished between what he calls *primary* and *secondary* refuge. Primary refuge refers to a vantage point that is itself enclosed or concealed. Clearly, the second dimension of this study involves primary refuge, and the very fact of its existence suggests that such an environmental affordance can have an important impact on an observer.

Figure 23.2. Scene from the Concealed Vantage Point category.

Second, trees are also quite important components of natural environments. This is hardly surprising. Earlier studies have also found the presence of trees to be a strong positive predictor of preference for such environments (Daniel and Boster, 1976; Gallagher, 1977, as described in R. Kaplan, 1983). Nature, represented mainly by trees, has been shown to be a highly valued component of urban environments as well (Herzog et al., 1982). The third dimension of this study strongly suggests that large older trees may be a special category of nature. Realizing this, planners may wish to consider preserving such trees whenever possible, especially in urban settings.

Prediction of preference

Role of content. The upper portion of Table 23.1 contains mean preference ratings as a function of content domain and viewing time. As is evident, Large

Figure 23.3. Scene from the Large Trees category.

Trees was the most preferred dimension (mean preference = 3.79), with Unconcealed and Concealed Vantage Points having similar ratings (mean preference = 3.27 and 3.39, respectively). The mean preferences for the three categories differed significantly ($F = 23.75$, $df = 2$ & 226, $p < .001$). Subsequent tests showed that Large Trees had a significantly higher rating than did the other two dimensions ($p < .05$), which did not differ from each other.

Clearly, content has an impact on preference. The specific pattern of results raises two issues. First, there is no difference in preference between concealed and unconcealed vantage points. Brush (1979) has reported a similar finding. Thus even though people react differently to the two types of environments by placing them in different categories, one type is not liked more than the other. Both types are moderately well liked (mean ratings slightly above the midpoint of the scale). With further knowledge of characteristics possessed by the most preferred scenes within each of these categories, planners would be better able to maximize preference in either type of setting. Suggestions concerning such distinguishing characteristics are offered in the section on predictor variables.

Table 23.1. *Mean ratings for each variable and content domain*

Variables	Content domain			
	Unconcealed Vantage Point	Concealed Vantage Point	Large Trees	Mean
Preference				
15 sec	3.32	3.47	3.95	3.58
200 msec	3.07	3.25	3.47	3.26
20 msec	3.31	3.23	3.53	3.36
Mean	3.27	3.39	3.79	
Identifiability	3.32	2.98	3.79	
Coherence	2.96	2.39	3.98	
Spaciousness	3.10	2.21	3.24	
Complexity	3.23	3.59	3.04	
Mystery	2.76	3.19	3.57	
Texture	2.64	2.36	3.69	

Second, large old trees are relatively well-liked natural elements. This finding reinforces the observation made earlier that preserving such trees ought to be given a high priority by urban planners. In this regard, a more specific suggestion can be made. The six most preferred scenes within this dimension included pathways (mean ratings above 4 on the 5-point scale), and in at least three of these scenes, the trees had a rather prominent role as pathway border elements. Hence, it may well be that large old trees create an especially pleasing effect as pathway border elements.

Since several of the scenes in the Large Trees category contained pathways, and these scenes were highly preferred, it is of interest to see whether the effects of large trees and pathways can be teased apart. Fortunately, there are some data that speak to this question, although conclusions must be tentative because of very small samples. The one scene in the Large Trees category that did *not* contain large trees did include a prominent pathway. Its mean preference rating in the 15-second condition was 4.25. The other five pathway scenes, which did contain large trees, had a mean preference rating of 4.14. The remaining four scenes, which contained trees but no pathways, had a mean preference rating of 3.64. The comparable means for all scenes in the Unconcealed and the Concealed Vantage Point categories were 3.32 and 3.47, respectively (see Table 23.1). Thus it might appear that large trees enhance preference slightly and pathways enhance preference substantially, with their combination no more preferred than pathways alone. Unfortunately for that theory, there were three

Figure 23.4. Pathway scene from the Unconcealed Vantage Point category.

scenes in the Unconcealed Vantage Point category that featured pathways but no large trees, and their mean preference rating in the 15-second condition was 3.32, exactly equal to the mean for the entire category. Hence, a tentative conclusion is that pathways in the general context of large trees are highly preferred, but pathways in the general context of unconcealed vantage points are not especially preferred. Obviously, further research is needed to isolate just which specific variables may be causing this difference.

Predictor variables. The predictor variables can be analyzed in two ways. One analysis examines the *means* of each of the predictors within the nature categories. The lower portion of Table 23.1 contains these means. The other analysis looks at the *relationship* between preference and the predictor variables within each nature category. For this analysis, the mean preference rating for each scene in the 15-second condition was correlated with the mean rating for each scene on each of the predictor variables. Since scenes were the units of analysis, sample

Figure 23.5. Pathway bordered by tall trees from the Large Trees category.

sizes are rather small for the Concealed Vantage Point and Large Trees categories. Results must be viewed with caution. Only correlations significant at p < .05 are reported.

For the Unconcealed Vantage Point category, the means for the predictor variables were all within a half-point of the middle of the rating scale. Within this category, three predictors were positively related to preference: identifiability (r = .55), coherence (r = .47), and spaciousness (r = .45). These are all variables that help one to organize and make sense of a setting (Kaplan and Kaplan, 1978, 1982). Their prominence as predictors suggests that when one is out in the open, there is a premium on being able to figure out where one is and where one could get to quickly. S. Kaplan (1973) offers some interesting evolutionary speculations on why this might be the case.

The Concealed Vantage Point category was much more volatile in its profile of mean ratings. These scenes were relatively high in complexity and quite low in coherence, spaciousness, and texture. Within the category, none of the predictors correlated significantly with preference. The pattern of predictor means indicates that these scenes contain a relatively large amount of unstructured information. In an urban setting, Herzog et al. (1982) found that scenes having this pattern (alleys and factories) were distinctly disliked. In the natural setting, these scenes, on the contrary, were not disliked. Clearly, the content of the scenes compensated for the unfavorable pattern of predictors. The source of the compensation would appear to be in what the scenes afford. They afford concealment, enclosure, what was referred to earlier as primary refuge. Such an affordance is apparently highly valued by an observer.

The Large Trees category had high ratings for identifiability, coherence, texture, and mystery. The first three properties help one make sense of the scene, while mystery fosters what Kaplan and Kaplan (1978, 1982) call involvement. In this case, the involvement consists of drawing the observer into the scene to discover what is only partially revealed from the current vantage point. Kaplan and Kaplan argue that scenes high in both making-sense and involvement properties will be most preferred, and that was certainly the case in this study.

Within the Large Trees category, there were three strong predictors of preference. Identifiability ($r = .79$) and mystery ($r = .91$) were positively related, while spaciousness ($r = -.69$) was negatively related. The six pathway scenes were primarily responsible for these correlations. As mentioned earlier, they were the most preferred scenes in the category. In addition, they were also relatively high in identifiability and mystery, but relatively low in spaciousness. Hence, large trees as pathway border elements may be pleasing in part because they are recognizable, familiar elements and because, when properly arranged, they add a strong sense of mystery to a scene. It appears that this arrangement works best in relatively smaller scale (lower spaciousness) settings.

Incidentally, the three pathway scenes in the Unconcealed Vantage Point category also had high mean ratings in identifiability and mystery. It appears that pathways may be perceived as familiar and relatively mysterious in general. Thus the preference difference between the pathways in the Large Trees and Unconcealed Vantage Point categories cannot be attributed to differences in identifiability or mystery.

Effects of viewing time

As Table 23.1 indicates, the overall preference means for the viewing times are very similar. Statistical analysis revealed no difference between the two short-duration conditions, but the average preference rating for the short durations combined was significantly lower than the mean in the 15-second condition ($t = 2.37$, $df = 113$, $p < .025$). This result should be viewed with caution, since the short-duration conditions were run two years after the 15-second groups. Still, it is interesting to note that with urban settings, Herzog et al. (1982) obtained just the opposite result, with short-duration viewing yielding higher mean preference ratings. Since urban scenes in general are less well liked than nature scenes, this set of findings might suggest that preference reactions are heightened with *longer* viewing time. However, R. Kaplan (1975) obtained exactly the opposite pattern of results: urban scenes liked less and natural scenes liked more with *brief* viewing times. She suggested that preference reactions might be heightened with *shorter* viewing times. It seems clear that viewing time can be an important variable, but more research will be necessary to understand its specific effects.

The pattern of preference means across the three content categories was the same for all three viewing times (i.e., the interaction of viewing time and content category was not significant). Large Trees was the favorite category, with the other two categories having lesser and roughly equal preference. Hence, for these field-and-forest settings, relative preferences do not change with brief as compared with extended viewing times.

Implications

The implications of this study are both general and specific. The general implication is that the methodology used here and in several other studies (see R. Kaplan, 1983; S. Kaplan, 1979b, for reviews) can be very useful in helping anyone understand how people view their environment and what they like or dislike about it. There is virtually unanimous agreement that planners should strive to understand the perception of those they serve. The methods used in this research could be very helpful in attaining that goal.

More specifically, the findings of this study suggest the following guidelines for planners:

1. People react differently to concealed and unconcealed vantage points, but neither type of setting is better liked. This means that both types of arrangement could be used to advantage by planners. They should note, however, that with unconcealed vantage points, it seems to be important that the setting be well structured spatially. In such settings, people apparently feel a need to determine

just where they are and how they could find their way to a place of safety quickly.

2. Large old trees represent an aspect of nature that is generally preferred. Preservation of such trees in urban settings deserves serious consideration.

3. Pathways seem to have a powerful influence in certain natural settings. When pathways are combined with large old trees in a relatively small-scale setting, it is possible to achieve a highly coherent and yet mysterious environment that is very highly liked. Whenever such a combination occurs in an urban environment, its preservation seems highly advisable.

4. Mystery continues to be an important variable in accounting for preferences in natural environments. In this study, its predictive power was especially potent in the Large Trees category. In general, the possible application of mystery both in using natural elements in urban settings and in arranging urban elements deserves attention and further exploration.

Acknowledgments

This research was supported in part by National Research Service Award #1 F32 MH05938–01 from the National Institute of Mental Health. Part of the work was completed while the author was on sabbatical leave at the Thurstone Psychometric Laboratory of the University of North Carolina at Chapel Hill. I thank my sponsor, Forrest Young, and the entire staff of the Thurstone Laboratory. I also thank Stephen and Rachel Kaplan, of the University of Michigan, for their continuing support and friendship and James Blakey, of Grand Valley State College, for photographic assistance. Requests for reprints should be sent to Thomas Herzog, Department of Psychology, Grand Valley State College, Allendale, Michigan 49401.

24 The emotional quality of scenes and observation points: a look at prospect and refuge

Jack L. Nasar, David Julian, Sarah Buchman,
David Humphreys, and Marianne Mrohaly

Introduction

One of the goals of planning professionals is to create and manage urban public settings that have aesthetic value to users. Toward that goal, researchers have examined the emotional quality of various properties of settings (see reviews by Wohlwill, 1976; Zube, Sell, and Taylor, 1982).

Appleton (1975b) argued that natural selection has led humans to prefer settings in which, without being seen ("refuge"), they can see a broad vista ("prospect"). Such preferences would have aided survival by affording a safe observation point and the capability to safely see, predict, and act in relation to potential predators, prey, and mates. A preference for more openness or spaciousness in scenes has appeared in a variety of contexts (Gärling, 1976a; Horayangkura, 1978; Nasar, 1981; Wohlwill, 1974). For protection, the evidence is less extensive. Law and Zube (1982) found preferred scenes as having natural framing elements (possible cues to refuge?) in the foreground.

This study examined both prospect and refuge. Individuals were expected to prefer more openness in a natural scene, and more enclosure in the observation point. Prospect and refuge were expected to have an interactive effect on response.

Methodology

Four conditions were considered: (1) Closed view/Protected observation point, (2) Open view/Protected observation point, (3) Closed view/Unprotected observation point, and (4) Open view/Unprotected observation point. The protected observation points were alcoves with concrete above and masonry behind and on

This paper was first published in *Landscape Planning* 10 (1983): 355–61.

Figure 24.1. Closed view and protected observation point.

Figure 24.2. Closed view and unprotected observation point.

Figure 24.3. Open view and protected observation point.

Figure 24.4. Open view and unprotected observation point.

one side, but not visible to the observer looking at the view. Vegetation obstructed the views; but in the open view, it was farther from the observer.

Other than the treatment variations, each condition had similar environmental and population characteristics. The investigators selected sites that they judged as similar in upkeep, diversity, noise, number of people, vehicles, vegetation, and man-made objects in view. These sites (Figures 24.1–24.4) were located less than 200 feet apart at two west entries to the Ohio Union on the Ohio State University Columbus campus. Notice that the view from the unprotected and the protected observation points is essentially the same: These observation points were located several feet apart.

There are many modes of environmental experience (included sitting, walking, and driving) and many kinds of variation in the openness of a viewing point or a view. While the narrow focus of this research (pedestrians' reactions to slight differences in the openness of a vegetative view and in the built enclosure behind them) may limit the applicability of the results, this study can nevertheless inform us about the appropriateness of Altman and Gauvain's (1981) dialectic approach to research and Appleton's (1975b) evolutionary perspective.

Assessment scales

From Kasmar's (1970) lexicon of environmental descriptors, twenty-eight affective scales were selected for consideration. Eight judges (graduate students in city and regional planning) assessed on a 7-point scale the appropriateness (very appropriate–very inappropriate) of each scale for assessing emotional responses to outdoor scenes. Those scales having the highest mean scores ($\bar{x} = 5.5$) were retained for use as 7-point scales. The resulting list included interesting–boring, ugly–beautiful, safe–unsafe, attractive–unattractive, dislike–like, repelling–inviting, relaxed–tense, excited–calm, and insecure–secure. To avoid response set and to mitigate order effects, the positive pole of each scale was assigned at random to 1 or 7, and multiple forms, each with the scales in different orders, were used.

Subjects

Sixty students (fifteen per site) took part in the study. Each was stopped one at a time at the sites on clear spring afternoons. Subjects were chosen such that each treatment condition had a balanced number of male interviews of males and of females, and female interviews of males and of females.

We experience our exterior surroundings intermittently over time. For instance, you might pass by the same area several times a week. By interviewing passers-by about their reactions to a view in sight, we believed we could get responses to such intermittent long-term exposure. In addition, *in situ* responses more likely apply to the natural experience with the environment than do responses to simulations, such as slides.

Procedure

Each subject was directed to the observation point, handed the instruction sheet and questionnaire, and asked to read the instructions. The interviewer indicated the view to be assessed and stepped out of the subject's field of vision.

Results

Principal axis factor analysis was used to condense the assessments for use in the experimental comparisons. A three-factor solution (accounting for 68% of the trace) was derived, in which the eigenvalue for each factor exceeded 1. Each factor accounted for at least 12% of the trace. The first factor, labeled Evaluation, had six scales with substantial loadings (those greater than .50) on it: dislike–like, repelling–inviting, ugly–beautiful, attractive–unattractive, interesting–boring, and relaxed–tense. The second factor, labeled Internal Arousal, had two scales with substantial loadings on it: relaxed–tense and excited–calm. The third factor, labeled Safety, had two scales with substantial loadings on it: safe–unsafe and insecure–secure. To represent these independent factors in the analysis of variance, for each one, the scale with the highest loading was retained: dislike–like, excited–calm, and safe–unsafe.

Males indicated a significantly higher level of safety ($s = 2.18$) than did females ($s = 2.90$) ($F = 7.94$, $df = 1$, $p < .01$). The open view was assessed as significantly more safe ($s = 2.2$) than was the closed one ($s = 3.0$) ($F = 8.18$, $df = 1$, $p < .01$). The observation point $\times$ view interaction (Figure 24.5) reveals that the difference in safety between the two views was more pronounced from the unprotected than from the protected observation point (effect approached significance, $F = 3.93$, $df = 1$, $p = .06$).

For dislike–like, the results (Figure 24.6) suggest that for the females, the protected observation position was preferred to the unprotected one, while for males the opposite was the case (effect approached significance, $F = 3.67$, $df = 1$, $p = .06$).

Discussion

The applications of the results may be limited to the specific conditions considered here: a specific variation of natural prospect and built refuge as it was experienced by a select set of adults (students residing in a city having a rela-

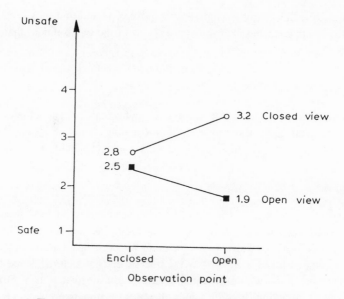

Figure 24.5. Interactive effect of view × refuge on evaluation.

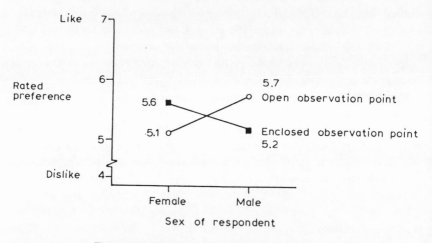

Figure 24.6. Interactive effect of refuge × sex on safety.

tively low density and flat terrain). However, the results provide limited support for the need for a dialectic research approach (Altman and Gauvain, 1981) and for the expected role of prospect and refuge (Appleton, 1975b). With regard to the dialectic, let us highlight again the evidence that even for the small variations examined here, in addition to the view, the observer's context (in this case, location and sex) seemed to influence emotional response. In agreement with Appleton (1975b), the open view was judged as safer than the closed one, and this effect was more pronounced from the unprotected observation point than from the protected one. However, in contrast to Appleton (1975b), this effect did not carry over to environmental preference, and males (unlike females) liked the setting with less refuge. Whyte (1980) also found similar differences between the sexes in the use of urban spaces. Perhaps evolution has influenced the sense of safety of such settings without affecting preference, at least for relatively safe settings. In any case, a better understanding of environmental aesthetics may require consideration of more than the characteristics of the view. Research might examine further the role of context in affect and rely less on simulations set in the laboratory. For design, it seems that for different observer contexts, different environmental configurations may be fitting. For campus settings like the one examined, reductions in the vegetative open spaces might reduce the perceived safety of those spaces.

25   Aesthetic preference for rural landscapes: some resident and visitor differences

Brian Orland

Introduction

The 1980 census revealed that the historic growth of urban areas at the expense of the rural may have reversed. Rural overpopulation and other factors, such as droughts in the Great Plains, had driven many to turn to urban areas in the search for work. Social and administrative concerns had generally been for the adjustment and adaptations of the rural migrant to the city, and for the erosion of the supposed economic and cultural advantages of urban life by pressure from increasing population (Taylor, 1968).

This situation has now changed. Planners are now faced with the pressures on rural areas brought about by an influx of newcomers, often with different needs and values from those of the existing rural residents. Few are moving to rural areas in search of work; in most cases, the motivation is for a somehow better life style. Some of these newcomers work in urban centers and commute to work. For others, the rural home is a vacation or retirement retreat. In planning for the development necessary to satisfy this increase in population, it will be necessary to weigh the values of the existing residents against those of the newcomers. This paper addresses the issue of how these two groups respond to the scenic beauty of their surroundings. The rural landscape is a mix of natural and human influences; newcomers might be expected to value the natural aspects but to dislike signs of development. In contrast, to existing residents, signs of development might signal economic prosperity, and familiar productive landscapes may represent a higher value for them than for the urban migrant.

Two experiments were conducted to investigate different aspects of the responses of urban and rural dwellers to rural scenery. In the first, rural-landscape scenes from an upland valley in central Arizona were viewed and rated by panels of respondents residing in urban Tucson. Results indicate that highly preferred "natural" aspects of the rural landscape can be affected negatively when viewed in the context of other scenes bearing signs of development. This finding has implications for the planning of rural communities where new residents are seeking to live in surroundings satisfying their image of rural life.

In the second procedure, residents of the same valley and further urban respondent groups viewed scenes of natural and human-influenced rural landscapes from a variety of places with which they were unfamiliar. Results indicate that the responses of rural and urban dwellers are not similar for all types of landscape setting. While responses to many scenes were similar, the respondents from upland Arizona responded much more favorably to scenes of grassland and range land, the types of productive landscape common around their homes. Respondent familiarity with the type of landscape depicted appears to influence preference ratings positively, indicating the necessity for considering the differing values of residents and newcomers in planning rural communities where long-time residents and retiring or vacationing newcomers will coexist.

Background

Migration to the rural landscape

Growth is now greater in nonmetropolitan districts than in metropolitan areas, and several incentives for the urban migrant have been identified; among them are escaping from city pressures, enjoying the recreational amenities of the countryside, and simply "going home" (Blackwood and Carpenter, 1978; Sofranko and Williams, 1980). This movement has significance for resource planners because although the problems of urban growth are still with us, there is a growing concern for maintaining the quality of the rural environment. For many, cultural and aesthetic values now appear to take precedence over the economic goals that motivated the migrations of earlier generations (Sofranko and Williams, 1980). Population increases and the visible consequences of economic development have brought changes in the landscape that will inevitably alter the scenic qualities of rural areas and have marked impacts on these values.

At a reduced scale, the Arizona landscape is under particularly intense pressure as a popular Sun Belt destination of migrants from the Frost Belt of the midwestern and northeastern United States. Urban and rural centers have expanded to house an influx of those who wish to enjoy the recreational and aesthetic benefits of the Southwest. Meanwhile, the homes and roadways to accommodate the migrants threaten the very things they came to enjoy. Although impacts of development are largely prevented in parks and on other public lands, rural areas enjoy little protection from the visual impacts of development and represent a critical meeting point of urban and rural, visitor and resident values.

Values of residents and migrants

The values and expectations of the rural resident and the urban migrant or visitor do not always match. In the case of the resident, Corlese and Jones (1977) found that established residents of Utah communities affected by energy projects felt that their home towns had changed with the influx of new residents. The atmosphere had become "less relaxed, friendly, traditional and harmonious . . . and more difficult and competitive" (p. 85). Sofranko and Williams (1980) demonstrated that this effect on host communities was far from intentional because the migrants tended to be seeking the life style cherished by the existing residents. For the visitors, McCannell (1976) highlighted the tourist's desire to see the real, traditional, untouched things. Even so, it appeared that residents of tourist destinations often only tolerated the tourist as a necessary evil.

The rural landscape is the setting where many of these conflicts take place and bears the marks of the two groups, the established and the newcomer. Decisions about how these landscapes should be developed are being made with little understanding of what people value in these settings and whether all the participants have similar perceptions of their surroundings.

The rural setting is a combination of natural features and the marks of civilization in the forms of cropping patterns, utility poles, farmsteads, and the like. While not as natural as the wilderness or forest, the rural landscape is far more natural in content than is the urban. There is a marked tendency for North American and European groups to prefer natural over built settings (Ulrich, 1983), but there is no clear point at which human intrusions become enough to cause the environment to be perceived as not natural and thus less valuable. Other writers note that "natural" is not a narrow definition, but may include crop lands and distant buildings as typical inclusions (Kaplan, Kaplan, and Wendt, 1972; Palmer, 1978). However, the vacation-home haven of the immigrant may be the unwarranted sign of overdevelopment to the resident; the commercial-timber stand of the resident may bring an unwanted symmetry to the views of the immigrant.

Many studies have reported high levels of agreement among individuals in their preferences for natural environments (e.g., Daniel and Boster, 1976; Shafer, Hamilton, and Schmidt, 1969; Zube, Pitt, and Anderson, 1975). Some limited differences have been found among the preferences of professionals such as landscape architects (Buhyoff, Wellman, Harvey, and Fraser, 1978) or foresters (Daniel and Boster, 1976), but for groups of the general public, consensus seems to be the rule (Ulrich, 1983). Against these studies, a number of scholars have suggested cultural background or historic knowledge of the viewer as a determinant of preference for a given landscape (e.g., Lowenthal, 1968; Tuan, 1973). However, as Ulrich (1983) points out, studies investigating cultural variations in response to natural environments have largely failed to link any difference in evaluations of the landscape to differences in cultural background of the study respondents.

Moreover, the cultural differences addressed in these studies have commonly been those of race or nationality. From the point of view of planning decisions regarding rural development, those sorts of differences are of limited interest, since it is rarely possible to predict or control the ethnic composition of a planning district. Nevertheless, it would be of great use to know if urban–rural differences (Sofranko and Williams, 1980; Taylor, 1968) are linked to individual preferences for the rural landscape, since it is inevitable that most migrants into rural areas will be from urban origins.

In this study, it was proposed that investigation of the responses of groups of urban and rural people to rural scenes not only could provide valuable information about the highly valued and less valued parts of those scenes, but also could give insights into the ways in which urban and rural "images" of the rural landscape may differ.

Study objectives

The rural Arizona landscape offered an opportunity to investigate rural–urban background where the line is still strongly marked between long-term rural residents and more recent arrivals looking to rural areas for vacations and retirement homes away from the city. It was proposed to:

(1) Investigate responses of urban Arizonans to predominantly "natural" and "human-influenced" scenes from a rural-vacation, retirement-retreat, weekend-home location. The hypothesis to be tested was that the preference of these respondents would be for the natural aspects of the rural landscape and that association of these natural aspects with human influences would diminish their value.

(2) Investigate responses of a further group of urban Arizonans, and of residents of the study location mentioned in (1), to "natural" and "human-influenced" scenes of unfamiliar rural settings. The hypothesis was that rural respondents' evaluations of familiar and productive rural landscapes would be more favorable than the responses of the urban respondents.

Study approach

A behavioral approach was used to quantify respondents' evaluations of the chosen rural landscapes. The Scenic Beauty Estimation method of Daniel and Boster (1976) lies within the psychophysical paradigm of landscape assessment and is based on the assumption that perceived landscape quality is determined primarily by the measurable physical characteristics of the landscape setting. Scenic Beauty Estimation is a technique for scaling subjective evaluations of landscape settings. The scaling procedure is similar to Thurstone's Law of Categorical Judgment. In the SBE procedure, color-slide surrogates for landscape settings are shown to panels of respondents who, typically, rate each slide scene on a 1 to 10 scale. The SBE program standardizes ratings and assigns aggregated scores either to stimuli or to respondents on the basis of the individual ratings.

Study location in Young, Arizona

The town of Young, Arizona, is in the Sierra Anchas about 60 miles east of Phoenix as the crow flies – many more miles and about four hours by very rough and largely unpaved road. Young has a resident population of around 300, living from ranching, forestry, and other seasonal labor. The people live mostly in an upland valley, separated from the outside world by the surrounding mountains. Pleasant Valley is at about 5,000 feet elevation, the surrounding mountains up to 9,000 feet, and the passes in and out of the valley at 7,000 to 8,000 feet. Young sprawls along the one road through the 20-square-mile valley bottom, which is mostly grazing land with some rare fields of crops such as corn or sorghum. Many inhabitants are descendants of the original settlers and have lived in the valley all their lives. In earlier years, the valley was extensively cultivated, but changes in the water table have reduced its usefulness as crop land. A small number of ranches survive, relying on grazing rights leased by the U.S. Forest Service – the owner of all the land around the rim of the valley.

Experiment 1: Urban respondent ratings of scenes from Young

Method

The focus of the first experiment was for residents of an urban area – in this case, students living in Tucson, Arizona – to evaluate scenes from the landscape surrounding Young. The residents of urban areas are the source of potential migrants to rural areas in Arizona whose perceptions of scenic quality will help to guide the development of landscape-management policy. Landscape scenes were represented by color-slide surrogates compiled using a random-sampling procedure.

Compilation of the photo sample. The landscape around Young was photo-sampled by taking photos from the road at .25-mile-interval stations along the 10 miles of roadway passing through the town. At each station, four photos were taken in four directions at 90 degrees to each other. The resulting color slides showed scenes that ranged in content from the grassland of the valley bottom

Figure 25.1. At higher elevations around the rim of the valley, the road is bordered by the ponderosa pines of the Tonto National Forest.

Figure 25.2. On lower slopes, a mix of small ponderosa pine and piñon juniper prevail on grazing land mostly owned by the U.S. Forest Service.

Figure 25.3. In the valley bottom, privately owned grassland is dotted with scrubby conifers and cholla cactus.

Figures 25.1–25.3. Predominantly natural landscapes around Young, Arizona.

with scattered dwellings to ponderosa pine forest in the surrounding hills. From these, a set of 110 "natural" scenes as well as a set of 50 "human-influenced" scenes were compiled. Two experimental stimulus sets were created from these. For the first, the 110 natural scenes composed the set. For the second, the 50 human-influenced scenes were substituted for 50 of the natural ones in the first set to create a human-influenced 110-slide set. Natural scenes included grassland and managed forest, but no buildings or roadside signs. Human influences ranged from fence posts and signs to dwellings and commercial buildings. In all cases, human influences were seen against a natural background – for example, ranch houses in grassland, signs and fences in forested areas.

Study participants. Urban respondents were drawn from lower division psychology classes at the University of Arizona. All were residents of Tucson at the time of the experiment. Students at the University of Arizona represent a wide array of backgrounds; many of the students are from the major urban centers of Arizona and adjoining states. Student respondents have been found to be suitable surrogates for the general public in this sort of experiment (Daniel and Boster, 1976).

Slide-show procedure. The two sets of color slides were shown to independant panels of respondents, one at a time, in a random order. Standard instructions directed respondents to assign a scenic-beauty rating to each scene on a scale from 1 (low scenic beauty) to 10 (high scenic beauty). The SBE computer program (Daniel and Boster, 1976) was used to generate standardized SBE scores for individual scenes.

Results

Predominantly "natural" and "human-influenced" scenes from Young were viewed by independent panels of urban respondents ($n = 25, 25$). It was expected that responses to natural scenes would reflect the consensus noted by others (e.g., Shafer et al., 1969; Zube et al., 1975). and would display the intuitively reasonable relationships between physical features of the landscape and patterns of response found in other studies (e.g., Daniel and Boster, 1976). Examination of the human-influenced slide scores might then reveal the effects of intrusions onto scenes otherwise comparable with counterparts in the natural set of scenes.

   ANOVA tests were performed on the raw ratings of each respondent group to establish minimum statistical criteria. Intraclass correlation coefficients were calculated to estimate agreement within respondent groups (Berk, 1979) and gave values of $r = .92$ and $.95$ for natural and human-influenced scenes, respectively. These values are well within the range found in this type of experiment (Daniel and Vining, 1983) and indicate high agreement within respondent groups.

Figure 25.4. Extensive development visible in the middle distance. This scene received a scenic-beauty score of −136.

Figure 25.5. Some development visible in the middle distance. This scene received a scenic-beauty score of −117.

Figure 25.6. Limited evidence of development in the far distance. This scene received a scenic-beauty score of −94.

Figures 25.4–25.6. A range of human-influenced scenes from the valley.

Table 25.1. *Experiment 1: ANOVA summaries*

Source	df	MS	F	p
Predominantly natural				
Observers	24	143.40	63.12	<.0001
Slides	134	27.77	12.23	<.0001
Observers × slides	3,216	2.27		
$r_g = .92$ $r_{icc} = .24$				
Human-influenced				
Observers	24	91.31	34.51	.000
Slides	134	50.08	18.93	.000
Observers × slides	3,216	2.65		
$r_g = .95$ $r_{icc} = .36$				

A priori examination of color slides by a panel of judges had established four categories of scenic content represented in the slides of natural settings: (1) mixed forest and woodland, (2) range land and grassland, (3) riparian woodland, and (4) woodland visually dominated by foreground road embankment. The background content of category 4 mostly matched the vegetative mixes of categories 1 and 2, but the scenes were visually dominated by wide expanses of roadway evident in the foreground; hence their separation into another category. In general, mixed forest and range land scored more highly than did the rather overgrown riparian scenes and scenes with extensive road embankment. Analysis of variance for natural and human-influenced slide ratings by area type revealed a significant ($p < .01$) difference between ratings received by range-land scenes in the two slide sets. When seen in the context of all natural scenes, range-land scenes were rated relatively high – values similar to those received by mixed-woodland scenes. In contrast, ratings of range-land scenes in the human-influenced context were rated much lower (Figure 25.1). In the latter slide set, as expected, range-land scenes received generally lower ratings than did other natural scenes. Again in that slide set, views with buildings and fences in ranch-land settings received scores similar to those received by range land alone.

Experiment 2: Urban and rural responses to an unfamiliar rural setting

Method

The second experiment sought to evaluate the responses of rural and urban residents to predominantly natural or human-influenced rural scenes. Respondent

Figure 25.7. Grazing land on lower slopes of the valley rim. This scene received a scenic-beauty score of -59 in the natural context, and of -155 in the human-influenced context.

Figure 25.8. Valley-bottom grassland. This scene received a scenic-beauty score of 2 in the natural context, and of -131 in the human-influenced context.

Figures 25.7–25.8. Typical grassland scenes that received relatively high scores when viewed among other natural scenes, and significantly lower scores when viewed as part of a set of human-influenced scenes.

values may differ for certain landscape types, having implications for the design of land-planning policies that respect the differing values of both existing rural residents and urban-background newcomers. Landscape scenes were represented by color slides selected from a variety of sources as surrogates for a range of landscape types.

Compilation of the photo sample. For this part of the study, slide sets compiled by Palmer (1980) were used. These slides were drawn from a variety of sources in the United States and Europe and thus represented scenes generally unfamiliar to Arizona respondents. The natural scenes were of a variety of evergreen and deciduous woodlands, some of the latter seen in winter without leaves. Human-

Table 25.2. *Experiment 2: ANOVA summaries/Tucson (University of Arizona) observers*

Source	df	MS	F	p
Forest scenes				
Observers	24	80.00	24.35	<.0001
Slides	79	32.06	9.76	<.0001
Observers × slides	1,896	3.29		
$r_g = .91$ $r_{icc} = .21$				
Countryside scenes				
Observers	24	54.71	17.39	<.0001
Slides	59	31.45	9.99	<.0001
Observers × slides	1,416	3.15		
$r_g = .91$ $r_{icc} = .22$				

influenced scenes included cultivated crop land, open grazing land, some buildings, and scenes of roads and fences in wooded settings.

Study participants. Urban respondents were once again drawn from lower division psychology classes at the University of Arizona and were residents of Tucson at the time of the experiment. Rural respondents for the second procedure were Friends of the Library Board of Young, Arizona, attracted by a small financial incentive.

Slide-show procedure. Color slides were again shown to panels of respondents, one at a time, in a random arrangement. As before, standard instructions directed respondents to assign a scenic-beauty rating to each scene on a scale from 1 (low scenic beauty) to 10 (high scenic beauty). The SBE program was used to generate standardized SBE scores for individual scenes.

Results

"Natural" and "human-influenced" scenes from the Palmer (1980) study were viewed by independent panels of rural respondents ($n = 30, 30$) and urban respondents ($n = 25, 25$). Intraclass correlation coefficients indicated good agreement within respondent groups, values for all four respondent × stimulus set cells being $r > .90$. A two-way analysis of variance of ratings by respondent origin and by stimulus set (natural or human-influenced) revealed no significant interaction effect but a main effect ($p = .033$), respondents' urban or rural background. These results were investigated by detailed examination of, first, natural scenes and responses and, then, human-influenced scenes and responses.

Table 25.3. *Experiment 2: ANOVA summaries/Young (Library Board) observers*

Source	df	MS	F	p
Forest scenes				
Observers	24	45.76	14.00	<.0001
Slides	79	45.42	13.90	<.0001
Observers × slides	1,896	3.27		
$r_g = .93$	$r_{icc} = .30$			
Countryside scenes				
Observers	24	55.42	17.79	<.0001
Slides	59	27.57	8.85	<.0001
Observers × slides	1,416	3.11		
$r_g = .90$	$r_{icc} = .20$			

Predominantly natural scenes. Once again, a priori examination of color slides had established categories of natural-scene content: (1) deciduous trees in winter, (2) thick forest with much undergrowth, and (3) open forest with clear understory. For both respondent groups, highest values were assigned to open forest, lower to thick forest, and lowest to deciduous trees. One-way ANOVAs were used to investigate the results. Differences among categories for each respondent group were significant ($p < .05$). Differences between respondent groups for each category were negligible (Table 25.2).

Human-influenced scenes. Examination of human-influenced color slides established four content categories: (1) water features; (2) woodland and forest; (3) developed, with farms, fences, and other built features; and (4) grassland and cultivated grazing. One-way ANOVAs investigated the results: For water, woodland, and developed scenes, differences between respondent groups' ratings were minimal. Generally, water scenes were more attractive than woodland, and developed scenes were rated still lower. For urban respondents, grassland scored significantly lower than did the other three categories ($p < .01, .05, .05$, respectively) and differently from the way it scored for rural respondents, ($p < .05$), who found grassland as attractive as woodland. Preferences noted for water scenes and for woodland over developed scenes follow the trend noted in Ulrich (1983) for preference of the natural over the man-made. However, there appears to be marked disagreement on the evaluation of grassland scenes that lack overt evidence of human influence (Table 25.3).

Discussion

The first experiment reported here addressed the question of whether human influences in a typical rural landscape had any effect on a possible visitor's evaluations of its natural features. The second investigated the question further by asking whether urban and rural people differed in their responses to different types of rural landscape, even where familiarity with the particular setting was not an issue.

In the first instance, the reported main effect of natural versus human-influenced slide-set context for scenes of grassland suggested that the presence of human influences did have an effect on the way in which urban residents evaluated certain features of the rural landscape. In this instance, grassland scenes viewed as part of the "natural" slide set appeared to be judged as natural features and thus were rated relatively highly. These high ratings follow the patterns of general preference for natural scenes noted by other investigators (e.g., Ulrich, 1983). However, it appears that those same grassland scenes when shown as part of the "human-influenced" set were somehow associated by the respondent group with the scenes of farm buildings that occurred in grassland settings and were thus judged as a human-influenced feature. Grassland scenes were thus rated rather low.

In short, it is suggested that for the urban respondent, the usually highly rated natural components of the rural landscape appear to be "devalued" by association with the less desirable visible signs of rural economic development.

The second experiment gave further support to these ideas. Significant differences in ratings by urban and rural respondent groups also arose in respondent ratings of grassland scenes. Again demonstrating a general preference for natural over human-influenced scenes, natural scenes received similar ratings from both urban and rural respondents. However, for grassland scenes that formed part of the human-influenced set, there was a main effect of respondent place of residence. Grasslands were rated low by urban respondents, and higher by rural. Some of the grasslands depicted bore overt signs of management activities, and urban respondent ratings followed the same pattern as in the first experiment, rating these scenes lower than other natural scenes, such as woodland. Rural respondents, however, rated grassland as attractive as woodland. Apparently, the grassland scenes were judged to be as attractive as the more "natural" woodland scenes, despite the evidence of human involvement.

In seeking an explanation for this urban–rural difference, it may help to consider the surroundings in which the rural respondents live. The upland-valley landscape around Young, Arizona, is predominantly open grassland, typical of the high plains of North America. Comparison of the composition of this landscape with that of the range-land and grassland scenes making up part of the human-influenced slide set in the second experiment revealed marked similarities – generally open views of pale, dry range land with backgrounds of coniferous forest.

The inevitable greater familiarity of rural respondents with this type of scene may have been a determinant of their responses, and it demands further investigation. While urban respondents rated grasslands seen in a human-influenced context similarly low in both the experiments reported here, rural respondents viewed the grassland as relatively more attractive, as attractive as they and the urban respondents judged the more natural scenes.

Conclusions

Two effects of importance to the planning of rural-landscape development have been noted in this study. First, the presence of human influences in otherwise "natural" landscapes may dispose viewers to regard the whole natural/human-influenced composite as human-influenced and hence perceive it as relatively less attractive than a purely natural landscape. This effect is congruent with the consistent preferences for natural versus built environments noted by many other investigators (e.g., Ulrich, 1983), but contradicts the views of other investigators who have suggested that "natural" scenes may include such human influences as cultivation and distant buildings without negatively affecting perceived scenic beauty (Kaplan et al., 1972; Palmer, 1978).

Second, responses to certain types of landscape settings are to some degree influenced by the place of residence of the respondent. In the instance reported here, grassland scenes received relatively higher ratings from rural respondents when compared with respondents' similarly high ratings of natural scenes. This finding is not consistent with other studies that have noted the widespread consensus of panels of the general public in evaluating scenes of natural and rural landscapes (Daniel and Vining, 1983; Ulrich, 1983).

These findings are important both to researchers and to landscape planners. In the planning of new development in rural areas, it may be important to consider carefully whether the encroachment of human influences – buildings, fences, and the like – is going to negatively influence the responses of newcomers and visitors and hence reduce the appeal of a new area to those people. This negative effect can clearly have great implications for economic development in rural areas. In a related area, tourism planning, the issue of improving the image of potential tourist destinations is taken very seriously.

It appears that residents of a given landscape type form a unique impression of how the value of that landscape relates to others. The residents of Young appear to have developed a value system that places grassland landscapes relatively higher on a scale of perceived scenic beauty than do the values of urban residents. This may be an effect of increased familiarity, of perceived economic benefits offered by the range-land scenes, or of a fondness for the setting developed through cultural associations. Whatever the reason, the residents of Young appear to have an image of what the grassland landscape means to them that differs from the image of the residents of Tucson.

Although the study of relationships between image and behavior is well developed within the fields of geography and recreation planning, recent reviews (i.e., Daniel and Vining, 1983; Zube, Sell and Taylor, 1982) have pointed to deficiencies in the ability of landscape assessment to offer guidance in situations where people's expectations of what a place should be like are influenced by the context in which the place is seen. The study reported here reinforces the need for increased efforts to assess the antecedents of environmental image, in terms of both the physical components of the landscape and the educational, cultural, and economic backgrounds of important subgroups within the general public. Better understanding of these factors in determining environmental preferences will assist in creating land-use and landscape-development policy for places such as the rural landscape, where people are meeting in new circumstances with widely different backgrounds and widely different expectations about their surroundings. Successful planning will occur only when these different needs and values are acknowledged and accommodated.

26 Familiarity and preference: a cross-cultural analysis

Rachel Kaplan and Eugene J. Herbert

People are said to like that which they know. In other words, preference is assumed to increase with familiarity. At the same time, there is the expression that "familiarity breeds contempt," suggesting that perhaps the relation between these two constructs is negative. One might know an area very well and not like it, and one might also like some settings that one had never seen before. Thus it would seem that there is truth to both positions. The effect of familiarity on preference has in fact been demonstrated to be complex, rather than necessarily positive or negative (Kaplan and Kaplan, 1982).

Cross-cultural studies provide an interesting opportunity to study these issues. Presumably, individuals from another culture would be less familiar with scenes taken in this country. While they may have been exposed to some visual images before, they are unlikely to have had the same degree of exposure as a group for whom such scenes are relatively local. One would expect, then, that the local preferences would be higher if familiarity has a positive influence and lower if preference is adversely affected by familiarity.

Several studies have examined visual preference in a cross-cultural context. Ulrich (1983) argues that such studies have shown more similarity than discrepancy in the perception and evaluation of the environment. Zube (1984) concludes that the similarity in cross-cultural evaluation is related to the similarity between the cultures. When these are diverse (e.g., with respect to life styles and economic indicators) and when the landscapes are strikingly contrasting, the differences in preference are greatest. In studies comparing more similar cultures, there is greater agreement. The initial report by Wheeler (1984) of ratings of a set of scenes by participants in various Western European countries as well as students at the University of Arizona supports Zube's conclusions. By and large, the similarity in judgments is considerable, although there is greater variation for some kinds of environments (e.g., dwellings) than for others (e.g., natural scenes). Interestingly, Nasar (1984) has reported that both American and Japanese students prefer foreign scenes to native ones.

Other factors may also affect the relationship between preference and famil-

379

iarity. In particular, it has been shown that some highly familiar settings are not preferred (R. Kaplan, 1977b). If the settings are generally low in preference, familiarity may actually serve to reduce the preference further ("breed contempt").

The scenes used in the current study are not of special places or vacation shots (as was the case in most of Wheeler's [1984] stimuli). They are of everyday, rural places. Although all were taken in one county in Michigan, many of them are typical of vast portions of the United States, and none of them is particularly striking or picturesque. The scenes are of natural areas, although some residences are visible in a few instances. It is far less likely that with scenes such as these, participants abroad would have had much direct familiarity.

The comparison (cross-cultural) group in this study consisted of individuals in Western Australia. In terms of Zube's (1984) analysis (and prior research with Australians [Zube and Mills, 1976]), we are dealing with two cultures that are quite similar in life style and economic issues. The vegetation in Western Australia is, however, distinctly different from that in rural Michigan. The most characteristic scenery would have fewer trees, and the predominant color of foliage is the gray-green of the eucalyptus, rather than the more saturated greens of deciduous trees.

In light of prior research that showed strong similarities between Australians and Americans, one might thus suspect minimal differences in preferences. At the same time, however, the two landscape types are distinctly different, and it would thus be reasonable that the preferences would differ. The familiarity argument would be supported by higher preference ratings by those more familiar with the setting (i.e., Americans). But given prior findings, the opposite of such results is also a possible outcome.

Method

The American sample (Herbert, 1981) consisted of 97 students at The University of Michigan; the Australian sample consisted of 122 students at the University of Western Australia. In both cases, participants were fulfilling a requirement in their psychology course. At both universities, the introductory psychology courses attract a broad range of college students who are not necessarily interested in further study in the field.

In both countries, participants viewed fifty-five slides of Oakland County, Michigan, and rated each scene using a 5-point preference scale ("How much do you like the scene"). Three practice slides were used to orient the participants to the task, and the rating form was numbered for sixty slides to avoid an "end effect." Each slide was shown for ten seconds. Participants were gathered in small groups, and two random orders were used in these slide presentations.

The fifty-five slides were selected to represent a diversity of landforms and land covers present in six rural townships in a county that includes portions of Detroit. The area includes scenery typical of much of the midwestern and eastern portions of the United States, consisting of abandoned fields, forests, and rural housing. There is little topographic variation in the region. None of the slides showed people or close-ups of particular houses or plants.

Results

Scene preference

The correlation of the preference ratings for the two samples for the fifty-five slides was .84. Across all scenes, the mean preferences were 3.29 ($SD = .48$) for the American sample and 3.08 ($SD = .52$) for the Australians ($t = 3.06$, $p < .01$).

These results suggest that on the whole, the Michigan students preferred these scenes, which were more familiar to them, but that the relative preferences were strikingly similar for the two samples. All fifteen scenes that the Australians rated as highly preferred (mean 3.5 or greater) were among the most preferred for the American sample as well. All nine least preferred scenes (means below 3.0) for the American sample were among the least preferred for the Australians.

None of the scenes was significantly preferred by the Australian students. There were eighteen scenes that the American sample rated at least 0.4 more favorably than the Australians. The scene with the greatest discrepancy (means of 4.1 and 3.3) depicted a large body of water with distant background forest and a cloudy sky. The other seventeen more favored scenes were of relatively rural settings, with open fields in the foreground and forests in the background.

Perceptual differences

Mean preference ratings in themselves, however, can tell only a part of the story. While the means are based on only the magnitude of the rating, the relationship among the ratings is also useful to explore. Thus the preference ratings can also be used in category-identifying procedures to extract patterns of relationships underlying these judgments. These patterns, or groupings, provide a means of examining perceptual categories implicit in the preference ratings (R. Kaplan, 1985).

The preference ratings from the American sample had, in fact, been examined using such an approach. Specifically, Herbert (1981) had carried out a nonmetric factor analysis (Smallest Space Analysis III, Lingoes, 1972).[1] Four groupings

Table 26.1. *Preference means for the scene groupings[a] based on American (UM) and Australian (UWA) data sets*

	UM-based analysis			UWA-based analysis		
	No. of scenes	UM	UWA	No. of scenes	UM	UWA
Residential	10	2.81 (4)	2.79 (3)	9	2.91 (4)	2.91 (3)
Vegetation	17	3.52 (2)	3.42 (2)	14	3.48 (2)	3.44 (2)
Pastoral/Open Rural	15	3.07 (3)	2.64 (4)	21	3.12 (3)	2.70 (4)
Manicured Landscapes/Picture Book	4	3.93 (1)	3.83 (1)	3	3.90 (1)	3.80 (1)

[a]Numbers in parentheses represent rank order of preferences within a column.

were identified in that study: Residential, Vegetation, Pastoral, and Manicured Landscapes.

The same analysis based on the preference ratings from the Australian sample yielded basically similar groupings. Table 26.1 summarizes these results. The mean preferences are provided for each of the four groupings when using the factorial solution based on the American sample as well as the Australian-based solution. In addition, each of these sets of means is rank ordered.[2]

The Residential grouping remained virtually unchanged. One scene, depicting a relatively more distant residential setting, did not load on this cluster for the Australian sample, but the remaining nine scenes were the same (Figure 26.1). The mean preference ratings (2.8) for these scenes were also identical for the two samples, reflecting a relatively low preference for these rural residences viewed in the context of otherwise more natural scenes.

Herbert (1981) described the original Vegetation grouping as "composed exclusively of nature items as portrayed by various admixtures of forest, water, and re-growth vegetation" (p. 71). The Vegetation grouping based on the Australian sample shared twelve of seventeen scenes in the original grouping (top row, Figure 26.2) and added two more scenes. The two additions were particularly grassy (e.g., lower left, Figure 26.2), and the five scenes that were dropped were distinctly less green than were the rest of the grouping (e.g., lower right, Figure 26.2). Thus while the two Vegetation groupings were very similar, the one based on the Australian sample included a stronger sense of lush, green landscapes and more forested settings. The mean ratings (3.4 and 3.5) for the two samples were essentially identical.

The Pastoral grouping identified by Herbert (1981) included fifteen scenes (top row, Figure 26.3). Two of these were dropped from this category in the Australian-based analysis. However, eight additional scenes were included in

Figure 26.1. Scenes from the Residential grouping.

Figure 26.2. Of these scenes from the Vegetation grouping, the two in the top row were common to both the American-based and the Australian-based factorial analyses. The ones in the bottom row were unique to each of these analyses, respectively.

Figure 26.3. Of these scenes from the Pastoral grouping, the two in the top row were common to both sets of analyses, while the bottom row represents the different perception of the Pastoral grouping to reflect a more open, rural landscape pattern on the part of the Australian sample. These scenes also represent the domain of greatest difference in preference between the two samples.

Figure 26.4. The original Manicured Landscapes grouping included the two scenes in the top row, as well as the one at lower left. The revised version of this category, based on the Australian data, included the two in the top row, as well as the one at lower right.

this grouping for the Australians (bottom row, Figure 26.3). The grouping with these additional scenes suggested open, rural land rather than a pastoral setting. The mean rating for these scenes was distinctly lower for the Australians (2.6) than for the Americans (3.1). This difference is true whether one uses the original Pastoral scene category or the expanded Open Rural grouping based on the Australian data. The seventeen scenes with discrepant mean differences mentioned earlier were included in the latter grouping.

The Manicured Landscapes grouping originally consisted of four scenes. Three of these are included in Figure 26.4 (top row and bottom left). The most comparable category based on the Australian data retained the two shown in the top row in Figure 26.4, and added an additional one (lower right, Figure 26.4). The revised grouping is difficult to describe. These are "picture book" landscapes, and the mean preference (3.9 and 3.8) for these was equally high for the two samples, when based on either the original or the revised category.

Discussion and conclusions

The most striking finding in these analyses is the strong resemblance between the two samples. There is strong similarity in the relative preferences for the scenes and in the perception of the types of environments, as reflected by the factorial solutions.

There are notable differences, however. The most striking of these is the relative ratings of the Pastoral/Open Rural grouping. Although these scenes were far preferred to the Residential grouping by the American students, for the Australian students, they were the least preferred of the four groupings (scored either in terms of the Pastoral or in terms of the Open Rural categorization).

In fact, the overall finding of significant differences in mean preference for the two samples are totally attributable to the differences in preference for the Pastoral/Open Rural scenes. The conclusion that Americans prefer their more familiar landscape is thus quite misleading. Results of this study suggest a much more complex conclusion.

It seems that while the Australian sample was not directly familiar with any of the scenes included in the study, the genre of the open, rural settings is one they recognize readily. This is probably the most common nonurban landscape in the area near Perth, Western Australia. In this case, a relative sense of familiarity with settings that are not highly preferred seemed to breed contempt.

There is no way to determine from these results whether the novelty of the more lush, forested settings served to augment the preferences of these landscapes for the Australian sample. In other words, the relative similarity in results between the two samples can be interpreted in terms of cultural similarity between the two countries; alternatively, however, these results may reflect increased preference for more novel settings and decreased preference for seemingly familiar settings.

It may at first glance appear that this pattern of results is discrepant with the findings of other studies in this area. After all, previous work suggests a considerably more straightforward set of relationships than those reported here. But previous studies have tended to rely on overall preference ratings and on correlations between ratings. Relying exclusively on either of these procedures would have led to an apparently straightforward conclusion in this case as well. Even studies that have been sensitive to content differences have approached content in terms of a priori experimenter-defined patterns. The results of the present study demonstrate that the very process of category definition depends on cultural and/or familiarity factors.[3]

It was only by the examination of different factorial solutions for the two groups that the interaction between content and participant population in the determination of preference became evident. Preference is indeed a complex area of human concern. Clearly, more research is needed to begin to untangle this subtle relationship. Equally clearly, it is important to employ methodologies that are sensitive not only to content, but also to content as implicitly defined by the preferences of the participants themselves.

Acknowledgments

This study was carried out while the second author was with the Forests Department, Western Australia. D. J. Illingworth's assistance made it possible to collect data at the University of Western Australia. We wish to thank T. J. Brown and S. Kaplan of the University of Michigan for their help and encouragement, especially in the earlier phases of the project. Support from the Urban Forestry Project through Cooperative Agreements 13–655 and 23–84–08, U.S. Forest Service, North Central Forest Experiment Station, is also gratefully acknowledged.

Notes

1. The use of factor analysis is recommended only if there are at least five times the number of observations as variables included in the analysis. It is not clear whether the same precaution is necessary for nonmetric procedures that do not rely on the actual magnitude of the correlations for further steps in the algorithm. In any event, the studies discussed here do not meet this criterion and the results should, therefore, be viewed with the same caution that is due any empirical work carried out on a limited basis.

2. It is hard to provide a direct comparison of the two nonmetric factor analyses. The same criteria were used for extracting the groupings and for including the items (eigenvalues greater than 1.00 and loadings of at least .40 on no more than a single grouping).

3. In a further examination of differences between Australian and American samples, this time using Western Australian landscapes, the perceptual categories were, once again, essential in revealing important cultural differences (R. Kaplan and Herbert, 1987).

Section III

Applications

Editor's introduction

If the design of public settings (such as streets, parks, or recreation areas) is to have visual appeal to the many and diverse passers-by and users, decision makers must integrate extant knowledge of environmental preferences into the design. Frequently, such information is overlooked. The first paper in this section (Lozano) develops some urban-design strategies that respond to empirical findings on visual preferences.

An important question, then, is how to translate such ideas into realities. Specifically, how does one bridge the gap between the development of design guidelines and their acceptance and application in public spaces through public policy? Individuals interested in environmental aesthetics can benefit from an improved understanding of the realities involved in translating proposals into public policy.

When there is one user or client and a short (one or two years) time to complete a project, the integration of the user's or client's aesthetic needs into the design is relatively straightforward. The papers in this section focus on the more complex situation in which there may be multiple users, multiple clients, and a long (sometimes open-ended) time frame. Obstacles to and opportunities for success in such situations are discussed. The emphasis is on application and process. Among the process issues addressed are how various cities and nations have implemented aesthetic controls (Preiser and Rohane), how courts and regulations at the national and state level affect aesthetic control (Pearlman), how one state dealt with aesthetic control (Ridout), how to frame recommendations (Sancar), and how to sustain implementation (Hurand).

Public policy is the product of a variety of forces (many of which are more powerful than empirical data or expert opinion). Various contexts require distinct kinds of linkages between design research and politics to bring recommendations into physical reality. The papers in this section indicate the need to extend research findings into the broader arena of public policy. Hurand argues for the need for greater sensitivity to local values in choosing strategies. Sancar views aesthetic decisions as part of a larger system and argues that for research to influence urban form, the researcher must understand and consider the externals.

393

The law is an important influence on urban form. Pearlman describes the legal framework within which aesthetic controls can be implemented, and Preiser and Rohane review the kinds of aesthetic controls that are currently being used.

The choice of method may vary with decision-making contexts. The papers discuss a variety of contexts: nations (Preiser and Rohane; Pearlman), states (Ridout; Pearlman), cities (Preiser and Rohane; Lozano), and rural towns (Hurand). The papers vary in their strategy of analysis: case study (Ridout), experiential perspective (Hurand), theory (Sancar), application of empirical data (Lozano), and review (Preiser and Rohane; Pearlman). However, they share a concern that researchers should consider public-policy issues in framing their research questions. Researchers may have to leave the comfort of the laboratory for the uncertain world of public policy if they want to see their findings applied in environmental design.

27   Visual needs in urban environments and physical planning

Eduardo E. Lozano

Introduction

The clarification of and agreement on visual objectives and their design implementation are, without doubt, among the weakest stages of the planning process. This design shortcoming is reflected in the product itself; few of the large-scale projects built in the last decades are able to evoke satisfactory aesthetic responses from the lay public or the critics, and it is possible to advance the hypothesis that this visual dissatisfaction is part of a major conflict between people and the new built environment. In contrast to the negative feelings evoked by contemporary urban design, there is widespread consensus on the positive visual qualities of many urban creations of the past, suggesting that the present inability to deal with the aesthetic issues must be traced either to major changes in the nature of cities or to major changes in the process of planning and design – and most likely to both factors.

Although the methodology to introduce visual inputs in the design process is poor and inadequate, the visual qualities of the built environment are extremely important. There is still a debate among social scientists on whether the environment bears important effects on humans and society, and it is possible to find researchers taking a whole range of different positions. Among the skeptics are key writers, such as Webber (1967), who emphasized the importance of non-geographical space in human relations, and Gans (1968), who stressed the importance of social systems, as against the built environment.

On the side of the believers of environmental effects, it is possible to find a growing body of studies in several social sciences, such as Michelson's *Man and His Urban Environment* (1970) in sociology, Hall's *Hidden Dimension* (1966) in anthropology, and Sommer's *Personal Space* (1969) in psychology, to mention only three of the outstanding authors in those fields.

Although there are no definitive research conclusions and the field of visual-

This paper was first published in *Town Planning Review* 45 (1974): 351–74.

environmental studies is still in a state of flux, there are growing indications that the environmental effects are "being stated in a more sophisticated way – as a complex, systemic, ecological interaction" (Rapoport, 1971, p. 6). The purely visual effects must be extended to explain, and be explained by, social, psychological, and behavioral effects, closely interrelated among them, and not quite sorted out, yet, by basic research.

There are other clues indicating that the visual-environmental effects are to be considered seriously; it is astonishing to realize that some of the most critical urban problems must be traced to these "soft" explanatory factors. One example is the destruction of part of the public-housing complex Pruitt-Igoe in St. Louis, Missouri, blown up by the local housing authority at a cost higher than the original construction budget, because the brutalizing conditions developed in the barrack-like environment of the project left no option (Rainwater, 1970; "St. Louis," 1972). This case is not an isolated phenomenon; on the contrary, it is rather a common and universal pattern: Public-housing projects of similar characteristics also have elicited negative reactions from people all over the world – many of them poor shantytowners who still prefer their neighborhood – shown in antisocial behavior and in refusal to move to the project. It is postulated that the effects of the built environment on people have to be interpreted in a wider context, considering the "relationship between man and the cultural dimension . . . in which both man and his environment participate in molding each other" (Hall, 1966, p. 4).

The complex relationship between the physical world of form and space and the different dimensions of humans – biological, psychological, social, and cultural – has been a most fertile area for research, leading to increasing interest on the part of social scientists and even to the generation of new interdisciplinary fields, such as environmental psychology. Most of this research work is characterized by an exploratory nature and a focused approach, selected probably in terms of the specific skills and interests of the research team; for those reasons, there is a wide gap between the research findings – many of them tentative and even conflicting – and operational knowledge for planners and designers.

This gap between basic research and professional practice may be a factor explaining why the published works with most impact on physical planners have not been produced by social scientists, but by planners and designers themselves. This statement needs qualification: Even though the works of some social scientists have reached widespread influence in the profession, their effects are felt at a social-advocacy level, with minimum impact on visual and environmental issues – good examples are found in the opposition of many professionals to urban renewal projects disrupting stable neighborhoods, based on arguments of social cohesion with exclusion of any visual consideration. One of the most influential works in the visual field has been Lynch's *Image of the City* (1960), which developed a systematic approach to note and study certain visual elements of the built environment, based on their perceived image by the people; the impact of Lynch's work is apparent in the number of planning studies that began to analyze visual structures through the identification of nodes, links, edges, and landmarks after its publication. Another influential work has been Venturi's *Complexity and Contradiction in Architecture* (1966), which brought a refreshing change from the canonical aesthetics of the modern movement.

One burden of the aesthetic criticism is the architectural tradition of criticism, which has been notoriously unable to generate common criteria and acceptable methodologies of analysis to evaluate built forms and spaces. In this respect, the lay public and the "educated" critics show little difference in their reliance on highly subjective concepts to illustrate an aesthetic reaction. Most of the assessment tends to remain at a gross level, failing to recognize the components of the problem, which is considered in its superficial totality, and often ending up in a simple polarity of positive or negative feelings – perhaps with some historical comparison to add perspective. Words such as *monotonous, regimented, chaotic,* and *confused* are used to synthetize a negative assessment; words such as *balanced, clear, interesting,* and *dramatic* are used in the case of a positive assessment – seldom with a definition of and a justification for the use of a particular adjective.

A more serious problem is that the common aesthetic criticism cannot be developed into meaningful operational criteria for planners and designers, because it fails to identify causative factors, resulting effects, and explanatory reasons for attaching positive or negative subjective reactions to a given environment. Unless a complete solution, found to be acceptable and desirable, is repeated completely, there is little that the designer can do, and this approach carries the danger of imposing a solution foreign to the new problem – that is, a degree of maladaption of the physical solution to the specific needs, requirements, and conditions.

Nevertheless, the intuitive and subjective reactions mentioned earlier are valuable clues because they provide an impressionistic idea of consistent human feelings – perhaps even needs – found in response to environmental stimuli, regardless of the lack of specification and definition.

The objective of this paper is to establish a link among some of the evidence generated by the experimental research in the natural and social sciences, the clues provided by the subjective and intuitive reaction of people in an environment, and the process of planning and design – which demands objective criteria and operational methodologies. The intention is to focus on what may appear to be significant visual issues, without any pretense at a comprehensive formulation, but stressing the importance of reaching conclusions of operational value for practitioners. First, a set of visual hypotheses is proposed, based on research evidence and oriented to postulates of value to the urban designer, the aim being to contribute to the development of more objective criteria for analysis (criticism and evaluation) and synthesis (planning and design). Second, the hypotheses are used in the study of selected urban areas, comparing their explanatory capacity vis-à-vis the current aesthetic criteria. Third, tentative planning criteria are presented, coherent with the accepted visual hypotheses.

The human need for complex visual inputs: a combination of orientation and variety

The search for the visual hypotheses proceeded almost heuristically at the beginning, until a pattern began to emerge and guide the successive search with increasing certainty and understanding of the explanatory factors; the cycle was usually from the identification of intuitive reactions in specific environments to the disentanglement of research findings in the social- and natural-science literature, analyzing specific visual-psychological, sociological, and cultural phenomena.

The two main hypotheses can be stated as follows:

First hypothesis. The human being has deep needs for a combination of different visual inputs from the environment. Some visual inputs must construct a simple (low) order, easily understandable, that would result in a continuity of fully anticipated experiences – fulfilling the orientation needs of the observer. Other visual inputs must construct a complex (high) order, only partially understandable, that would result in a sequence of partially or fully unanticipated experiences – fulfilling the variety and surprise needs of the observer. Thus the environment must generate a set of visual inputs of varying complexity defined by different levels of visual orders.

Second hypothesis. The different visual inputs are not conflicting or exclusionary; on the contrary, they are complementary and have to be combined in the same environment. The absence of one type of visual input handicaps the effects of the other visual input: Environments organized on a very low (simplistic) order exclusively would not result in a satisfactory orientation, but in disorientation – due to lack of visual "clues" and/or to sensorial rejection of a monotonous image; environments organized on a very high (complex) order exclusively would not result in a satisfactory variety, but in confusion – due to lack of visual "commonalities" linking the succession of experiences and/or to sensorial rejection of a chaotic image. Thus both types of visual inputs need each other for a satisfactory performance, actually reinforcing mutually in their visual roles.

It is striking that most contemporary urban designs are not based on a combination of visual inputs at different complexity levels, but tend to be influenced by a kind of pendulum law, stressing either an exclusive low-order organization (modern-movement purism) or a pseudo-complex organization (superimposed picturesquism).

In order to qualify our hypotheses, it is necessary to recognize the limits of their validity. The relationship between the physical world and people is established through the visual psychological processes of perception and cognition. However, there is a sociocultural dimension that has to be added, resulting in the "subjective and cultural relativity of perception and cognition" (Rapoport, 1971, p. 4), as a function of the specific cultural or subcultural group that is being considered. This is a critical stage because "[s]elective screening of sensory data admits some things while filtering others, so that experience as it is perceived through one set of culturally patterned sensory screens is quite different from experience perceived through another" (Hall, 1968, p. 27). The cultural factor bears the final responsibility in linking the visual inputs from the environment with the experiences stored in the human memory, and thus with human reaction and behavior.

Also, in analyzing the visual aspects of an active environment, it is difficult to separate them from its organizational aspects – such as activities, population, and movements. Indeed, it may be desirable not to separate them, since the interaction between form and content is of extreme interest to designers. "In order to be of maximum utility, awareness of the physical form of the environment and knowledge of its activity characteristics should be complimentary and reversible" (Steinitz, 1968, p. 235).

Orientation

Orientation can be defined as the awareness experienced by an observer of his own location in a given environment, involving consciousness of relative direction of and distance to specific orienting-landmarks, which determines his relative position in an understandable and highly predictive spatial organization.

The sense of orientation should be within the capacity of people who are relatively familiar with a place, as well as of people who have never been in the particular place but who should be able to understand the visual messages based on previous experiences with similar places – in other words, sharing a culture. Orientation, then, is the result of a successful match of visual clues generated by the environment, and perceived and understood by the observer, with cognitive structure and/or images stored in the observer's memory.

The importance of orientation is stressed by social scientists; Hall (1966) wrote that "man's feeling about being properly oriented in space runs deep. Such knowledge is ultimately linked to survival and sanity. To be disoriented in space is to be psychotic'' (p. 105). The ancestral need for orientation was (and is for primitive cultures) essential for daily survival of humans in the natural environment, and this psychological demand is still felt even without a direct danger; one of the most fearful experiences is to find oneself lost in a place that was assumed to be known.

The contemporary urban life continually reinforces the need for orientation. The obvious example is the potential level of personal danger found in many areas of the metropolis, which can be compared with the survival needs in a wild environment; in this regard, certainty of one's location and of the capacity to sort out the minimum-risk paths is essential. But the examples need not rely on such an extreme case of personal threat; ample need for orientation can be found in many other instances of the urban life: the decision to exit a highway while driving at high speed, the frustrating search for an address in suburbia, or the finding of a place to obtain specific goods or services. However, the core of the argument for orientation should not be the functional need to reach particular places, but has to be found at the psychological level, in the sense of reassurance and enjoyment experienced by a well-oriented observer and, in the opposite case, in the distress felt by a disoriented person.

Since the process of experiencing orientation is the result of a match between perceived visual clues and cognitive structures in the observer's memory, it is clear that orientation increases with the personal experience of environments. This dynamic feedback between observer and environment underlies some visual-psychological processes; in the analysis of "meaning," Steinitz (1968) includes "the knowledge latent in environmental forms and activities to which people are exposed" – in our words, the visual messages, or inputs, to be perceived; "the knowledge gained as people learn the characteristics of their environment" – that is, the assimilation and memorization of the visual inputs in a cognitive structure; and "the knowledge upon which are based the plans of action used by people to satisfy their various individual and social purposes" (p. 233) – that is, the match of perceived inputs within the frame of the cognitive structure, which permits the observer the orientation synthesis.

The issue of "meaning" and congruence between form and activity is related to the debate about whether buildings should "express" their use. Steinitz (1968) indicates three levels of congruence: type, intensity, and significance of places. From the point of view of orientation, the design problem is to develop "expressions" broad enough to encompass minor changes in users and hierarchy and meaningful enough to permit a sensorial (not literal) identification. This process of orientation, as mentioned, includes cultural factors; the visual inputs, clues to reconstruct a whole with the help of the cognitive structure, are understandable only to those familiar with the cultural code used. What may be a clear orientation message to a member of the local culture, could be just a cipher to a member of a foreign culture – and it is possible that the clue may even be missed to this newcomer. This is of extreme importance to urban designers, because it indicates the benefit of developing coherent reinforcements of the cultural code understood and used by the local population. This is found in well-established cultures, in which an economy of visual means results in a wealth of significant meaning. Designers intending to organize new visual systems should make special efforts to repeat the main orientation messages using several (often redundant) visual inputs; and it may be important to include part of the older visual system to provide a transitional bridge that may help the population in developing the "new memory" and the "new culture." Particularly disturbing could be the times of crises in which the old system is deteriorating and the new system is not yet well shaped; at those times, incoherence and internal conflicts may easily develop, and it can be argued that crises define our present urban period, explaining the inconsistencies found and the inability of designers to deal with them. One example of this problem is found in an analysis of congruence between form and activity, in which lack of congruence is said to deprive the individual of the ability to generalize from past experience, forcing him to a constant and difficult learning process without familiar clues (Steinitz, 1968). The psychological stress in such cases is due to the need to continually revise the cognitive structures or even to abandon them without a suitable replacement; the individual's confidence in the stored information would be so low as to introduce doubts even in reference to valid images.

The cultural inputs in the orientation process are also important wherever urban areas tend to be composed of groups belonging to different – and in many cases mutually foreign – subcultures. It will be necessary to construct visual messages accessible to all members of the population, if a common sense of orientation is to be regained; but at the same time, it may be worthwhile to reserve orientation inputs accessible to only the members of the local subculture, to increase their sense of belonging to their place – visual inputs that would be perceived in terms of local variety by outsiders. In this way, metropolitan (universal) and local orientation domains could be developed, using the different levels of cultural accessibility of the urban groups.

Variety

The word *variety* is used as a shorthand for a set of environmental qualities, all of which correspond to a higher (more complex) order that defies full understanding in anticipation of the complete experience of the environmental sequence.

Variety can be defined as the characteristic of an environment made up of sets of similar but not equal elements that belong to a common and recognizable taxonomy (typology), perceived by the observer in terms of the rhythmical differences appearing within the commonalities unifying the set.

This definition may be applicable to the cases of analogy (functional similarity) and homology (formal similarity). As mentioned before, it is not desirable to separate the functional and the formal aspects of the environment; however, there may be cases in which incongruence between the characteristics of the physical form and of the activities may appear to be a sign of the transitional crises that our cities are experiencing.

Diversity is another related quality that has to be considered. The definition of diversity is an expansion of the definition of variety, in order to include the characteristics of an environment made up of sets of related elements that belong to a more universal taxonomy and show less commonalities than in the case of variety.

Diversity implies higher levels of differentiation than variety and may be applicable only to cases of analogy. For example, a row of residential brownstones exhibiting minor architectural-detail differences is a case of variety within a common typology and homogeneous use, but diversity should be applied to qualify all the set of different building types devoted to housing in a given area.

It must be clear that the differentiation between variety and diversity is a matter of agreement, since both qualities are actually points along a continuum.

The quality of environmental variety is closely associated with the experiencing of surprise, since the complete sequence cannot be apprehended fully in anticipation. The degree of surprise felt by the observer is a function of the rhythm of variety and the range of change – as well as the relative familiarity with the place. If the changes occur following a rhythm not discovered by the observer, he can be left totally unprepared to expect the next stage of visual inputs, constituting a surprising sequence of events – their weight in the total visual experience depending on their relative degree of change. The changes, however, could occur following a rhythm understandable to the observer, in which case a situation of partial expectation would be developed, constituting an anticipatory sequence of events of climactic or cyclic type. Even in the case of observers familiar with the place, in which the sequence of varied visual inputs lacks an intellectual surprise, the eye still is easily attracted by the changing environment, resulting in a confirmation (and enjoyment, in many cases) of the memorized experience.

The importance of variety has been recognized by social scientists, and many of the research reports focus on the problems of environments without variety, stressing the negative effects of monotony. There are, also, findings asserting positively the human need for variety. Novelty and variety appeared as qualities enjoyed by people, necessary for psychological development and sensorial stimulation, indicated in the human's choice for changing and interesting environments (White, 1959). Rapoport and Kantor (1967) summarized other findings: Complex environments, presumably more varied, are preferred over simpler ones, presumably less varied. Adults, specifically, choose variability and uncertainty – potential surprise? – in their visual and auditory stimulation.

As mentioned, however, the most detailed arguments for variety are found in research probing into the negative effects of the absence of it.

Consequences of monotony (lack of variety)

The exclusive stress on simple visual organizations, at the expense of other, more complex visual inputs, would not lead to a higher level of orientation; on the contrary, most likely it would result in a decrease of orientation. If monotony is defined as the quality of environments that lack visual variation, it can be said that monotonous environments lead to disorientation for two main reasons.

First, monotonous environments result in sensorial rejection to a degree that they are partially unperceived. The study by Evans and Piggins (1966) explains this phenomenon:

All biological systems, particularly those equipped with specialized receptors and complex nervous integration centers, respond primarily to a changing environment . . . the eye reacts with exceptional speed to a flickering target viewed peripherally and is sometimes surprisingly inattentive to familiar more centrally placed objects. This preference for the new or the changing is an essential mechanism of any system which is to survive for long in the physical world as we know it, for a "steady" environment seldom, if ever, occurs in nature. What would happen to the physical system if it were confronted with an unchanging [environment]? The target in view soon becomes perceptually unstable and finally actually disappears from view, [although] disappearances are by no means permanent, and the part of the image which vanishes generally returns to view within seconds, after which some other part or parts disappear. (pp. 293–98)

This indicates that an environment organized exclusively on a low-level order – the traditional aesthetic of modern urbanism – would not be properly seen, in the sense of a conscious perception. Monotony, then, can be considered a sensorial deprivation, since human senses tend to reject such visual conditions. One of the most common experiences in monotonous environments is to pass by a place, regularly, without being aware of the physical elements in it; when a functional need forces an observer to perceive part of the physical world, there is often a feeling of mild astonishment at "discovering" part of what should have been familiar a long time ago.

The sensorial deprivation caused by monotony not only reduces the potential for orientation, but also can result in other negative effects for the observer, if he is immersed in the monotonous domain for considerable time. According to Hall (1966): "As he moves through space, man depends on the visual messages received from his body to stabilize his visual world. Without such body feedback, a great many people lose contact with reality and hallucinate'' (p. 66). If the sensorial deprivation corresponds to the visual field, the results could be especially damaging, since vision is the main guidance mechanism we use in moving through space. Several other psychological studies reported perceptual disturbances, such as hallucinations, when people were subjected to monotonous, unchanging environments (Rapoport and Kantor, 1967).

The second reason explaining disorientation in monotonous environments is their lack of minimum differentiation.

Although in order to orient the observer, environments have to be simple and easily understandable, they also need to offer subtle gradations (oriented differences) to provide clues to direction and distances and/or to important changes in the form of orienting-landmarks. This means that in simplistic organizations, where a low-level order is prevalent throughout, the visual conditions for orientation are absent. An example of the orientation advantages of gradations and differentiation is a comparison between a Manhattan-type grid, where main avenues and transversal streets are clearly indicated, and a completely uniform grid. While the first one is simple and understandable, it is also able to communicate more information about the observer's location in it than is the second one.

It is apparent, then, that in order to provide the orientation information satisfactorily, urban designs should balance the need for immediate perception and cognition – simple order within the common cultural code – and a minimum differentiation to provide direction and distance clues, avoiding simplistic visual organizations that could result in banal (sensorially rejected) and/or disorienting environments. Thus the complexity of an environment combining different visual orders could strengthen the perception of the simple order; an example is where environmental variety is organized on a climactic end – that is, expectations are built up in the direction of a main space or building – which could also double as orientation clues, showing that the same visual inputs could play several roles.

Consequences of confusion (lack of orientation)

The exclusive stress on complex visual organizations, at the expense of other, simpler visual inputs, would not lead to a higher level of variety; on the contrary, most likely it would result in a decrease of variety. If confusion is defined as the quality of environments that lack visual orientation, it can be said that confused environments lead to undifferentiation.

In confused environments, formed by a continuous flow of diverse visual inputs without any perceived link, the observer is bombarded with disjointed visual inputs until he is saturated. His predictability is drastically reduced, since there is no understandable rhythm, and expectation is drowned in a fast build-up

to nothing. The saturation experience means that the observer loses the enjoyment of variety, and his senses become insensitive to this succession without
purpose. One example is the typical highway-strip development, in which a
variety of buildings, uses, advertising signs, and lights compete to attract the
driver's attention and patronage; however, being unable to search, compare, and
choose while keeping a minimum speed, the driver behaves in two ways: selecting randomly or following the first familiar sign that is found in the midst of a
confusion of lights, poles, and billboards. In other words, the slightest bit of
orientation immediately attracts the attention, relieved from the relentless undifferentiated confusion. This, of course, is not unknown to entrepreneurs, who
value highly the franchise of a national chain. Visually, the apparently various
signs and lights actually cancel one another.

Combination of orientation and variety

It is apparent not only that the human being needs visual inputs leading to a
satisfactory sense of orientation and variety, but also that both visual experiences
must be present in the same environment because they are mutually reinforcing –
the absence of one handicaps the other.

The need to combine orientation and variety in the human habitat has been
recognized in different ways. Whitehead (1925) contributed one of the earliest
notions, in writing on form in the most general sense:

A rhythm involves a pattern, and to that extent is always self-identical. But no rhythm can
be a mere pattern; for the rhythmic quality depends equally upon the differences involved
in each exhibition of the pattern. The essence of rhythm is the fusion of sameness and
novelty, so that the whole never loses the essential unity of the pattern, while the parts
exhibit the contrast arising from the novelty of their detail. A mere recurrence kills rhythm
as surely as does a mere confusion of detail. (p. 198)

Writing on information theory, Wiener (1967) presented a statement strikingly
parallel to Whitehead's – a legitimate comparison, since visual inputs are information units:

Messages are themselves a form of pattern and organization. Indeed, it is possible to treat
sets of messages as having an entropy like sets of states of the extended world. Just as
entropy is a measure of disorganization, the information carried by a set of messages is a
measure of organization. In fact, it is possible to interpret the information carried by a
message as essentially the negative of its entropy, and the negative logarithm of its
probability. That is, the more probable the message, the less information it gives. Clichés,
for example, are less illuminating than great poems. (p. 31)

Highly probable visual messages in the built environment are generated by
simple, understandable, and anticipated low-order systems and lead to increasing
levels of orientation until, beyond certain range, they result in a decrease of

information and become banal – a visual cliché. The optimal combination of visual messages must include a share of redundant (repetitive, anticipated) messages, to provide orientation, and a share of highly informative (new) messages, to provide variety.

Another version of the need for combination is provided by Cullen (1961), who wrote that the essential element is visual variety within a pattern, recognizing that absence of the first would result in chaos, while absence of the second would result in monotony. In the article by Rapoport and Kantor (1967), there is a quotation that relates to the proposed hypothesis: Hebb found that the sustained interest of a perceiver needs a stimulus field with some familiarity, and some novelty, and that if the novelty is absent, there is a corresponding lack of interest, since interest requires the unfamiliar factor to be learned.

Rapoport and Kantor (1967) also provide an interesting comparison with our hypotheses, since they recognize the problems of an environment organized around a single type of visual input:

the range of perceptual input has been defined between sensory deprivation (monotony) where there is an excess of order and a poverty of elements to observe and organize, and sensory satiation (chaos) where there is a lack of order and an overwhelming amount of elements to observe, impossible to organize. (p. 211)

Their proposal is to increase the level of "ambiguity" in urban environments, defined as "any visual nuance however slight which gives alternative reactions to the same building or urban group" (p. 210). Ambiguity would help to reach an optimal perceptual rate between simplicity and chaos, boredom and confusion, but this notion is rather different from our hypothesis and requires an explanation.

The argument used by the authors is that environments need "complexity" and that this is better achieved through "ambiguity"; there are several reasons for questioning this assertion. First, their interpretation of complexity seems to include only the complex subsystems in the visual environment, and does not account, explicitly at least, for a set of visual systems, ranging from complex to simple, as is our proposal. Second, it can be questioned if complexity is to be achieved through ambiguity; it appears that the reverse is closer to the design process, in which it is possible to organize complex physical systems, which, in turn, would generate variable levels of ambiguity in the different observers – that is, complexity is an operational variable within the designer's capacity to manipulate; ambiguity is the result of the interaction between observer and environment, changing with each observer. It must be remembered that among the few designers who purposefully attempted ambiguity were the Mannerist architects of the seventeenth century, and in most cases they were restricted to suggesting structural solutions in conflict with the gravity-logic solution. Third, ambiguity is not the only way of attracting the eye and the human interest; it is one of several. Consider also the slow process of discovery, the climactic expectation, the trial-and-error interpretation, or the veiled mystery among other possibilities. Ambiguity is a legitimate option to allow people's active dialectic with the environment in the generation of their own interpretations. However, if carried as the sole source of visual interest, it could be negative, such as when the psychological pressures it could generate conflict with the sense of orientation.

Our hypothesis, thus, is summarized in a need for a combination of visual inputs, the extremes of which can be defined in several ways: redundancy and new information, low order and high order, immediate understanding and incomplete understanding, orientation and variety. The functioning of the human mind, which always recognizes the simplest organizations first – according to the Gestalt psychologists (Arnheim, 1966a) – would permit the low-order subsystems of the built environment to satisfy the need for orientation, regardless of the complexities of higher order subsystems.

Brief analysis of selected case studies

The proposed hypotheses are used as criteria to analyze, visually, a set of case studies.

The first case study is the **Medieval Town** (Figure 27.1).

In most medieval towns, it is possible to find a set of orienting landmarks, such as defensive walls, plazas and squares with outstanding civic and religious buildings, and main streets organized in different patterns, which are easily identifiable due to their larger scale and recognizable form. This set of orienting landmarks keeps a constant relationship with the total urban structure: Walls are at the perimeter encircling the town; the main streets originate at the town gates and lead to main spaces or buildings; and so on. This permits an observer with a personal knowledge of the town or an observer with a previous experience in the same type of town (i.e., sharing the cultural code) to relate the visual inputs to his cognitive mental image, and thus to find orientation in the place.

Variety is found in the maze of secondary streets and alleys that form the interstices between the main arteries, resulting in a sequence of more or less surprising experiences for the pedestrian, several options in choosing paths, and alternative interpretations. However, an orienting landmark is always within a short distance from any place in the town, no area in the maze being more than a few minutes' walk from a visual reference. That means that it is possible to be immersed in the most surprising, intriguing, and even mysterious environment, and be aware at the same time that orientation can be restored within a short time – making the experience more pleasurable. The orientation inputs are never offered simplistically, but require some identification and localization from the

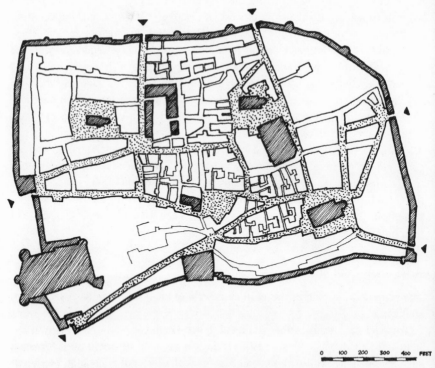

Figure 27.1. Plan of Tübingen. (From Howard Saalman, *Medieval Cities* [New York: Braziller, 1968] and K. Weidle, *Die Entstehung von Alt-Tübingen* [1954])

observer. Orientation and variety are closely interrelated in a process of mutual definition and support.

In the illustrated example, Tübingen (Figure 27.1), it is apparent that the dispersion of public open spaces and the network of main streets reduce the labyrinthine interstices to manageable dimensions. In most districts, the point farthest from one orienting landmark is between 150 and 250 feet away; the extreme case, of a pedestrian walking in the least favorable direction in an elongated area, is 300 feet.

Whenever towns grew beyond the relatively small size of most medieval centers, in the cases of successful commercial cities, vertical orienting landmarks appeared, providing clear three-dimensional patterns of reference. Those vertical elements took the form of towers, spires, domes, and campaniles, in most cases permitted by the greater economic affluence of the commercial city. An observer within a larger maze-like district would not need to physically walk to a reference, but need only raise his eyes to see a three-dimensional orienting landmark.

Figure 27.2. A view of dwellings at
Tübingen. (From Saalman [1968])

At a smaller scale (Figure 27.2), the interrelationship of orientation and variety is also apparent. The urban dwelling units belong to common typologies, with subtle changes in lot size and building height corresponding to the unit's location in the urban structure, providing directional and intensity orientation. However, every building is slightly different, being a variation on a common theme that results from individual interpretations within the sequential, marginal, and disjointed process of adding buildings to the medieval town with traditional craftsmanship.

A contrast to this visual organization is given by the **Renaissance and Baroque Squares** often built within the medieval towns, which although different in their "solution," are also satisfying (Figure 27.3).

Analyzed by themselves, these squares are organized on highly redundant visual information, and are so heavily dependent on the repetition of exactly equal elements on a regular geometric pattern that it is sometimes impossible to distinguish the component units behind the unified façades. However, the total impression is not one of monotony, the reason being that the squares become the exceptional element in the urban pattern, perceived as a variety input at the largest scale and as an orienting landmark.

There is no need for small-scale variety within the square, since the whole complex plays the role of a gigantic variation on the intricate pattern around it. Furthermore, the strong visual identity obtained through its regularity helps in developing clear mental images in the observers, permitting the square's role of an orientator-landmark.

The illustrated example, Park Crescent in London (Figure 27.3), summarizes

Figure 27.3. Park Crescent, London.
(From Leonardo Benevolo, *The Origins of Modern Town Planning* [Cambridge, Mass.: MIT Press, 1967] and Paul Zucker, *Town and Square* [Cambridge, Mass.: MIT Press, 1970])

the principle of a gigantic and ordered exception within the urban pattern. To further support the proposed interpretation, it is only necessary to imagine the crescent repeated many times one after another, an image of unbearable monotony and boredom that destroys all the positive qualities found in the unique crescent.

Before dealing with present urban cases, it is valuable to mention the rows of nineteenth-century town houses in the United States (Figure 27.4). The urban pattern carried a high level of redundant information, based on an easily understandable order of main and secondary streets, defining blocks regularly subdivided. Within this pattern, each dwelling developed a slight variation on the common town-house topology – due to individual wishes or conscious design decisions – variations that often were no more than rhythmical changes in the aggregate form, but that effectively raised the complexity of the physical order and introduced a subtle variety. With limited visual devices, these neighborhoods became the most outstanding urban creation of this country.

The contemporary urban design is made up of at least three main branches: the Modern Movement, a New Picturesquism, and the Suburban Development, to choose only those that have an influence in the physical environment of today.

The Modern Movement in architecture has had a profound influence on most design areas, and involved a sweeping simplification in the understanding of the built environment. At the same time that academic formal conventions were

Figure 27.4. New York town houses.
(From Vincent Scully, *American Ar-*
chitecture and Urbanism [New York:
Praeger, 1969])

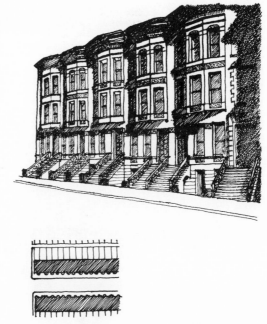

destroyed and new aesthetics created, the movement also impoverished the notions of human psychological needs and ignored all but the most simplistic visual
considerations. It is beginning to be apparent that the early visions of the Modern
Movement are responsible, to a large degree, for the recurrent failures of contemporary urban design.

The urbanism of Le Corbusier and the Bauhaus, two of the best-known influences, is one of absolute dominance of redundant information, extremely low-
order visual organization, and absence of variety at any level, resulting in an
oppressive feeling of monotony, coupled with disorientation. The visual clichés
(paraphrasing Wiener [1967]) originated in the mechanical repetition of identical
elements at all scales, all over the site, and the lack of complexity in the visual
inputs are expressions of the characteristics of the movement: narrowly defined
functionalism, and obsession with purism and clarity. In a world of abstract
utopia, dwelling units, buildings, and patterns are homogenized, and users and
activities are averaged.

There is a historical precedent for this impoverished design attitude, a line of
utopian proposals originated since the Renaissance that reflects the same detach-
ment from human needs and differences, aimed at the creation of abstract geo-
metrics without any relationship with a true urban community.

Figure 27.5. *Above:* Le Corbusier's
Plan Voisin. (From Scully [1969])
Below: New York public housing.
(From Scully [1969])

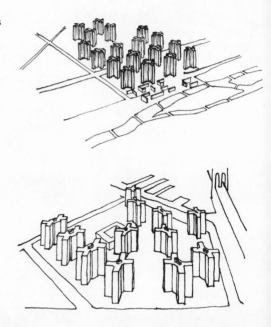

   The illustrated example, Plan Voisin for Paris (1925) (Figure 27.5), is pictured
by Scully (1969) next to a public-housing project built in New York in 1948; this
points to the most critical issue of the Modern Movement influence. Although
the top architects in the movement built very few large-scale projects (and their
smaller architectural designs often show more sophistication than do the grand
urbanist schemes), their "urban solutions" became a legacy that passed to the
mainstream of the design profession, synthetized in an easy-to-administrate reci-
pe of low-order organization, regimentation, and mechanical repetition. The
difference between Le Corbusier's urban proposals and our worst public-housing
failures is only of details, not of basic concepts.
   The most serious indictment of the Modern Movement is that the physical
typologies created by the "masters" degraded very rapidly when implemented
by lesser designers. Contrast this with the ability of the medieval master masons,
for instance, to build whole towns based on variations of common types, at a
consistent visual-quality level. In this respect, the Modern Movement was the
culmination of the Beaux-Arts artist–architect tradition whose work was unique
and not reproducible. However, even the quality found in the masters' work
could not avoid the basic conflict between the purist, detached environment they
created and the people living in it. We should be grateful that only one Unité
d'Habitation was built at any single site, a controversial monument rather than an
intolerable bore of a row of slabs. The dynamiting of the Pruitt-Igoe public-

housing complex in St. Louis blew up more than just a few buildings; it destroyed the Ville Radieuse utopia.

In reaction to the limitations of the Modern Movement schemes, a New Picturesquism trend began to emerge during the 1960s. The threat of monotony has been answered by the introduction of often-superficial design approaches that can be encompassed under the label of picturesque. One approach, which can remain as an exception, is exemplified by Habitat (1967), a housing project openly recalling images of Mediterranean villages and Indian pueblos, supported by expensive engineering, designed with high-pitched variation at the unit and aggregation levels, and characterized by very poor orientation. This probably is the climax of a movement combining historicism and high-cost technology within a formalist intention.

The mainstream of the New Picturesquism, however, can be found in the new towns that have been built since the early 1960s, which show the recurrent use of similar design approaches and visual results. The common "solution" to avoid monotony is the purposefully winding roads, which seriously reduce orientation at the large scale, and the strong volumetric articulations of the residential rows of town houses and garden apartments (the high-rise slabs have disappeared, except in public-housing or in central-city projects). This architectural push-and-pull, together with the changing lines of pitched roofs, exposed side walls, and jogged fences, aimed at obtaining a "moved" effect. Nevertheless, these designs would be rejected by a master craftsman because of their lack of good building sense and maladaptation to present-day technology and production systems.

In reality, the New Picturesquism attempts to re-create its own pseudo version of an old-town atmosphere, an image that is vitiated by a lack of understanding of the morphological roots of variety. Variety results from the nature of the problem and the environment and cannot be replaced by a willful formal play at the risk of becoming a confusion of details.

The New Picturesquism does not improve the visual orientation. The already mentioned winding pattern of roads makes it practically impossible to develop a cognitive image of the urban pattern; the usual repetition of quadrants or "villages" (the building block of most new towns) does not allow visual sequences of structured paths. The only orienting landmarks are the highway and the shopping center, which are practically invisible from most parts of the community because they seldom have a vertical reference. Perhaps the only orientation assistance found in most new towns has to be traced to their extreme simplicity in the movement pattern – sooner or later, a driver would reach the main circulation loop.

The illustrated example is Reston, Virginia, which has been hailed as the new town with highest quality design and which, in some respects, presents the only

Figure 27.6. Plan of Reston. (From
Reston documents)

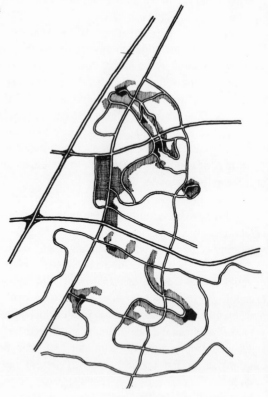

departures from the typical solution. The circulation system (Figure 27.6), composed of winding roads and culs-de-sac, results in an extremely confusing pattern
with negligible orientation value – lower than the typical loops that can be found
in Columbia, Maryland, for instance. The positive elements of Reston are the
relative degree of continuity of some high-density areas and local centers, which
can generate recognizable paths, in contrast with the isolation of Columbia's
"villages," as well as the different designs of each neighborhood, which can
provide some orientation clues after a preliminary experience. This latter characteristic, nevertheless, is questionable because the visual unity of the new town is
sacrificed in order to obtain "identifiable" neighborhoods, and the visual differences are originated simply on the will of the individual architects.

Reston (Figure 27.7) exhibits the typical volumetric push-and-pull of the
multifamily residential blocks plus superficial changes of color, texture, and roof
profiles. In the first neighborhood built, there was an attempt at reproducing a
"square" – a grand composition of a formal space half enclosed by buildings,

Figure 27.7. Reston's first village.
(From Reston documents)

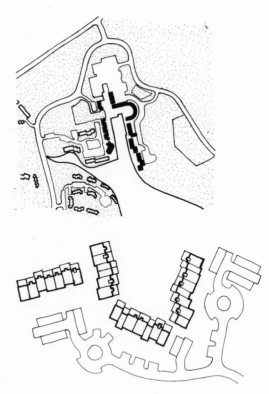

with a high-rise tower (built to achieve a "sense of urbanity," in the words of the developer) and opening to a body of water. This grand space, which has a scale suitable for a location in a large, dense district, as discussed in the case of the Renaissance and Baroque squares, is, however, bounded by buildings one dwelling-unit thick, behind which large parking lots and undefined open space extend. The square is, then, more of a stage – similar to Palladio's Teatro Olimpico in Vicenza – than a real urban space. It is an imaginary world created by using modern clichés and historical images, to simulate urbanity and variety within a suburban homogeneous new town – just as Disneyland simulates exotic or historical environments.

There are a few cases of contemporary urban design that should be discussed to prevent an overcritical assessment of today's state of the art. The illustrated example is Sert's married-student housing at Harvard University (1964) (Figure 27.8), a significant residential design in which a few unit types are combined in different ways, resulting in a variety of buildings, façade treatments (due to orientation requirements), and spaces. This variety is a legitimate answer to the

Figure 27.8. Sert's married student
housing at Harvard University.
(From Knud Bastlund, *José Luis Sert*
[New York: Praeger, 1967])

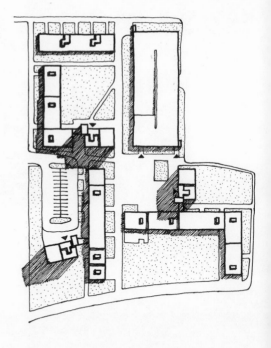

different wishes of families in terms of life styles. Orientation is provided through the spatial relationships of the three towers, which define the central open space; through the distinctive parking-structure form (although the automobile access is lacking orientation partly due to exogenous constraints of the site); and through the visual expression of elevators, which indicate high-rise entrances.

In terms of impact on our urban areas, the most important influence is clearly the Suburban Development. Unfortunately, it has not resulted in satisfactory visual environments, according to the criteria proposed in this study (Figure 27.9). The street system is mainly a set of winding loops that prevent understanding and orientation. The built system shows little differentiation or hierarchies. Nearness to particular elements is not reflected, detracting from both orientation and variety. There are no vertical orientator-landmarks to guide in the vast expanses of tract land; as a result, only residents can be oriented within a short distance from their houses and on the paths they usually follow.

The dwelling units, corresponding essentially to the same type, are superficially ornamented with different style disguises, which result in a confusion of details rather than in variety. Suburbia exhibits a poor combination of disorientation and confusion of details.

Figure 27.9. Suburbia (Deltona, Florida). (From Deltona documents)

Conclusions

If the proposed hypothesis of the human needs for complex visual inputs in a combination of orientation and variety is accepted, then it is apparent that most of today's urban designs have been visually unsuccessful. As was mentioned in the introduction, radical changes in the nature of cities – development speed and size – and in the nature of the planning process – institutionalization and centralization – have resulted in a gap of understanding and ability. The study of historical cases is valuable in assessing the proposed hypothesis and in exploring alternative design approaches and solutions, but those findings should be placed in the perspective of the present cultural conditions to have operational value. Two main conclusions have been arrived at.

First, *urban hierarchies are assigned visual roles,* an extremely important operational concept. The built environment can be interpreted as a system of subsystems of changing complexity and scale, on a number of hierarchical levels.

A structural hierarchy may be defined as a spatial system displaying a sequence of separable levels of sufficiently stable structures or processes, each level being paired with at least one other level; a unit at one level being composed of parts, each of which is a unit at the next level and so on. (Law Whyte, 1972, p. 611)

Each level of the physical hierarchies can be assigned a specific visual role, in terms of conveying messages of orientation or unity (low order) and of variety or rhythm (high order).

In the medieval town, the lower hierarchies of the street system were loaded with variety inputs, while the higher ones were loaded with orientation inputs. In the nineteenth-century town-house neighborhood, the lower (detail) hierarchies of the dwelling unit conveyed the variety inputs, while the higher hierarchies conveyed unity (the typology) and orientation (the street system). The important fact is that the visual roles always find a hierarchical level to be expressed, although the specific design solution may vary.

It is interesting to notice that although humans apprehend the visual context as a totality, since "the characteristic quality of the whole can be dependent on the universal interaction of literally all its parts" (Lorenz, 1972, p. 157), perception would follow the principle of similarity, as proposed by Wertheimer, tending toward connection and fusion according to relationships of size, location, form, color, and direction and starting with the simplest organizations (Arnheim, 1966a). That is, the low-order visual messages of orientation would be perceived more rapidly than the variety messages, establishing a time priority for the set of complex visual inputs.

Second, *an urban hierarchy can play a multiple visual role.* That is, the same physical element could be perceived as a landmark and as a variety input, depending on the relationship established by the observer with the rest of the physical environment – a case of positive ambiguity.

The Renaissance and Baroque squares at their highest hierarchy of the total complex, are both a gigantic variation on the surrounding pattern and an important orienting landmark within the city. The changes in lot size and building height (as well as use) found in many urban areas corresponding to the subdivision hierarchy are both subtle variations and subtle orientators, indicating the direction of highest urban activities and density.

After stating the two main conclusions obtained from the case studies, it is important to hint at some design implications within today's context, suggestive of alternative paths for future exploration.

Designed variety and technological constraints

An immediate approach is to design consciously to obtain a varied environment, but this rapidly raises the question of potential conflict with the (assumed) standardization constraints imposed by technology and/or the total production system, as shown by increasing costs. One solution to this dilemma is to maintain standard dwelling units and to combine them in different ways, resulting in a variety of buildings. That is Sert's solution. However, the assumptions of technological constraints are true only under given circumstances. Advanced technological methods, using automated plants, would permit the fabrication of building components with a range of sizes, forms, openings, and surface and

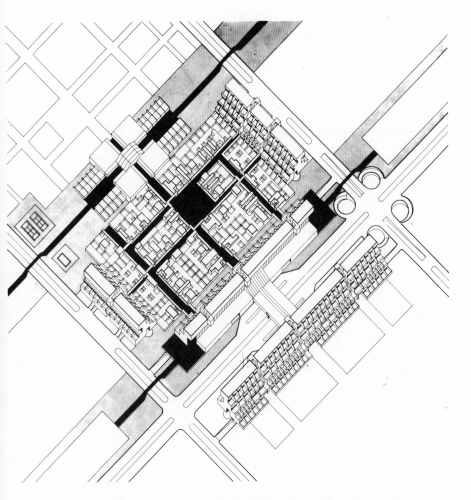

Figure 27.10. New community proposed in New Jersey. (From author's documents)

structural characteristics as long as a large production level were ensured. It would be possible to combine those components to produce a large variety of dwelling units, which would also widen the market's choice sufficiently to warrant a large production level. An example of this approach (Figure 27.10) is a new community designed by the author in which a variety of dwelling units (variations of a typology only) resulted from the combination of standard components. (In this case, the units could also be expanded in different ways, increasing the variation at any given time.)

Programming variety

Another possible approach is to generate variety at the program level. One of the consequences of the radical change in the nature of cities and planning is that large areas are planned simultaneously for a number of anonymous users, described through market analysis and social surveys. A more refined evaluation of the data, weighing their distribution more than their mean, would help in programming for the potential range of needs and wishes of the future population, resulting in a better fit with the real-world complexities and, eventually, in a more varied social and physical environment.

At this point, it is clear that economic and social segregation are factors contributing to the generation of homogeneous and monotonous environments, since they result in an absence of program variety. This issue has been expanded by Sennett (1970), leading us to believe that the visual simplification criticized in monotonous environments is a parallel to the social simplification of urban life.

The traditional land-use segregation, which forces single-use zones throughout the urban area, should be equally revised, since it also detracts from program variety. It originated in the purist aims of the Modern Movement, as shown in the Charte d'Athens, and fitted well the control aims of the planning bureaucracy. However, with the exception of clearly incompatible uses (pollution, noise, vibration, and other nuisances), there is no reason to eliminate interaction, overlap, sharing, and symbiosis, which are important factors in a program variety.

Variety through planning abstention

To a degree, variety must be self-originated in the urban community, and to think of planning the complete range of unexpected and spontaneous activities and physical elements that appear in cities, characterizing real urban life, is a mirage and a contradiction. Deliberate planning abstention is an absolute must, abandoning a false "comprehensiveness" that is both impossible and undesirable. Rather, the aim should be the creation of a catalyst that encourages the local forces to shape their own community. An open-ended design will permit the selection among several potential alternatives, leading to variety rooted in the local interpretation of needs and problems.

Designed orientation and the effects of speed

Since orientation is basically related to the human being in (potential) movement, the introduction of radically different speeds in urban areas after the technological advances of the past decades has created a growing dichotomy between the fast (motorized) and the slow (pedestrian) individual. It is true that the redundant orientation messages sent to a driver must be radically different in scale, size, interval, and content from those set to a pedestrian. However, as vehicular speed decreases, the integration of the driver with the pedestrian orientation system should increase, through explicit and intelligible visual presentation of urban activities, buildings, and people. Unfortunately, mode segregation, almost without speed consideration, has been another belief of modern urbanism, which leads to a dichotomy in visual experiences and a break in the orientation sequence (the parking-lot stage). It is interesting that the Parisian boulevards, for instance, are one of the most vital urban scapes, with people on foot and in (necessarily slow) vehicles finding symbiotic satisfaction and a balanced combination of orientation and variety.

In conclusion, the proposed hypothesis of a human need for complex visual inputs from the environment in a combination of orientation/unification and variety/rhythm appears confirmed by research findings, appears present in urban areas that, by consensus, are considered visually successful, and is absent in areas where the intuitive reaction is negative. There is a need for further research in the sociobehavioral and visual-psychological fields, but it is also evident that there is an urgent need, as well, for investigation of alternative design approaches for urban areas, and for reassessment of planning practices that have outlived their validity.

28   A survey of aesthetic controls in English-speaking countries

Wolfgang F. E. Preiser and Kevin P. Rohane

Objectives

This project was undertaken to study the range and types of aesthetic controls for the built environment found in an international sample of communities.

The absence of such controls in the southwestern United States, for example, has resulted in deteriorating aesthetic quality at a time when rapid growth could and should be channeled properly to maintain and to improve the quality of the built environment.

The background to this project was the authors' interest and experience in studies aimed at assessing the general public's perception of visual aesthetic quality. In the "Stronghurst study" (Harrington and Preiser, 1980), aesthetic compatibility was at issue. A sound, older residential neighborhood was confronted with new, industrial development at its perimeter, a type of development that would significantly alter the aesthetic quality of and views from that neighborhood. Three issues were investigated: (1) Is the general public capable of judgment and agreement concerning aesthetic issues? (2) Can the general public identify and reasonably judge visual aesthetic subcategories, such as massing, scale, ratio of window openings to wall surfaces, height, setbacks, landscaping, building color, and materials? (3) Can visual aesthetic guidelines be derived from empirical studies that may aid in conflict situations and arbitration concerning visual aesthetic compatibility of proposed developments with existing ones? The answer to all three questions was affirmative.

The "Sydney aesthetics study" (Preiser and Hall, 1980) consisted of showing slides of "believable" buildings to groups of respondents and systematically varying the description of building function. The result was that perceived building function biases aesthetic judgments of buildings. In this case, functions given were a child-day-care center, a doctor's or lawyer's office, and a drug-rehabilitation center. Further, perceived distance from or frequency of exposure to the object shown biased the respondent's aesthetic judgments; that is, if the proposed building was across the street from the respondent, it was rated much less favorably than if it was .5 mile or farther away. The same is true for frequency of

422

exposure; when the building was by-passed once a month, it was rated more favorably than if it was seen every day.

The need for an expanded scope of this type of study, which would cover a true cross section of communities from very small to very large, is underscored by the fact that at this time there is a wide discrepancy between the visual aesthetic quality experienced in communities that have controls and design review boards, such as Boulder, Colorado, and those that have none, such as Houston and Amarillo, Texas. With the goal of upgrading the quality of the built environment in mind, this study was undertaken to demonstrate that models exist for aesthetic controls and administrative mechanisms that could be adapted to those communities currently lacking such controls.

Methodology

The primary goal of the broad, initial survey was to identify the range of aesthetic issues addressed by regulatory bodies and documents. Of particular interest was to identify what was being regulated from an architectural and aesthetic quality point of view.

An international group of sixty-four city officials was solicited for copies of published regulations designed to control aesthetic issues within their jurisdictions. Communities in certain German-speaking countries and in Sweden were contacted, in addition to communities in English-speaking countries – that is, the United Kingdom, Australia, New Zealand, Canada, and the United States. The sample size of seven communities in Canada, fourteen in the United Kingdom, and fourteen in Australia and New Zealand refers to the fact that the largest cities in these countries, which have relatively small populations, were selected for participation in the survey. Thus a rather comprehensive coverage was achieved in Canada, the United Kingdom, Australia, and New Zealand. In the United States, where the sample size was twenty-nine, the participating communities were not necessarily the largest cities, but sometimes communities of smaller size that were known to have aesthetic controls in place. Therefore, the emphasis in this study was not on a truly representative cross section of responding communities. Because of the inadequate responses from non-English-speaking countries, the analysis of survey responses was restricted to the four English-speaking regions mentioned earlier.

All the aesthetic-control regulations reviewed were created and enforced by political entities – that is, city-planning boards, councils, or design review committees. Also, the regulations reviewed in this study were community based. No regulations that were regional or national in scope were identified or evaluated.

Contact was made with a relevant public official in each of the identified communities to solicit his or her cooperation in this study. A request was made for any documentation or regulatory mechanisms that they utilized to regulate aesthetic concerns in their respective communities. The response yielded information from sixty-four communities ranging in type from richly illustrated, multicolor guideline booklets to simple, narrative building-code manuals.

Summary of findings

The analysis of architectural-control mechanisms in four English-speaking regions – the United States, Canada, the United Kingdom, and Australia/New Zealand – showed a great disparity in the regulation of visual aesthetic concerns. Depending on the age and location of a given community, aesthetic controls varied significantly. They ranged from nonexistent controls in Houston and Amarillo, Texas, to a large number of regulated items in the United Kingdom and its former colonies. In general, many fewer items tended to be regulated in the United States when compared with the other English-speaking regions.

In the United States, aesthetic controls were concerned mostly with setbacks, signage, compatibility with predominant uses in a given area, heights or number of stories, treatment of mechanical equipment, and fences, walls, and screens.

In addition to the above-referenced items, a majority of communities in Australia and New Zealand, and fewer in Canada and the United Kingdom, were concerned with the following controllable items:

Admission of sunlight or daylight into buildings; spacing between structures; window placement for visual privacy; setbacks for visual privacy; transition buffers between differently zoned areas; pollution – that is, noise, smell, dust, vibration, and smoke; required amount of landscaping and conservation of existing vegetation; open space, plot ratio, density, or building coverage; off-street parking; vehicular traffic.

In Canada only, the following concerns were frequently addressed:

Exterior lighting; public art; private landscapes and site drainage; building-site width; color, texture, materials, surfaces, and finishes; play equipment; street furniture; on-site storage; and ancillary and accessory buildings.

In the United Kingdom, aesthetic controls cover the following additional areas of concern:

Volume, massing, form, bulk, or shape; articulation of façade, modulation or rhythm of voids, solids and fenestration; extensions or alterations of existing buildings; landscaping and buildings on historic sites; design subordination to valued buildings or other valued landmarks.

Review of results

The results from this study need to be qualified in light of the fact that the sample sizes of cities and communities in the participating countries were rather small.

Thus the use of statistical-analysis techniques was considered inappropriate. Therefore, only simple tabulations of frequencies of occurrence of certain types of regulations were assembled. The significance of the findings lies more in the qualitative nature and degree of what is regulated in different parts of the English-speaking world than in the numbers that show how many communities are regulating which items at the present time.

Although no specific hypotheses were stated at the initiation of the study, the authors did expect to find certain phenomena: (1) Countries belonging to the British Commonwealth would be greatly influenced by the long tradition of aesthetic controls in the United Kingdom; (2) older communities in the United States would be more regulated than newer ones, especially those in the western United States; and (3) the constitutions of the participating countries would be reflected in a basic way concerning individual versus communal rights and, therefore, the willingness to accept regulations and aesthetic controls. This, in turn, would be reflected in the visual aesthetic quality and unity of the different communities that participated in this study.

Aesthetic controls that were evaluated ranged from stringent, issue-specific regulations – for example, height restrictions – to the opposite extreme of generic, qualitative controls mandating "high" local design standards as determined by design review committees.

Controls varied widely among cities as to their individual scope and level of detail and as to what was regulated, although there was a strong agreement about the general topics addressed in Table 28.1.

According to Table 28.1, more than fifty of the sixty-four community samples regulated signage, height, the general compatibility of buildings, as well as off-street parking and loading zones. The next most frequently mentioned control issues were the conservation of existing vegetation, requirements for specific types of landscaping, requirements for open space, review of volume and massing, and review of materials and textures. About half the communities regulated color and finishes, style and character of buildings, as well as sizes of building lots.

The second analysis of the survey results was concerned with groups of communities rather than with individual cities. They were analyzed according to their political boundaries, categorized by Australia/New Zealand (AUS/NZ), the United Kingdom (UK), Canada (CAN), and the United States (US). These abbreviations for the four English-speaking regions will be used in the interest of brevity.

The comparison of aesthetic regulations across several English-speaking countries yielded a relatively short, finite list of aesthetic issues addressed by controls. In the following discussion, some of the significant findings of the study are highlighted.

A number of issues were common to most architectural controls. The intent of

Table 28.1. *Frequency of aesthetic-control issues*

Item	Regulated aesthetic issue	No. of communities citing issue
1	Signage restrictions	56
2	Height restrictions	54
3	General compatibility of building	52
4	Off-street parking and loading zones	50
5	Conservation of existing vegetation	48
6	Specific landscape elements required	45
7	Percentage of open space required	44
8	Volume and massing review	44
9	Materials and texture review	44
10	Color and finish review	36
11	Style and character review	34
12	Maximum and minimum size of buildings or lot	34
13	Visual air-pollution regulations	30
14	Underground-utility requirements	30
15	Percentage of site to be landscaped	29
16	Silhouette or profile review	27
17	Façade-articulation review	26
18	Public art required or encouraged	11

Total sample size = 64.

the majority of these was to ensure compatibility and aesthetic harmony of the appearance of future development with the existing architectural quality by maintaining homogeneity. The variables that contribute to visual compatibility exemplify this concern by required reviews of these building characteristics:

> Height
> Materials and texture
> Style and character
> Conservation of existing vegetation
> Volume and massing
> Colors and finishes
> Façade articulation
> Minimum-size restrictions

A related common intent of regulation was to control visual clutter. Such variables as signage controls, silhouette and profile review, underground-utility requirements, and pollution controls are included in this category. Table 28.2 summarizes the topics that are addressed by aesthetic controls, broken down by regions.

The regulated items listed in Table 28.1 were grouped into three basic catego-

Table 28.2. *Aesthetic controls by topic and region*

Item	Regulated item	Frequency (no. of codes)	AUS/NZ (%)	UK (%)	CAN (%)	US (%)
			($n = 14$)	($n = 14$)	($n = 7$)	($n = 29$)
Site and landscaping related						
1	Signs	56	93	79	86	86
2	Off-street parking and loading zones	50	100	86	86	66
3	Conservation of existing vegetation	48	64	86	71	72
4	Specific landscape elements required	45	64	86	71	72
5	Percentage of open space required	44	100	50	86	85
6	Underground utilities required	30	57	43	71	38
7	Percentage of site to be landscaped	29	86	7	71	38
8	Public art required	11	14	0	43	21
Buildings						
9	General compatibility	52	79	86	86	76
10	Style and character	34	57	57	57	48
11	Volume or massing	44	79	64	100	59
12	Silhouette or profile	27	43	50	43	38
13	Height	54	100	71	86	83
14	Color and finish	36	36	64	86	55
15	Materials and texture	44	57	79	100	62
16	Façade articulation	26	29	64	57	31
17	Maximum or minimum size of building or lot	34	79	14	86	48
18	Maintenance of a high design standard	24	36	71	43	21

Total sample size = 64.

ries: those contributing to pollution, those concerning site and landscaping, and those concerning buildings. This breakdown is represented in Table 28.2, which shows aesthetic controls organized by topic and region and tabulated by the frequency of items mentioned as well as the percentages of the total sample.

Site and landscaping issues

The use of signs, either free-standing or on the sides of buildings, was the most regulated site issue (88%) among all the codes. Except in 15% of the codes, the majority of communities in AUS/NZ, CAN, and the US addressed signs, while 20% of the UK codes did *not* address signs.

Off-street parking and the placement of loading zones were addressed by 86–100% of the codes in AUS/NZ, the UK, and CAN, while only 66% of those reviewed in the US regulated this issue.

The conservation of existing vegetation was a major component of regulations in 86% of the codes in the UK. Up to 72% of the codes had similar specific requirements in AUS/NZ, CAN, and the US.

A certain percentage of open space per site was required by codes in at least 86% of the communities reviewed in the US, AUS/NZ, and CAN, while in the UK, only 50% of the codes addressed the open-space issue.

The requirement for underground utilities was most common in CAN (71%), while in decreasing order, AUS/NZ (57%), the UK (43%), and the US (38%) chose to regulate the elimination of overhead utilities.

The requirement for a specific portion of a site to be landscaped was found in 86% of the AUS/NZ codes, but in only 7% of the codes in the UK. Landscaping was addressed in 38% of the US codes, while 71% of the CAN codes required a specific proportion of a site to be landscaped.

Public art as a part of site development was not addressed by any code in the UK, while 15% of the AUS/NZ codes, 21% of the US codes, and 43% of the CAN codes support artwork as an integral part of site design.

Building issues

The two most common issues found in the code regulations concerned general compatibility (83%) and specific height of structures (86%), while the upholding of "high" design standards was a stipulation in a small, but significant, number of the codes sampled.

The general compatibility of a building with its milieu was most certainly a universal objective of all the code restrictions. However, some communities selected specific issues to regulate. The style or character of a building was subject to regulation in 54% of all the codes reviewed. A building's massing or volume was addressed by all of the CAN codes and most of the AUS/NZ communities, while about 59% of the US and 64% of the UK sample dealt with this issue.

A building's silhouette was to be reviewed for compatibility by 43% of all the codes, while building height was regulated by 100% of the AUS/NZ and by 85% of the CAN and US codes. Only 71% of the UK communities chose to regulate building height.

The color and finish of a building was specifically regulated by 86% of the CAN and only 36% of the AUS/NZ codes. In the US and the UK, 55% and 64% of the codes, respectively, controlled the color and finish of buildings. However, more codes restricted the use of materials and textures: 100% in CAN, 79% in the UK, 62% in the US, and 51% in AUS/NZ.

The issue of façade articulation was mandated for review by 64% of the codes in the UK and 57% of the CAN codes, while in AUS/NZ and the US, only 29% and 31% of the codes, respectively, evaluated façade articulation.

The imposition of a maximum or minimum building or lot size was stated most commonly in CAN and AUS/NZ (86% and 79%, respectively) and least commonly in the UK (7%). In the US, 48% of the codes addressed a maximum or minimum lot size.

The upholding of high design standards was mandated by a significant number of communities – especially in the UK (71%), and to a lesser degree in CAN (43%), AUS/NZ (36%), and the US (21%).

A small number of codes in the sample addressed the issue of historic preservation of sites, buildings, and/or landscaped elements. The codes that required design subordination of proposed buildings located in historic zones were most commonly found in the UK and CAN, but also found in a few US and AUS/NZ communities.

Synopsis

1. The most *comprehensive aesthetic controls* were in use in the following communities: Brighton, UK; Brisbane, AUS; Calgary, CAN; Canberra, AUS; Dunedin, NZ; Halifax, CAN; Palo Alto, California (US); Santa Barbara, California (US); Southend-on-Sea, UK; Vancouver, CAN; and Wellington, NZ.

2. Particularly *detailed aesthetic controls* were adopted by in the following communities: Calgary, CAN; Columbia, South Carolina (US); Coral Gables, Florida (US); Darwin, AUS; El Paso, Texas (US); Halifax, CAN; Las Vegas, Nevada (US); New Plymouth, NZ; Oklahoma City, Oklahoma (US); and Phoenix, Arizona (US).

3. *Best formats of presentation* and clarity of written communication of aesthetic controls were found in the following communities: Adelaide, AUS; Boston, Massachusetts (US); Brighton, UK; Cambridge, Massachusetts (US); Carlisle, UK; Darwin, AUS; Dover, UK; Glenville, Illinois (US); Halifax, CAN; Highland Park, Illinois (US); Milton Keynes, UK; Regina, CAN; Reston, Virginia (US); San Francisco, California (US); Southend-on-Sea, UK; and Vancouver, CAN.

Research questions

Why is there a need for aesthetic regulation? There is no need to regulate that which always occurs within acceptable limits. Aesthetic regulations may limit the amount of visual variation or change within a community, and they often legislate against specific conditions that have been found undesirable. The purpose of aesthetic controls can be to regulate that which, if left unregulated, would result in the loss or deterioration of existing conditions.

The American legal precedent supports the right to legislate aesthetic standards based on the rights of property owners to preserve and maintain property values. The creation of historical preservation zones or districts, as well as controls and requirements for general compatibility reviews, serve this purpose.

Is this type of regulation intended as a countermeasure to the unresponsiveness of national architectural firms and developers, which are often unsympathetic to local conditions? On the contrary, aesthetic controls can be used to encourage new development with desirable features and attributes – for example, mandates for a specific portion of a site to be left open or more specifically landscaped, underground utility requirements, and bonuses for public artwork.

Where did the various regulatory agencies acquire the basic "issues to be regulated"? In general, it appears that each agency created documents to regulate a set of aesthetic issues that was uniquely important to the local community. Regulation of aesthetics does not occur at state or national levels – it is a uniquely local issue – and it is quite interesting to observe the commonality of issues that are addressed.

In this study, it was shown that a distinct commonality among regulated issues does exist. Is this a symptom of the international style in architecture? Or is it a cross-cultural continuity of design aesthetics? Could it simply be cross-fertilization due to networking regulators? Perhaps there are basic issues of good design that are truly universal. Is it feasible to create an international master set of aesthetic variables that regulate the development of communities? It would certainly not be a universal instrument, but a comprehensive menu from which each community could select appropriate sections and degrees of regulation.

Is there a correlation between the quantifiable aesthetic controls and the age of the community or country? Do the more pioneer-oriented cultures have less regulation because of their youth or because they have no identifiable character to preserve?

Another related question concerns a basic issue – the relationship of the structure of society and the ability to legislate and enforce architectural and aesthetic controls. Do countries with autocratic systems of government, such as eastern bloc countries, in fact have the ability to exercise more effective controls? And does this control produce a greater degree of aesthetic quality?

In conclusion, this and other studies concerning architectural and aesthetic controls point to important implications for city planners and urban designers. It is possible to regulate architectural and aesthetic quality to a certain extent and with varying success in different parts of the world. As the results of this study show, the regulation of architectural and aesthetic design quality relates to the interface of the private and public domains and is not easy to accomplish unless a certain willingness to be regulated exists in the population. Therefore, it can be said that a certain preparedness to act in the public's best interest is needed to create and support the enforcement of architectural and aesthetic controls.

Acknowledgments

Thanks are owed to the planning officers and planning departments of those communities that participated in this survey of aesthetic controls. We are especially grateful to the National Endowment for the Arts, which partially supported this effort (Grant No. 91–4282–164).

Appendix: selected representative excerpts from aesthetic controls

Buffers

Only those uses which are ancillary to and compatible with dwellings will be permitted within residential areas. These uses will be confined to those which do not attract substantial amounts of traffic into residential areas. (Perth, Australia: City Scheme Report)

Visual screen

The C-4 District shall be permanently screened from abutting or adjoining properties zoned for dwellings by a screening wall, fence, or other suitable enclosure of at least six feet in height. If the City Planning Department determines that a 10 foot buffer area is required between a residential zone and a parking lot, said area shall be planted with grass . . . (City of Colorado Springs, Colorado: Zoning Ordinances 640–640.1)

Site and neighborhood compatibility

The site shall be planned to accomplish a desirable transition with the streetscape, and to provide for adequate planting, safe pedestrian movement, and parking areas. (Highland Park, Illinois: Appearance Code)

Landscaped area (suburban residential)

At least one live deciduous tree, one inch in caliper measured one foot from the ground, shall be planted on the tract that is being developed as an NBZ [Neighborhood Business Zone] for each 25 feet or portion thereof of street frontage within the NBZ . . . (City of Colorado Springs, Colorado: Zoning Ordinance)

Image

Individual buildings should be well designed of themselves and relate together so as to form an integral part of the overall design [of the development]. A variety of building lines and heights should be used to create interesting enclosed spaces within the housing area, and to give each part of the housing area a distinct identity. (Cardiff, Wales: 8 – Radyr Court, Housing Area "AB" Planning Brief)

First floors

New buildings should maintain the grain of existing development in views across the City as a whole. For example, an area that can be seen from distant viewpoints and is composed of small terraces rising in steps house by house should not be the subject of development which introduces large single buildings. The terrace form of building should be maintained, and the relationship of buildings to the underlying land form should be continued. (Bath, England: 28(i) – Policies for the Conservative Area)

Visual clutter

Buildings should be sited on their lots in a manner which exposes qualitative design features to pedestrian traffic and obscures mechanical equipment, venting, storage areas, trash and garbage collection areas, etc. (Halifax, Nova Scotia: 17 – Proposed Planning Criteria Statement for Brunswick Street Area)

Prevention of ribbon (residential) development

Architectural design of the project shall be of such character so as to be compatible with the area and to harmonize with natural surroundings and be in keeping with local materials, colors, and design features in order to preserve and improve the appearance and beauty of the Waikiki area. (Honolulu, Hawaii: 5 – Bill #144 [1975] Draft Number 5)

Administrative structure

The City Council or the City Planning Commissioner may request that the El Paso Chapter of the American Institute of Architects, The El Paso Museum of Art, or similar agencies provide representatives in an advisory capacity regarding the suitability of uses and the compatibility of the proposed architectural changes [in Historic Zones]. (El Paso, Texas: 131 – Zoning Ordinance for the City of El Paso)

Information to accompany application

Legal description
Area of the site
Length of the frontage
Owner's name
Occupier's name
Existing use
Existing structures
Proposed dimensions, stories
Site plan
Number of motor vehicles

Number of persons engaged in proposed use
Nature of proposed machinery (if any)
Mechanical power load
Other – shadow diagrams, sight-line diagrams, elevations, environmental-impact statement

Control of Development (in Conservancy District) submittals: Detailed plans and drawings, elevations of building in its setting, existing site conditions, buildings, boundary walls, topography, trees, and other vegetation. Details of building materials to be used. (Dover, England: Conservation Policy Statement – Kent County Council)

29 Scenic-beauty issues in public policy making

Mollie Ridout

The landscape has been shaped by complex factors, both physical and cultural. One social process that has left its mark on the land is public policy (i.e., government action). Although the making of public policy is not always apparent to the general public, its effects are apparent. For example, government decisions to set aside public land, to construct interstate highways, or to extend sewer lines into rural areas have their effects: areas of natural scenery, isolated nodes of commercial development, sprawling subdivisions. Each decision contributes to shaping the world we see, as well as the way we live.

Visual quality, like clean air and fertile soils, is one of the nation's natural resources (Council on Environmental Quality, 1970). The value that our society places on visual beauty can be seen in the setting aside of national parks and in the immense numbers of people who visit these parks every year. While the national policy of scenic protection focuses primarily on spectacular settings, such as Yellowstone or the Grand Canyon, local governments and private citizens indicate their concern for the visual quality of the everyday environment by establishing architectural review boards, downtown improvement associations, garden clubs, and other such organizations. The economic expression of this concern for beautiful surroundings is clear if we compare the real-estate value of a property with a magnificent ocean view and a home with a view of the house next door. The economic incentive for visual quality also leads towns and cities to refurbish or thoroughly redesign their downtown commercial districts. On a broader scale, scenic beauty provides a critical resource base for the tourism industry, a major component of the economy in a number of states.

However, the development and population growth of the past decades have exerted enormous pressure on the visual environment. Jacob H. Beuscher describes the most serious problem in natural resources as

an increasing demand that open spaces, outdoor amenities, and aesthetically pleasing surroundings be preserved. To achieve these ends there must be orderly growth, prepared for through planning in part by the use of a variety of regulatory tools used singly and in combination. (Thomas, 1972, p. 22)

434

This is a clear call for policy that can protect and manage the natural and cultural elements that we value for their appearance and that constitute the visual resource. The National Environmental Policy Act provided a broad directive for federal agencies to "insure that presently unquantified environmental amenities," including aesthetics, "be given appropriate consideration" (National Environmental Policy Act, 1969). Unlike policies that are concerned with physical objects, policy to protect the visual quality of the landscape concerns what some decision makers consider to be an "intangible." Decision makers frequently regard this resource as too subjective for regulation or management. They may feel that regulation of visual quality could give rise to undue litigation, that it has already been taken care of in the course of managing other resources, that it is a frivolous concern or a matter of taste, or that its proper consideration requires no special action. These arguments all come to bear on the outcome of policy making for the visual resource.

Integral to the professional concern of the environmental-design professions is visual quality of the environment. Thus public policy, because it can affect visual quality in a pervasive and widespread fashion, is an appropriate concern for these professionals. Participation in the policy-making process can be improved only by more information: an understanding of how policy is created, what actors are involved, what issues may be raised by those actors, and how the policy-making process can be influenced.

In this study, a particular attempt to create policy for scenic beauty is described. Factors that influenced the outcome of the policy-making attempt are discussed, and some conclusions are drawn about what actions may be most effective for those who wish to enter the policy-making arena on behalf of the visual resource. The study focuses on a case that occurred at the state-agency level of policy making in Wisconsin. Some of its characteristics are unique to the specific situation. However, the participants raised issues that are characteristic of decision makers' concerns about visual quality. In the field of visual-resource studies, there has been little attention, as yet, to understanding how the special nature of a purported "intangible" like scenic beauty might influence efforts to create public policy. In this circumstance, a descriptive study of a specific case can be useful to generate information and suggest guidelines for future action. Documentation of the case was gathered in structured interviews with participants, tape recordings of agency and committee meetings, and reports and correspondence generated in response to the policy-making process. These data were submitted to a content analysis. The results are discussed within a framework of theoretical concepts drawn from the literature of planning, political science, and social psychology.

A history of NR 1.96 and the Scenic Beauty Committee

Beginning in 1980, a group of citizens attempted to convince the Wisconsin Department of Natural Resources (DNR) to adopt a policy whereby scenic beauty would be considered in a routine manner during the agency's regulatory and management activities. These citizens, because of both personal and professional interests in the visual quality of the environment, had observed and sometimes participated in the DNR's decision-making process and were concerned over a lack of consistency in the agency's treatment of the resource. One member of this group was the Wisconsin Public Intervenor, an assistant attorney general in the Wisconsin Department of Justice who is empowered to intervene in certain proceedings of the DNR in order to protect public rights in natural resources. He brought to the group his professional experience as an environmental advocate. He was joined by four professors from the University of Wisconsin–Madison who were engaged in research on landscape and natural-resource issues. The group represented a complementary array of knowledge and skills. Their collective experience included serving as an expert witness in an adversarial setting, implementing U.S. Forest Service visual-assessment techniques, and participating in state-level policy making as chairman of the Natural Resources Board, the DNR's policy-making body.

During the summer of 1980, they acted on their concern about the agency's consideration of scenic beauty by drafting a proposed administrative rule, which they titled a "Petition for the Creation of Scenic Beauty Rules." An administrative rule is a regulation or policy statement issued by a governmental agency in response to statutory laws. The purpose of such a rule is to make specific the general directive of statutes. The incorporation of scenic-beauty concerns into a broad rule would provide the legal means to challenge any negligence on the part of the agency.

The proposed rule, designated NR 1.96, would have required the DNR to preserve, protect, and restore legislatively and judicially declared rights in scenic beauty. Specifically, the agency would have to do so in five routine activities: in considering permits, licenses, and approvals; in undertaking procedures pursuant to the Wisconsin Environmental Policy Act (i.e., the preparation of environmental-impact assessments); when managing DNR property; when purchasing lands and easements; and when developing rules and legislation. The rule would also have required the department to apply visual management principles routinely in all actions.

The petition was signed by thirteen citizens from throughout the state. The petitioners approached the DNR staff, holding a meeting during the fall of 1980 with the director of the agency's Bureau of Environmental Impact. This staff member reported to the Natural Resources Board, the agency's policy-making body, that

in the absence of a convincing demonstration that there are meaningful methods to evaluate aesthetic impacts and because of the lack of direct or indirect authority, in most cases, to apply such an analysis to Department regulatory and management decisions, I strongly urge the National Resources Board to deny the petition as requested. (DNR, 1981a, p. 1)

The Public Intervenor decided not to take the petition to the board's February meeting, as he had planned. Instead, a small group of petitioners visited individual members of the board to discuss the proposed rule. They emphasized the importance of natural-resource aesthetics as a "new challenge" and the role of scenic beauty as a resource base for Wisconsin's tourism industry. They pointed out the fact that the National Environmental Policy Act requires appropriate consideration of "presently unquantified amenities," including scenic beauty, in decision making. They cited the existing authority for considering scenic beauty, including statutes, administrative rules, and the policies of some bureaus within the agency. They pointed out that the DNR had no comprehensive scenic-beauty policy, no bureau or staff member who specifically took responsibility for dealing with scenic-beauty questions, and only limited procedures for implementation. They stated that this condition resulted in a case-by-case review of scenic-beauty issues, with inconsistencies in the agency's decisions and no guidance for the staff members who must make such decisions. They listed cases that they had observed in which scenic beauty had been dealt with inconsistently. They asked the board members to support a cooperative effort between the petitioners and staff, aimed at developing the agency's awareness of scenic-beauty issues and skills for managing and protecting the resource. Interviews with participants in these meetings, including the chairman of the Natural Resources Board, indicated that the majority of the board members responded positively to this presentation.

As a result, the director of the Bureau of Environmental Impact was instructed to work out a compromise draft of the administrative rule that would be acceptable to both the staff and the petitioners.

The compromise language, drafted on September 25, 1981, made a general policy statement that "scenic beauty be protected, maintained and improved in actions over which the Department has management or regulatory authority." But rather than requiring consideration in the five areas of activity listed in the original petition, the new draft merely stated that the scenic-beauty policy extended to those activities "where permitted by law." It furthermore made it clear that the department was not required to develop written analysis or documentation if it was not otherwise required by law.

The director of the Bureau of Environmental Impact responded to the new rule draft with another memorandum, stating that "this proposed rule will highlight and formalize the generally held feeling within the Department that scenic beauty is important and should be protected" (DNR, 1981b, p. 1). Hearings were held on December 14, 1981, in Madison and on December 15, 1981, in Rhinelander. They opened the policy process to a number of new participants. Present at the hearings were representatives of the Wisconsin Paper Council, Wisconsin Agribusiness Council, Wisconsin Association of Manufacturers and Commerce, and Associated General Contractors of Greater Milwaukee, all opposing the rule. Appearing in favor of the rule were four of the petitioners, Wisconsin's Environmental Decade, the League of Women Voters, and five interested citizens. Six paper companies, five county forestry committees, and the Secretary of the Wisconsin Department of Agriculture, Trade, and Consumer Protection all submitted written testimony against the creation of NR 1.96. The petitioners submitted lengthy documents supporting their position with research findings and refuting the objections lodged by the opposing parties.

The issues raised by these participants related to three general questions: (1) whether such a rule as NR 1.96 was necessary, (2) whether it was within the authority of the department to administer, and (3) whether scenic beauty is a feasible subject of regulation.

On the basis of the hearing testimony, the director of the Bureau of Environmental Impact issued another memorandum, stating that "it is the staff view that the rule is unnecessary, confusing to the public and should not be promulgated" (DNR, 1982a, p. 1).

His argument for not promulgating the rule was based on an identical line of reasoning developed by the Wisconsin Paper Council in its written testimony. This argument held that NR 1.96 did not comply with the statutory definition of a rule because it did not implement, interpret, or make specific the legislation that is administered by the department. This reasoning is an interesting contrast to that in the earlier memorandums. In February 1981, he described NR 1.96 as a rule, although one that the agency did not want; then in October 1981, he described it as a rule that the agency did want. Furthermore, the Legislative Council Rule Clearinghouse had already approved the substantive content and technical accuracy of the proposed rule and stated that statutory authority for the rule did not exist.

The director concluded by recommending that the board adopt a resolution in support of scenic beauty and direct the staff to include scenic beauty, where appropriate, when promulgating new or revised administrative rules. A resolution is a statement of policy that, since it does not have the force of law that a rule has, cannot be used to enforce implementation.

In March, the Natural Resources Board rejected the broad policy for scenic beauty as embodied in NR 1.96. However, the chairman was interested in the possibility of considering scenic beauty on a program-by-program basis. He asked the DNR staff to evaluate the department's authority to consider scenic beauty and the current status of the department's consideration of scenic beauty, and to identify any need for further department consideration of scenic beauty.

This investigation (DNR, 1982b) found express authority to consider scenic beauty in nineteen statutes and implied authority in forty-three statutes and in twenty-two administrative code provisions that dealt with scenic beauty. The report also outlined several opportunities for adding scenic-beauty consideration to existing programs, including the Wisconsin Environmental Policy Act, approvals of waste-water-treatment systems, water regulation and zoning, the solid-waste program and the forestry program.

In response, the board members decided to inquire further into the matter of scenic-beauty consideration. In April, they voted unanimously to appoint the Ad Hoc Scenic Beauty Advisory Committee. This nine-member citizen committee was charged with evaluating DNR authority to consider scenic-beauty concerns in department regulatory and management decisions and with making recommendations to the board on possible administrative rule changes, legislative initiatives, and internal department decision-making procedures that would be necessary to ensure appropriate consideration of scenic-beauty concerns.

The committee met seven times from July through December 1982. The nine members represented diverse interests, including the Wisconsin Paper Council, Wisconsin Agribusiness Council, Wisconsin Association of Manufacturers and Commerce, solid-waste industry, League of Women Voters, Audubon Society, Wisconsin Wildlife Federation, and original petitioners. The director of the Bureau of Environmental Impact served as a nonvoting chairperson. In retrospect, all members of the committee have agreed that the committee failed to achieve its purpose. Disagreement about the issues interfered with the fact-finding process, and consensus was not possible. In January 1983, the committee presented two reports to the board.

The Board voted 6 to 1 to accept the majority report. This report, prepared by the committee chairman, was supported by five of the committee members, including four who represented industrial or business interests. It recommended that scenic-beauty consideration be included in the *Master Planning Handbook,* which guides the management of department property, in the evaluation of land for purchase or redesignation, and in the environmental-impact process. However, it left implementation to the discretion of staff. The report also supported changes in statutes and administrative rules recommended by staff in its April 1982 report. Finally, the report stated that while the survey techniques used by the professors in their research on scenic-beauty management may be useful, the techniques might be used only in situations where scenic beauty was the primary management objective. It also recommended that "scenic beauty quantification techniques cannot be applied to Department regulatory decisionmaking" (DNR, 1983, p. 6). This meant that even where appropriate or required, techniques to determine the visual impacts of or public response to a proposed activity could not be used by department staff.

The minority report, signed by four members of the committee, endorsed most of the recommendations contained in the majority report. However, it specifically took a stand against two of the recommendations: that which limited the use of scenic-beauty survey techniques to situations where it was the primary management objective, and that which did not allow scenic-beauty quantification techniques to be applied in department decision making. The minority members objected on the grounds that there are situations in which staff are required by law to quantify scenic beauty – for example, in issuing permits for dam construction; that these recommendations contradict the National Environmental Policy Act of 1969, which endorses "methods and procedures . . . which will insure that presently unquantified environmental amenities and values may be given appropriate consideration" (NEPA, 1969); and that they would set back the consideration of scenic beauty as well as fail to recognize current research and management of the resource. The report also called the board's attention to more than fifty specific scenic-beauty recommendations that included statutory, rule-making, and manual changes. Of these, forty received no action at all from the committee (Tlusty et al., 1983).

The chairman of the board informally told the signers of the minority report that the board was open to considering a condensed version of their report, with the possibility of developing handbooks or manuals for use within the department. The minority members did not respond to this offer.

Acceptance of NR 1.96 would have had sweeping effects. The proposed rule would have been an explicit mandate to an agency with direct or indirect impacts on the lives of 4.5 million people and 34 million acres of land to consider scenic beauty in many of its actions. NR 1.96 not only would have affected the public land managed by the DNR, but also would have carried scenic-beauty consideration to the private sector through DNR decisions on permits for bridges, dams, waste-treatment systems, and numerous other projects. The defeat of the proposed rule delays the explicit recognition of scenic beauty as a resource with the stature of air, water, soil, and wildlife. Nonetheless, the net result of the policy-making effort is still significant. A body of state-level policy makers was willing to hear and respond to the scenic-beauty issue over a period of two and a half years, to initiate information gathering on the subject, and to remain open to further proposals despite staff opposition. Such a channel provides crucial access to the decision-making process. Increased staff awareness of scenic-beauty issues and management techniques is an inevitable result of such an extended campaign, although one that is difficult to measure. Sensitivity to the issue can be assessed only as the agency continues to make management and permit decisions and amends or promulgates administrative rules.

What lessons can people who are concerned with the protection of visual quality draw from this case? The petitioners for NR 1.96 became aware of a number of factors in the policy arena that would influence the outcome of their endeavor. Some of these factors are typical of the policy-making process in general. General theories about the nature of special-interest groups and public-interest groups and about the elements of successful advocacy are found to have bearing in the case of NR 1.96. Other factors in this case are peculiar to the scenic-beauty issue. Objections about the ''subjectivity'' of the resource and the use of quantitative methods were prominent in the discussion of participants. Compounding the situation was a political climate that favored reducing the rule-making agenda and placing more decisions in local hands.

Advocacy and the policy arena

Interest groups are a major source of nongovernmental demands on decision makers. When the forestry and paper industries, agribusiness and manufacturers became involved in the scenic-beauty case, they made it clear that certain types of interest groups wield a great deal of power. Special-interest groups such as these arise out of the need of individuals to organize and take collective action if they are to make an impact on public policies and decisions. These groups have highly concentrated benefits to be derived from particular governmental decisions.

Because of their business orientation, none of these groups welcomed regulation. They saw scenic-beauty consideration as difficult to regulate, with the potential for interfering with business development and operation. The forestry and paper industries objected the most, fearing that such regulation would curtail clear-cutting practices in northern Wisconsin.

The issue of benefits distinguishes special interests from the public interest. The public interest involves benefits that are so diffuse and widespread, it is not worthwhile to the individual to take action. This is an example of ''market failure,'' where actual demand for a collective good does not get transmitted accurately to the supplier (Ranney and Nasoff, 1972). The scenic-beauty petitioners embody the idea of market failure in that they represented no identifiable constituency, although they expounded general values that drew vague endorsement from almost everyone.

The difference between special interests and the public interest has an important effect on how these groups relate to a regulating agency like the DNR. Special-interest groups, on the basis of expecting highly concentrated benefits from certain agency decisions, find it worthwhile to invest in representation in the form of attorneys and lobbyists. Also because agency regulation can affect their economic success, these groups are inclined to develop a long-term, stable relationship with the agency. This relationship is fortified by the group's technical expertise, on which the agency, lacking resources to do its own research and testing, may come to rely. The nature of the special-interest group leads to frequent, enduring, well-defined and mutually beneficial contacts with the agency (Ranney and Nasoff, 1972).

In contrast, the public interest, with diffused costs, is not as easily organized as special-interest groups. The lack of payoff for the individual means that the public interest is not represented as persistently as special interests. The cost of information and short-term interest in localized problems minimizes the individual's participation. The lack of organization and representation, lack of expertise, and lack of long-term, stable relationships with agency personnel detract from the influence of the public interest.

This imbalance in the ability of different groups to gain access to decision making has important implications for public policy. The agency tends to develop policies that are responsive to the concerns of special-interest groups, since those are groups that engage their attention most. In this case, there appears to have been a shared understanding between the special interests of the paper industry and the agency. The director of the Bureau of Environmental Impact relied heavily on the reasoning of the Wisconsin Paper Council's representative in preparing the March 1982 memorandum, which rejected NR 1.96, even to the point of directly quoting the council's position paper on the rule.

In the scenic-beauty case, however, the Public Intervenor had an official role that had affected departmental decision making in the past. In addition, the professors and the representatives of the League of Women Voters and the Audubon Society had had experience with the agency in policy-making or planning procedures. The petitioners had gained access to top-level decision makers when they presented their ideas to the members of the Natural Resources Board.

Given this starting point for the scenic-beauty advocates, what conditions led to the defeat of NR 1.96? Access to key decision-making points is important, but there are other elements that support successful advocacy. In addition to (1) access, there are (2) mastery of skills and (3) constituent support (Trubek and Trubek, 1979).

The petitioning groups offered a mixture of skills, with the Public Intervenor bringing professional advocacy skills to the issue, while other petitioners brought technical expertise. However, both agency officials and interest-group representatives found it difficult to understand the advocates' ideas. The majority felt that specific examples of situations where scenic beauty had not been properly considered and examples of how the resource could be evaluated and managed would have clarified the proposals that the advocates were making. The advocate needs to be able to speak to the decision makers' situation, supplying technical information that answers their needs in terms that are congruent with their experience. Trubek and Trubek (1979) point out that a mastery of technical issues and an ability to present them in a flexible format are two of the strengths of a skilled advocate.

Another important skill is the ability to participate in the give-and-take of negotiation. Interest-group representatives must have a shared concern and a real interest in finding a resolution to the problem. They must be willing to enter into a true exchange of ideas in order to let other parties know where they stand (Dawson, 1981; Derouin, 1981). They must be willing to listen to and respect the position of other representatives, and willing to accept the fact that the resolution may be a middle ground that cannot be predicted or determined in advance.

A number of participants criticized the advocates for having taken an aggressive, inflexible stand on their issue. The industrial representatives who were members of the Ad Hoc Scenic Beauty Advisory Committee had a past record of negotiating successfully on environmental issues. Several elements of successful negotiation were missing in this case – the discovery of a common interest or an area of agreement that could serve as the basis for developing further agreements, and the frequent, sustained contacts that could lead to a negotiating attitude, a willingness to participate in the give-and-take of working out a solution.

Advocates have to develop this flexible attitude within a carefully chosen arena, however. The advocate's experience, skills, and understanding of the special-interest groups must be brought to bear in analyzing the situation in which the negotiations will be carried out. It may be necessary to refuse to participate in an arena where the advocate will be at an obvious disadvantage. The Scenic Beauty Advisory Committee appears to have been such an arena. Its membership included a large number of business and industrial representatives, at least one of whom had already demonstrated strong opposition to the issue. It was chaired by an agency representative who had been inconsistent, unhelpful, and usually in opposition to the advocates throughout the policy process. Advocates have to be aware of the arenas in which their skills make them most effective and capitalize on those skills, maintaining access to the decision-making process at points where they can be most effective.

A third element of successful advocacy is the ability to demonstrate constituent support (Trubek and Trubek, 1979). The issue of a constituency for scenic beauty was raised by several participants during the Scenic Beauty Advisory Committee meetings. Industrial representatives who challenged the need for scenic-beauty consideration pointed out the advocates' lack of a constituency as proof that such consideration was not needed.

The scenic-beauty advocates had chosen early in the process not to develop a constituency, because they felt that the issue could be handled best by providing well-researched information to the agency and avoiding the emotional atmosphere of a confrontation between constituents and staff. Given the special-interest representatives' skepticism about whether a problem really exists, a constituency is a useful tool. Constituents emphasize the existence of a problem and lend credibility to the advocate's claim that it is a significant problem. A constituency helps create the pressure that makes it necessary for all participants to commit themselves to finding a solution. This pressure was clearly lacking in the scenic-beauty case.

The difficulties that the scenic-beauty advocates encountered in the policy arena are consistent with theory and research findings concerning policy implementation. The subtext to their quandary is the "fact/value dilemma" (Rein, 1976; Simon, 1945). If facts, the empirical products of social-science research, are to contribute to the development of public policy, they must connect logically with the values, the ends or objectives, that inform the policy-making process. Success depends on these values being shared, and "the political process is the primary means by which shared values are created" (Rein, 1976, p. 95). Lindblom and Cohen (1979) say that scientific knowledge is only supplemental to the "ordinary knowledge" (originating in common sense, causal empiricism, or thoughtful speculation) on which decision makers rely. They recommend "interactive" methods of problem solving as being more productive than analytic methods alone.

Baum (1980a, 1980b) has commented on the need for political interaction in developing policy. Although his study focused on planners, the planners' self-definition as intellectual problem solvers, attempting to bring about broad, large-scale changes in the physical or social environment, suggests that they play a role analogous to that of the scenic-beauty advocates. Baum's research identifies a disjunction between the planners' perception of their own goals and expertise and their power to bring about implementation. He draws the conclusion that this problem stems from the planners' failure to appreciate the role of the organization as a political milieu. To be effective, it is necessary "to see problems as implicated in a web of perceived interests and to think of remedies in terms of strategies for organizing coalitions of those interests for particular courses of action" (Baum, 1980a, p. 493).

Meltsner (1979) presents some guidelines for operating in the political milieu, which include

1. defining relevant policy space and issue area
2. identifying interested actors with their motivations, beliefs, and resources
3. locating crucial sites of decisions
4. looking for likely areas of consensus and conflict affecting possible exchanges on issues.

Meltsner suggests preparing for an advocacy role by drafting alternative scenarios based on information developed under each of the categories. Krumholz, Cogger, and Linner (1975) give examples of the efforts by the Cleveland City Planning Commission to shape local policy. Their experiences lead them to conclude that effectiveness depends on a willingness to take an activist role and engage in protracted, vocal participation and on an ability to offer something that decision makers want and can relate to.

Forester (1980, 1982) likewise gives practical recommendations that rely on interaction among people rather than "the power of documents" and isolated technical work. Forester's work has focused on communication, the control of information, as a key factor in effective political participation. Developing good communication skills depends on recognizing the extent to which participants are vulnerable to the management of comprehension, trust, consent, and knowledge. He concludes that effectiveness depends on the ability to

1. cultivate liaisons and contacts
2. listen carefully to gauge concerns and interests of all participants
3. develop skills to work with groups and conflict situations rather than expecting progress to stem mainly from isolated technical work
4. encourage participation and activism of interested citizen groups
5. anticipate external pressure and look for countervailing pressure (Forester, 1980).

In summary, some factors that affected the scenic-beauty case are typical of the advocacy arena and must be addressed if the advocacy effort is to be successful. Advocates must understand the political and interactive nature of their task. Success depends on the development of an effective relationship with the agency in order to counterbalance the natural advantage that special-interest groups have. The effort to develop such a relationship should be focused on key decision makers, and an effort must be made to identify their concerns and beliefs. The advocate needs to identify the arenas in which decisions will be made and capitalize on interactions that have proved successful, such as one-to-one meetings in this case. Negotiation is the primary mode for building shared values and creating a receptive audience for policy proposals. Successful negotiation depends on a willingness to look for areas of agreement and on an understanding that the solution may lie in a new and unpredicted middle ground. However, negotiation needs to be carried out in carefully chosen arenas where the advocates can capitalize on their skills. Where the advocates have difficulty convincing others that their issue is serious, a vocal constituency and carefully chosen and graphically presented examples of the problem are essential tools. Interested citizen groups need to be encouraged to take an active role, and external pressure, such as that applied by the paper and forestry interests in the scenic-beauty issue, have to be anticipated. Rather than attempting to avoid this kind of pressure altogether, the advocates need to identify countervailing pressure – other interest groups that can be brought into the arena.

Advocacy and the visual resource

Other factors that affected the outcome of the case were peculiar to the scenic-beauty issue. In hearings, meetings, and staff reports, participants made it clear that they considered scenic beauty to be a subjective entity and found it difficult to accept the idea of measuring and regulating the resource. Staff members trained in other fields, primarily the biological sciences, did not find scenic beauty amenable to traditional procedures of quantification and regulation. The advocates' educational material did not alter these beliefs. It seems highly probable that such concerns will be expressed in any case of policy making for the visual resource.

Participants expressed difficulty in understanding what problems existed with scenic-beauty consideration and how these problems could be rectified in the agency's day-to-day decision making. They repeatedly called for specific examples of mismanagement and illustrations of alternative solutions to such problems. They also asked for examples of how agency staff could make judgments about and prescribe management of the resource. The advocates' inability to address concerns such as these seriously undermined the credibility of their effort. The ability to speak to the decision makers' situation, supplying technical information that answers their needs in terms that are congruent with their experience, is an essential ingredient in effective advocacy. The question of how scenic beauty could be quantified was raised so frequently during the case that it was incumbent on the advocate to provide specific examples of how quantification could be carried out within the framework of the agency's routine activities. If it is not possible to do so, the advocates will have to demonstrate that quantification is unnecessary or that it can be circumvented by some other evaluative technique.

The advocates have raised two points in response to these issues. They suggest that objections on the part of the agency were not straightforward expressions of the agency's concerns, but were excuses to cover the fact that a key agency official had personal differences with the advocates and was sympathetic to the forestry and paper industries, which opposed the regulation of scenic beauty. The advocates also question whether developing a set of examples could convince industrial representatives that the agency has not considered scenic beauty. It is in the interest of these representatives not to acknowledge that a problem exists, regardless of whether they believe it or not. Although these points are speculative and are peculiar to this case, they cannot be discounted as a possibility in future cases. To the extent that these points may be true, they serve to reinforce the idea that maximizing the advocates' access to sympathetic decision makers and establishing a nonadversarial negotiating relationship with the special-interest groups would be profitable endeavors. By acquiring knowledge about potential participants and analyzing their positions, it may be possible to reduce the influence of such adversaries by finding a way to enlist their cooperation or by looking for alternative channels of access to the decision-making process.

An unavoidable problem in this case was the political climate at the national and state level engendered by President Reagan's "new federalism." This climate was carried into the agency when Governor Dreyfus appointed several Natural Resources Board members who favored reducing the rule-making agenda and placing more decisions in local hands (Jordahl, 1983). Such a political atmosphere obviously bodes ill for the creation of broad policy for the scenic-beauty resource and suggests that a delay in renewing the policy-making effort may be appropriate.

Another factor also suggests the wisdom of waiting for a more opportune time. One of the advocates has described a progression in creating new policy for the visual resource. The process begins with awareness of the issue and proceeds to a program for mitigating adverse impacts on visual quality before it approaches the level of broad management objectives that recognize scenic beauty as a resource on the same level as soil, water, or wildlife. Although a broad policy that encompasses statewide consideration of scenic beauty on a level with other natural resources is the most desirable outcome of an advocacy effort, the events of the NR 1.96 case suggest that this policy process is still primarily at the awareness stage. Some procedures are in place for mitigating adverse impacts, and through management plans for several of Wisconsin's rivers, scenic-beauty policy is being attempted on a geographically limited scale. But it appears that for the time being, those who will figure in the development of scenic-beauty policy must have their awareness and understanding of the issue heightened with specific examples of inadequate protection. An evaluation of how the petition for NR 1.96 has increased the DNR's awareness is yet to come. This is the time to let the seeds of that awareness grow and to prompt that growth by introducing specific cases that require agency consideration. With the accumulation of these cases and the advent of more accommodating political conditions, the appropriate time for initiating broad policy for scenic beauty will arrive.

Recommendations

Although it is not possible to make unimpeachable generalizations from a single case study, some peculiar aspects of the scenic-beauty issue make it highly inviting to at least attempt to draw general observations out of the case that could have application in other situations. The study of the visual resource is a relatively new undertaking, and knowledge about the resource is seldom transferred to those who make political decisions. Therefore, the existence of even a small amount of case-specific information directed toward policy making for the visual resource may be significant for the professional who wishes to enter that arena. The following recommendations are offered as a summary of lessons that might be drawn from the evidence presented by the case of NR 1.96 and the Ad Hoc Scenic Beauty Advisory Committee. They are presented in the hope that they may guide future advocates for visual-resource quality in preparing and carrying out a foray into new territory.

- Be prepared to deal with special-interest groups. Know who they are and what stake they have in the regulation of visual-resource quality.
- Be prepared to convince them that a problem exists by using specific examples with graphic illustrations of existing situations where there have been adverse impacts, and simulations of proposed solutions.
- Be prepared to enter a negotiating situation in which flexibility and willingness to exchange information are important. Be prepared to look for areas of agreement and to establish frequent, sustained contacts with other participants.
- Choose the arenas in which you will negotiate. Capitalize on those arenas where your skills have the most effect. Refuse to enter an arena where the odds are obviously against you.
- If other participants are not responsive to broad policy recommendations, be prepared to negotiate at a level of site-specific cases.
- Develop a constituency. Target groups with appropriate interests, and build a base of support through the group's opinion leaders. Educate the group on your issue, and provide an agenda for action on their part.
- Discover which officials are sympathetic to your issue, and capitalize on one-to-one contacts with them. Concentrate on key decision makers who have the power to force a resolution of the issue. Discover which officials are not sympathetic, and look for alternative channels of access to the agency.
- Pitch the information that you deliver to the needs of the agency. Frame recommendations in terms that are compatible with the agency's routine activities and that are understandable in terms of day-to-day decision making.
- Develop examples that demonstrate the existence of a performance gap in the agency's treatment of the resource and that demonstrate alternative consequences to typical management and regulatory decisions.

Acknowledgment

I wish to acknowledge the assistance of Dr. Richard E. Chenoweth, Department of Landscape Architecture, University of Wisconsin–Madison, in defining the research topic and developing the research methodology.

30   Coping with aesthetics and community design in rural communities

Fred A. Hurand

Introduction

The aesthetic of small rural communities, as with any other community, is a function of their historical growth and development. While one would expect that the concepts, ideas, and methods of planning and design that operate in major urban centers could easily be converted and applied to these small urban spaces, in reality, this is not true. Scale is not the most dramatic difference. Rather, it is the dynamics of change that make comparisons difficult. Unlike their larger urban counterparts, these towns have not developed into large, dynamically changing organisms. Rather, their scale and size have been arrested at some earlier stage of development, leaving a simple comprehensible environment that is not only aesthetically but also, to many, socially pleasing. The physical form of these towns reflects their homogeneity of life style, attitudes, and values. Stability or decline is a better description of their dynamics. Thus although there are indeed shifts in scale, there are also major differences in problem focus that challenge traditional planning and design methods.

This paper is not intended to present new research. Nor is it intended to develop new theory. It is a collection of thoughts about the effectiveness of current research and planning tools that may be applicable in rural settings. These thoughts result from this author's personal experiences working with smaller rural towns on the Columbia Plateau of eastern Washington. Working with communities to solve local problems produced a skepticism about many of the methods utilized to assess the aesthetic qualities of environments. Few research techniques and planning methods seemed applicable. Working with a community is an important concept. It assumes an interactive affair with citizens of the community. In fact, the process must be transactive, whereby information, problems discovery, and solutions are shared between the planner and the community (Friedman, 1973). This concept casts some shadow of doubt on the efficacy of many of the more traditional assessments of aesthetic quality. More about this later. First, the stage must be set.

449

The forces of stability and decline

The setting of these reflections is the vast prairies and deserts of the Columbia Plateau in eastern Washington. Towns are a direct result of the economic forces that dominate any region. In this area, dryland farming is the heart of a productive agricultural system that shaped the location and character of the towns within the region. The discovery of the utility of this vast natural resource led to rapid colonization of the area from the 1870s to 1910. The first significant communities were located around the major segments of the Columbia and Snake rivers, the core of the major transportation network for distributing wheat to a larger market area. When the Northern Pacific Railroad crossed the Rockies in the 1880s, the focus of town development shifted to the rail network, which opened a larger region for agricultural development. Thus the fortunes of these towns were now linked not only to natural resources, but also to the entrepreneurial interests of the railroads. To improve its profits within the region, the Northern Pacific promoted the Pacific Northwest in the Midwest, inducing greater migration to the area. Seizing on this growth, local land developers created towns throughout the region, and communities competed with one another for larger numbers of the incoming migrants (Meining, 1968).

Most growth has limits. As the agricultural industry expanded, as crop practices improved, and as mechanization arrived, dramatic changes occurred in these communities. By the late 1920s, community growth peaked and declined (Table 30.1). Ranchers no longer needed large pools of seasonal labor. Workers migrated away from the major farm centers to larger urban areas. Thus began a protracted period of decline for these towns, ending with a relatively stable population that persists to the present.

There are specific economic realities of living in these communities that affect the aesthetic appearance. First, the economic catchment area is limited to the residents of the town and of the immediate agricultural area. The town is the primary source of goods and services for this area. Early in the community's life, the additional town and catchment-area population required a larger share of goods and services than does the current population. Commercial space and structures expanded to fill this need. Outmigration created excess space. The result is underutilized or vacant commercial buildings.

This problem was exacerbated by improved highway transportation, coupled with increased mobility associated with the automobile, producing a shift in the location of purchases for specific commodities. In the past, these minor centers were the nuclei of shopping for inhabitants of the town and the immediate rural catchment area. The ability to travel greater distances in order to make major purchases in larger urban centers, where prices and selection were more competitive, created a concomitant contraction of services within these towns. Table

Table 30.1. *Representative decennial censuses*

Town	Year and population							
	1910	1920	1930	1940	1950	1960	1970	1980
Almira	368	450	339	466	395	414	376	349
Creston	308	317	216	281	268	317	325	318
Garfield	932	776	703	674	647	607	610	599
Ritzville	1,859	1,900	1,777	1,748	2,145	2,173	1,876	1,800

Source: Office of Financial Management, State of Washington, *The Pocket Data Book* (1980).

30.2 shows regional shopping patterns for two of the larger communities within this region. This shifting of buying locations is not unique to this area. Pollman and Wishnick (1980) report similar findings in Wisconsin. Their study concluded that the two main reasons for seeking commodities outside the immediate town were a wider assortment of merchandise and lower prices.

There are other problems associated with the economics of a small town. For example, health services are limited or nonexistent. Doctors have found it unprofitable to rely on a limited economic base for survival. Thus they have left the community for a more lucrative location. Speculative housing starts are marginal. The costs of building housing exceeds the market value for used housing, especially when there is little potential for selling a speculative unit.

These economic factors also affect local government. Local government depends on local taxes to support service delivery and capital construction. Since many of these towns were built early in the century, their capital infrastructure is antiquated and, in some cases, dysfunctional. Minor improvements through the years have not kept pace with deterioration. Available local economic resources limit the ability to pay for substantial improvements, leaving the towns dependent on external funding through grants from state and federal agencies. Since these resources are also limited, major improvements are delayed. This can place the community in jeopardy, with agencies demanding improvements because of environmental hazards. When the situation becomes intractable, the community begins to look for some form of growth but has very little potential for receiving it. Valuable resources are expended in attempts to attract new industry and business, with virtually no payoff.

Stability can also yield positive economic benefits. During the economic recession of the early 1980s, employment in this area remained relatively stable. The three major agricultural counties in this region had an average unemployment rate of 6.2%, much lower than the double-digit unemployment of the state. One county had the lowest unemployment in the state, 4% (''Unemployment

Table 30.2. *Regional shopping patterns*

| | Location of percentage of purchases by commodity | | |
Town and commodity	Local	Spokane	Other
Ritzville			
Hardware	86	9	5
Farm supplies	67	6	27
Major appliances	55	39	6
Clothing	41	52	7
Wilbur			
Hardware	43	28	19
Farm supplies	59	21	20
Major appliances	41	48	11
Clothing	18	70	12

Note: Spokane is the major regional urban center and is within sixty to ninety minutes' driving time.
Source: Surveys conducted by the Department of Urban and Regional Planning for the Ritzville and Wilbur Planning Commissions in their comprehensive planning processes, 1978 and 1979, respectively.

rates,'' 1982, p. 13). In other words, the region is able to withstand larger shifts in the national economy that affect other areas. As a contrast, those rural Washington counties that are adjacent to the agricultural counties and relied on the forest industry during this period experienced unemployment in excess of 20% and have been less resilient in rebounding from this recession.

The aesthetic form of rural towns in the region

What effect do these factors of growth, decline, and stability have on the aesthetic form of these communities? Traditional aesthetic research is limited in its ability to explain the forces that shape the aesthetic form. Much of the behavioral research used to assess aesthetics focuses on human reaction to existing forms. Some researchers have also used the results of this research to guide planning decisions, particularly in natural settings (see Smardon, Palmer, and Felleman, 1986). Although it is important to assess these reactions in order to develop fundamental relationships between perceived pleasure or displeasure and the environment and its forms, action to shape these forms also depends on knowing how to intervene in the forces that create these forms.

A better understanding of these forces can be gained from a framework derived from Spreiregen (1965) and Wolfe (1975). The forms that ultimately appear in an environment begin with the underlying natural features of a place. These features are then transformed by a series of human activities that mold and shape the environment. The taxonomy of these human-made determinants of urban form include private entrepreneurial decisions, land values, tax structures, life styles, public plans and policies, regulations, capital-improvements budgets, and circulation plans. Although these determinants of form may be obvious, they are, in fact, the variables that influence aesthetic outcomes and the cluster of factors that the planner or designer has the ability to influence and, thereby, shape the future character of the environment.

Several of these determinants of form have particular application to the forces that shaped the aesthetic form of small towns on the Columbia Plateau. The period of greatest growth within these towns parallels the height of homesteading in the West. This national development policy encouraged migration to the region and, thus, the growth of its communities. Linked to this policy was the national system of land subdivision, organized around the township and range grid-coordinate system established by the Northwest Ordinance of 1787. This system subdivided the West into 6-mile-square grids, regardless of the underlying terrain. These grid squares were further subdivided into square-mile sections, facilitating land transfers and becoming the framework for town sites throughout the region. Land developers acquired these sections of land and inserted yet another uniform grid pattern of blocks and streets, which served as the matrix on which the human-made environment was created. Given that agricultural lands were valued for their resources, the land occupied by the towns was normally the least productive, lying in bottom lands and ravines or on hillsides.

This gridiron matrix provides a convenient means for reading and comprehending the structure of these towns. It also provides a continuity of form from community to community. This continuity is also inherent in the spatial distribution of land uses. Bisecting each town is a major highway that serves as the spine for commercial development. This development is rarely more than a block in depth and extends in varying lengths along the highway. Somewhere within a block or two of this main axis is the railroad and the focus of the town's major economic activity, the grain elevator. The elevator is also the dominant spatial form of the town, towering several stories above all the other structures within the town. Extending from these major activity centers are residential areas of the town, which decrease in density toward the periphery of the community. At its edge is a mixture of residential and farm activities that provides a useful transition to the rural agricultural areas, creating harmony between the larger landscape and these towns.

The commercial areas represent each town's economic role in the region and express the forces of growth, decline, and stability. For those towns that have remained the major foci of various subregions, their relative size and the attractiveness associated with their catchment areas maintain a moderate level of business activity. The smaller communities have been less fortunate, and their business centers have contracted to meet smaller demands. Regardless of size, each town has its proportionate share of vacant land and buildings, standing as a testament to its economic importance.

The aesthetic results are an interesting mixture. The primary buildings within these commercial centers were constructed during each town's period of greatest growth. Some have acquired modern appliqués to their façades. Others stand vacant and decaying. The result is an elegant foundation defaced by obtrusive interventions and deterioration. Thus the forces that would normally ensure a vital, changing aesthetic have been removed. Decline and stability are written in the faces of these buildings.

Standing in quiet contrast are the residential areas. These areas also reflect the town's stages of development, but represent a wider temporal range of design and construction. Each era of vernacular residential architecture is represented within the larger fabric of the subdivision system. Although there is a mixture of residential types, the dominant form is the single-family detached home. Unlike larger urban centers, where residential land uses are separated and segmented, in these rural communities, apartments and mobile homes are interspersed with the single-family units. The density of homes within the original platted core is greater than that within newer developments on the periphery, expressing the original and current subdivision-platting systems.

The public buildings within these towns serve as symbolic foci. The major public facility is the school, which serves as the social nucleus of the community, if not of the greater region surrounding the town. The school is generally located within the residential area, along with the churches that represent a variety of denominations. The city hall is normally a modest structure within the commercial center of the town. For those communities fortunate enough to be county seats, the courthouse becomes a dominant feature. The capital infrastructure is less visible, but one of the major problem areas for the community. Aging water and sewer systems suffer from lack of capital for repairs.

The resultant aesthetic form is not unlike that of other rural agricultural areas of America. Melnick (1977) identifies a similar pattern in a region of Kansas. He isolates several factors that provide a common heritage for these Kansas towns: an initial period of major growth, an organic growth pattern, ties to the immediate agricultural region, a split between the agricultural and the commercial business areas, automobile orientation, uniform building heights, common building materials, cafés within the commercial cores that serve as local meeting places, a bank in each commercial district, gasoline service stations on the edges of the commercial core, and newer residential districts standing in contrast to the older residential areas proximate to the commercial core (Melnick, 1977, p. 6). Most of these descriptors would characterize the towns on the Columbia Plateau.

The consistency and coherence of the pattern and form of the towns within this region are symbolic of the common underlying heritage. They were created in the same era of town building, flourished on the resources of this era, and then contracted to fit more recent economic realities. The resultant appearance and form are a unique combination of pleasant small-town living tarnished by minor decay.

Planning and design in stable rural towns

Large growing communities must cope with problems associated with an expanding and changing environment. The focus is on planning long-range futures and setting plans and policies to ensure that growth is directed into proper areas, that this growth does not exceed the community's capacity to provide services to these areas, and that the resultant aesthetic environment meets community standards. What happens when there is no growth or there is even decline? Are there no problems to cope with?

There are many community problems in small rural towns. However, the attention shifts from the long-range to the immediate. Decline creates a different set of problems. Foremost of these is the lack of sufficient economic base to support local business and the public services demanded by the population. For example, as the towns on the Columbia Plateau grew, capital infrastructures were created to cope with increasing demand. As they contracted, little was done to maintain these systems. Now ancient water and sewer systems suffer from infiltration. Available economic resources limit the community's ability to pay for the repair of these facilities. Problems exist, but the methods to cope with them are far different from those available to communities with greater resources.

Aesthetic concerns begin to lose their prominence in an environment with more immediate and pressing problems, but they are not neglected. Community pride is linked with appearance. The deterioration of the town's commercial center is always a visible and constant reminder of its inability to change its economic role. Appearance becomes a symbol of the town's health. Regardless, these small rural towns represent a significant aesthetic resource of America. Given the number of these communities across the Columbia Plateau and throughout the United States, we cannot afford to lose this resource (Wildrick, 1978, p. 318). But if we are to save these aesthetic resources for towns with limited financial resources, how are we to do it?

Small towns do have resources that can help create an atmosphere for revitalization. The most valuable of these resources is self-help. Small rural communities rely on self-determination coupled with community self-help. Collective action is much easier in places with homogeneous values and attitudes and a strong sense of local identity, the mainstays of small rural towns. This has been demonstrated by their ability to stage annual fairs and rodeos and, in some cases, even to undertake civic-improvement projects (Denman, 1981). The focus is on small achievable projects. These efforts can be converted into effective local action to preserve local businesses and aesthetic resources. This requires a different approach to planning and design, one that identifies individual problems rather than comprehensive solutions. It also requires the planner to interact effectively with the community to identify these problems and to develop the means to solve them. As Toner (1979) notes, the citizens of small towns can clearly define their problems but may have little interest in general comprehensive planning. These two elements, coupled with the ability of these communities to effectively act cooperatively, can be the keys to planning in these communities.

It is particularly important to use a transactive planning model (Friedman, 1973) in small towns. Dialogue and mutual learning ensure a common understanding of problems and local values (Nellis, 1980), not only between the planner and the citizens, but also among the residents themselves. This attitude should be carried throughout the process to include the development of solutions as well as implementation programs. The issue of planning with, rather than for, poses some serious questions about tools for assessing and managing aesthetic resources within these communities. A review of the literature in aesthetic research and visual-resource management shows an apparent lack of interactive mechanisms by which solutions evolve from a participatory approach to solving aesthetic problems. Most research attempts to define theoretical concepts and linkages between the attributes of the environment and the cognitive processes of internalizing information (i.e., S. Kaplan, 1975; Wohlwill, 1976). Although this is useful knowledge, the translation of this knowledge into action presumes that the only contribution of the citizens in a setting is their role as respondents. A similar argument can be made about tools used in visual-resource management. Most of these methods are utilized by trained resource managers and are applied in settings where the public has little interaction with the decision makers. This is especially true with the U.S. Forest Service and other federal agencies involved in visual-resource management (see Smardon et al., 1986).

Returning to the issue of small-town aesthetics and design, a variety of interactive techniques have been developed to isolate problems and suggest solutions. Wildrick (1978) suggests an adaptation of the approach of the American Institute of Architects Regional/Urban Design Assistance Team. He states that bringing a team of designers to town may be too costly. His solution is to limit this visit to one or two professionals in a much more restricted time frame. The professional's responsibility is to develop, along with the community, a simplified visual survey designed to identify critical problems that can then be utilized as a foundation for developing solutions at a later date. Kasprisin and Stouder (1983a, 1983b) have elaborated on this model and developed what they refer to as a small-town design studio. This method employs the services of a designer and a community-development specialist. A three-day design charette is established in a drop-in center within the town. People may come in and suggest problems and solutions. The designer acts as pencil for the community by expressing suggestions in plan and three-dimensional form. At the end of the workshop, the community has a working document that can be used to develop implementation programs for community improvements. Sanoff (1981) describes a technique that utilizes a drawing of street façades with missing buildings. Participants in his workshops slide a card with a variety of alternative forms and façades past a hole in the drawing, which reveals how each alternative would appear in relationship to the whole street.

All these techniques encourage interaction between the designer and the citizens of the community. Each method also points to specific problems and solutions. Attention can then be given to these problems and solutions, dividing them into those that are immediate and achievable and those that may need more resources programmed over a longer period of time. Cooperative community action can then be applied first to the immediate and achievable and then to longer-range issues. It is important that action is taken on achievable elements. This takes advantage of community awareness and initiative. It also creates visible results. Masterson (1981) and Sieberling and Masterson (1981) describe two projects, one in Iowa and one in North Carolina, that produced design results utilizing the initiative and resources of active community groups, such as business associations and church groups.

Longer-range solutions can be tied to community capital-improvements programming. Scheduling design improvements that coincide with improvements to the local infrastructure can achieve visible results without excessive impacts on public budgets. This capital web of public improvements can become an impetus and framework for private redevelopment. These improvements can be focused on specific problems and scheduled to ensure that the costs match projected local public income, with each scheduled improvement taking another small step toward specified design objectives.

Conclusion

The aesthetics and ambience of small-town America are valuable resources. The maintenance of these resources is important. For those rural small towns that have less potential for growth than those that are more advantageously located, special care is needed to ensure their longevity. Although they may not have the potential for growth, they do have inherent potential for change and revitalization. Their resources are limited; therefore, it is essential that any action be positive and effective.

Planning for change in these communities must include the citizens of the community. Since financial resources are limited, the collective action of the residents is the most valuable resource available to the community. Their involvement ensures that any solutions will be intimately linked to their perception of their community and that they will be involved in the implementation of programs designed to achieve these solutions. Any plan must include specific achievable objectives that have the greatest potential for completion. Collective action must be attached to these specific objectives by assigning tasks to individuals and groups that can act on them. The completion of each task builds greater confidence in tackling larger, more complex problems. The result is a more effective strategy for community revitalization that leads to the preservation of a valuable national resource – the rural small town.

31 Toward theory generation in landscape aesthetics

Fahriye Hazer Sancar

The ancient Greek philosophers classified human pursuits into four exclusive and exhaustive categories: truth (the scientific), plenty (the political-economic), virtue or fairness (the ethical-moral), and beauty (the aesthetic). Today, there exists little evidence that contemporary people have greater ability to produce or enjoy beauty than their predecessors (Ackoff, 1975). The planning and management fields, as well as the fields of human perception and behavior, have devoted considerable attention to studying and interrelating the first three pursuits. Aesthetic issues in planning, however, have not enjoyed the same popularity. Eventually, the idea that aesthetics is antithetic or hostile to the other pursuits has gained a widespread and almost instinctive acceptance (Ackoff, 1975).

This claim can readily be substantiated by a quick survey of the mainstream planning approaches represented in popular textbooks. The normative-rational models of planners, designed to provide logical decision processes, have ignored the issue of aesthetics altogether in favor of the first three pursuits. In consequence, the normative planning techniques and models in public-policy planning have not benefited the social scientist interested in incorporating aesthetics into decision making. Having been excluded from a rigorous modeling framework, the aesthetic dimension has been treated with the overriding objective to generate "defensible," a posteriori evidence to define intrinsic aesthetic qualities for visual-impact assessment of environmental developments. This latter scenario has been a reaction to the planners' omission, promoting an adversarial position in support of the belief that the aesthetic pursuits of humans are indeed in direct conflict with the remaining three. In turn, social scientists have sought agreement on aesthetic preferences via public surveys in which all other relevant issues – economic, ecological, political – are excluded. As such, survey information has been of little value to planners in incorporating aesthetic issues into public decision making.

Incorporation of aesthetic concerns into public policy has been primarily

This paper was first published in *Landscape Journal* 4 (1985): 116–24.

through environmental-impact statements or cost–benefit analyses undertaken by various agencies as a response to the National Environmental Policy Act (NEPA) in 1969. Since then, several attempts have been made to evaluate the aesthetic aspects of proposed projects concurrently with technical, social, and economic factors in such fields as land-use planning, forestry, water resources, highway planning, park services, and public-utility planning (Redding, 1973).

Aesthetics, however, does not appear to be one of the concerns in the state-of-the-art environmental models or in methodological studies and applications of public-policy planning. Brock (1977), for example, in his article "Desiderata of Normative Public Policy Models," suggested a set of desired norms that should be considered in modeling, and then reduced these to three categories that parallel those of the ancient Greeks – with the exception of the aesthetic norm. Examples of public-policy models in which various aesthetic issues have been similarly neglected in urban, regional, water-resources, recreation, transportation, energy, and economic decision making have been given by DasGupta, Sen, and Margolin (1972), Gass and Sisson (1975), and Cohon (1978).

It is interesting that in decision models used for product design, marketing, or predicting consumer choice, the concept of "style" or aesthetics becomes the major issue. Because of pressing competition, investment, and profit motives, the salability of a project as dictated by its aesthetic attractiveness to the consumer (i.e., the general public) is of primary importance. Hence, in evaluating a new car design or in choosing a car from numerous alternatives, aesthetics are explicitly considered. In contrast, when public-policy issues, such as the selection of a power-plant or waste-disposal site or the choice of alternatives for resource management, are debated, aesthetics are hardly a concern (e.g., Sage, 1977, pp. 360, 410).

There have been few exceptions to this general trend in the policy-planning area where the aesthetics issue has been explicitly mentioned to demonstrate the all-inclusiveness of the methods used (Keeney and Raiffa, 1976). Edwards (1977) has reported a decision model for regional development in which the aesthetic impact of various alternatives is evaluated. Nijkamp (1979) has provided an illustration of multicriteria environmental design in which the concept of attractiveness of an area, as measured by an environmental index, has been used. In another study (Sancar, 1977, 1980), aesthetic ramifications of physical-environmental changes have been quantified by a similar "environmental attractiveness index," determined by user preferences in a dynamic model of downtown development. Thus even though it is technically possible and relatively easy to incorporate the aesthetic dimension into normative decision models, such efforts have been rare.

The objective of this paper is to examine, with a critical eye, this dilemma and the quality in planning and landscape-aesthetic-research fields and to propose a resolvent approach for aesthetic evaluation when considering the planning, design, and management of landscapes. In doing so, the policy relevance of the mainstream landscape-assessment approaches, or, more specifically, of the psychophysical approach, will first be described. The shortcomings of the latter, from policy, planning, and management points of view, are identified with regard to preference evaluation, consensus generation, identification of stakeholders, and respondent participation. Then its theoretical relevance within the framework of existing perspectives to theory-generation and landscape-perception paradigms is evaluated; from this, it is concluded that it does not provide an appropriate basis for theory construction in landscape aesthetics.

The concept of "grounded theory" and a set of requirements for theory generation (Glaser and Strauss, 1967) are then introduced to construct an appropriate perspective that integrates theory generation with the actual planning and design of landscapes. Within this perspective, an approach based on a reflective-dialectical process using normative decision models is proposed. The general rules of this approach and its treatment of the human actors, the situational model, and the design and use of the situational model via participant interaction are given.

Policy relevance of aesthetic-assessment research

The psychophysical approach

In contrast to mainstream planning approaches, there is a substantial effort in the area of visual assessment to incorporate the aesthetic dimension into the planning, design, and management of landscape resources (R. Kaplan, 1979a, 1979b; Kaplan, Bufford, Lewis, Kaplan, and Horsbrugh, 1978). The research in this area is led mainly by social scientists in environmental psychology, rural sociology, and geography. Consequently, social-science methods are applied to construct descriptive (i.e., empirical) models to represent people's responses to the visual-aesthetic aspects of the environment. The main concern in these studies is to identify the manipulable landscape or environmental features and to assess their values (Gobster, 1983; Pardee, 1983). These values can then be used to develop "defensible" manuals or guidelines based on "publicly ascertainable" facts for decision making within the legislative process or for environmental-impact reports (Bufford, 1973; Cook, 1983). This process, which emphasizes problem-focused research and addresses the development of methods for such research, now appears to have emerged as a dominant paradigm and research direction. Zube, Sell, and Taylor (1982), in their article "Landscape Perception: Research, Application and Theory," refer to this paradigm as the psychophysical approach.

In this approach, aesthetic assessment is done by evaluating the preferences of the general public, via survey methods, when considering landscapes and management options for landscapes (R. Kaplan, 1979b; Zube, 1976). The survey results are analyzed to determine the consensus on the aesthetic acceptability of the available alternatives or of various objective landscape features. The degree of consensus regarding an alternative is the evidence for the defensibility of the related aesthetic values in a court of law or in public debate.

Policy-planning implications of the psychophysical approach

The relevance of the psychophysical approach, as described above, to policy planning may be discussed in relation to the type and nature of the information made available through the process and whether or not this information is appropriate and/or useful in providing warrant or justification for policy. One may question the appropriateness of using information on preferences or values rather than other types of behavioral data. The answer to this question is indeed positive. It has been successfully argued elsewhere (Paris and Reynolds, 1983; Vickers, 1965) that the information of primary importance for policy consists of human values, rather than "objective facts" about human behavior, although this does not necessarily mean that the latter is irrelevant. Since preferences are value judgments, they appear to be most appropriate for policy issues concerning aesthetic matters.

As a separate issue, one may also question the appropriateness and utility of public preferences, given the way they are acquired and evaluated. It is this question that needs to be addressed, since there are numerous points for criticism in the procedural aspects of the psychophysical approach when evaluated from a normative policy perspective.

Data evaluation and consensus generation

An important issue is the evaluation of the preference data to search for overall or for localized consensus. This process involves the use of an aggregation method so that the information is meaningful for policy purposes. As such, it faces the same difficulties as in the construction of a "social welfare function" (Luce and Raiffa, 1957), or a "public interest criterion" (Klosterman, 1976). In social-welfare-function procedure, the difficulty lies in defining a "fair" method for interpersonal comparisons of individual states of mind and aggregating these in such a manner that the preferences indeed do serve the best interests of society. Such an aggregation has proved to be a logical impossibility (Arrow, 1951).

The public-interest criterion, in that it is normative, has been presented as an alternative to the former and to the prevailing modes of rational or incremental planning. That is to say, it is based on the interests of the general public and is argued to be scientific because the criterion would be verified to be so. It is implied that available policy alternatives should be analyzed by comparison and by invoking the empirically established needs of different segments of the population. The planner then advocates policies or actions that are in the interest of the general public. The deficiency in the public-interest-criterion approach is that the method used to define the criterion has never been clear.

In earlier applications (Daniel and Boster, 1976; Shafer and Mietz, 1969), the psychophysical approach has attempted to show regularities in public preferences on visual characteristics, regardless of the policy context or of different interests. More recent applications (Dickhut, 1983; Zube, 1973; Zube and Pitt, 1981) have paralleled the "public interest" approach. The problems experienced in both the social-welfare function and the public-interest criterion can also be identified with the psychophysical approach insofar as policy relevance is concerned. These issues have been typically ignored in the literature, and preferences have been statistically summarized and presented in a purely descriptive fashion with the intention of providing warrant for policies. Since it is not logically possible to derive an "ought to" from an "is" (Bernstein, 1978), this practice has promoted further encouragement of present adversarial legislative processes, in which the creative, the informational, and the preferential/judgmental aspects of decisions are treated indiscriminately and simultaneously in a confusing way (Howard, 1975).

Identification of stakeholders

Another disadvantage of the psychophysical approach is that it encourages regulatory or administrative agencies and other public decision organizations to concentrate only on specific actions or outcomes (such as defining use zones or color schemes) rather than encouraging a more desirable attitude of considering the actual decision-making process itself. Interest groups are referred to as the "general public" instead of devising means to identify the legitimate range of decision takers or stakeholders. Furthermore, attention is focused on particular alternatives rather than on designing effective methods to obtain and evaluate the preferences as an integral part of the policy-planning process.

Respondent participation

The final issue concerns the differences between the implicit assumptions about the nature of preferences in the psychophysical approach and those in a policy-making context. In the former approach, the "general public" voices a one-time-only, single, and "unbiased" response on an aesthetic issue via surveys reflecting none of the other relevant dimensions of the policy context. This reflects the social-scientific desire to obtain law-like generalizations concerning preferences, whereas the type of information required in the policy context is normative (Paris and Reynolds, 1983). That is, unlike scientific generalizations, these may be changed by choice. Therefore, conditions that make such choices both possible and explicit need to be integrated into the policy-making process. To achieve this, all relevant aspects of the specific policy-planning situation need to be presented so that the respondents can make trade-offs among competing dimensions, such as equity, efficiency, freedom, and so on; or, they may be asked to identify their aesthetic preferences independent of these other dimensions when such assumptions of independence are realistic. The respondents should be aware of a reference point, or "the base rate" (Tversky, 1977a), which implicitly influences their preferences and may consist of the present state of affairs, some future expectation, or an ideal state of affairs. The base-rate information also should include the probabilities of outcomes (or impacts) and the confidence one may place on this information. The respondents also should be given the opportunity for critical self-reflection by allowing them to compare their preferences with those of others.

There is strong evidence in the policy and planning literature, as well as in the literature on social and organizational change and in the cognitive sciences, that all these issues need to be adequately addressed in order to legitimately incorporate preference data into public decision making (Edwards, 1981; Fischoff, Lichtenstein, Slovic, Derby, and Keeney, 1981; Linstone and Turoff, 1975; Studer, 1982; Tversky and Kahneman, 1981). Unlike the psychophysical approach, public preferences, if they are to be relevant and useful in a policy context, should be allowed to change and evolve in a social-learning process that is aimed at a discursively achieved rather than a statistically summarized consensus.

Theoretical relevance of aesthetic-assessment research

A legitimate argument against the criticisms given in the preceding section could be that anything short of a full-fledged theory about landscape aesthetic perception and preferences will be inadequate in dealing with the complexities of such decision issues. It can further be claimed that studies, as done in the area of visual assessment of landscapes, are directed at contributing to the development of exactly such a theory. Eventually, there will be a suitable number of case studies in which the relationship among landscape properties, human purposes (i.e., goals of the users and observers), and the interaction outcomes (e.g., benefits, satisfaction, impact, aesthetic value, economic utility) would be fully explicated. The above arguments may be evaluated by the progress there has been in constructing a theory of landscape aesthetics that is based on problem-focused research in aesthetic assessment; this, in turn, largely depends on the capabilities of the epistemological perspectives available for providing such a basis.

Perspectives in theory generation

The available perspectives for constructing a viable theory in the social sciences may be grouped under three major approaches (Dunn and Swierczek, 1977). These are the universalistic, the situational, and the integrative perspectives. The universalistic perspective assigns priority to the abstraction, formalization, and generalization of relations using a hypothetic-deductive strategy. The situational perspective emphasizes the generation of contextually relevant information for the planning and management in specific settings. Ideally, this perspective assigns priority to extending the theories developed in the universalistic perspective to specific situations. In contrast to these two, the integrative perspective focuses on the "generation of grounded theory," based largely on an inductive process (Glaser and Strauss, 1967). It is possible to relate these perspectives to the existing body of literature in landscape perception and preferences and to evaluate their contributions to the development of theory. This will be done by referring to the two most recent reviews of the state of the art – by Zube et al. (1982) and by Priestley (1983) – published in the two leading journals for landscape planning and research.

Landscape-perception paradigms

Zube et al. (1982) have identified four general paradigms of landscape perception research: the expert, psychophysical, cognitive, and experiential. The expert paradigm is based on the evaluation of landscape quality by observers trained in the fields of art and design or in resource management. More recently, this paradigm has been either combined with or replaced by the psychophysical paradigm, in which the assessment of landscape aesthetic qualities is achieved through testing the general public or making evaluations of selected populations. Both approaches are problem focused and are primarily concerned with specific landscape context, such as recreational rivers, ski resorts, scenic highways, and so forth. This paradigm reflects the situational perspective in terms of theory development.

The cognitive and experiential approaches are more strongly concerned with understanding the human evaluative processes, and have more direct interest in developing a theory of landscape perception. The cognitive approach emphasizes the mental process of perceiving, and seeks to understand not the immediate landscape properties, but the predispositions or interventions in the thought process that lead to aesthetic appreciation. The primary interest is in discovering the universal characteristics of both perceptual and evaluative processes, and their relationships to the universal characteristics of landscapes. This paradigm clearly represents the universalistic perspective of theory development.

The experiential paradigm considers landscape values that develop over time through the interaction of people and landscapes, whereby both shape and are shaped by the process of interaction. The majority of studies embracing this approach have focused on literature and art as creative expressions rather than on living environments or landscapes; researchers generally have used unstructured phenomenological exploration exclusively. This approach also is universalistic in its aims, but differs from the cognitive approach in terms of the research methods used. The research methods used in the cognitive approach are aimed at rigorous verification, whereas the studies done in the experiential approach do not allow for systematic generalizations.

None of the four paradigms identified by Zube et al. (1982) may be associated with an integrative approach to theory development. Having reviewed the body of research in all four paradigms, they identify a "conspicuous theoretical void in the majority of the research" (Zube et al., 1982, p. 25); this conclusion also is supported by Priestley (1983), who shows that there is no evidence of a convergence toward a unified theory of landscape perception.

Requirements for theory generation in landscape aesthetics

Zube et al. (1982) draw a set of minimum considerations that must be taken into account in any adequate study of landscape perception and that are based on the recognition that the central issue is the interaction between humans and landscapes. Landscapes permit movement and exploration of the situation, forcing the observer to be a participant. Landscape perception always involves action, control, manipulation; in other words, it calls forth purposeful actions. Landscapes are always encountered as part of a social activity. At any time, they provide more information than can be used (they are redundant), which implies that specific information relevant to the context – the purpose, the activity – is picked up by the observer. This information is received through multiple senses and comes both from the outside as well as from within the focus of attention.

An important consequence of these considerations is that any process of inquiry in landscape perception should allow for the participation of the observers in purposeful activities relevant to the landscape in question. Such inquiry ideally needs to be grounded in actual situations – that is, in practice – dealing with issues concerning the enjoyment of landscapes and the making of decisions. Furthermore, the decision situations also should meet the conditions as outlined in the previous section on the policy relevance of aesthetic-assessment research, for theoretical relevance. Zube et al.'s (1982) considerations and the above interpretation suggest the need for an integrative approach to fill the theoretical void in landscape-aesthetics research.

The main characteristic of such an approach has been identified (Glaser and Strauss, 1967) and applied more recently to theory development in the field of organizational change and development (Dunn and Swierczek, 1977). The integrative approach introduces the concept of "grounded theory" and is based on the premise that the adequacy of a theory cannot be divorced from the process by which it is generated. When the process of generation focuses primarily on verification of a preconceived theory, the emergence of the theory itself is taken for granted. Since there is a clear theoretical void in the area of aesthetic research, it is the emergence of theory that should receive attention. This may be achieved through a strategy of comparative analysis, flexible use of both qualitative and quantitative data, and secondary analysis of substantive data (Glaser and Strauss, 1967). The fundamental characteristics of the theory would be derived from those cases where (1) the conditions are reliably represented (internal validity), (2) the conditions typify those found in other situations (external validity), (3) the findings contribute to the generation of concepts by comparing the information obtained through different methods (reflexivity), and (4) the understanding of consensus is promoted among groups with conflicting frames of reference (translatability) (Dunn and Swierczek, 1977). The integrative approach to generating a grounded theory provides the conceptual framework that allows the use of information for theoretical development in the area of landscape perception that can be gained only through landscape-policy planning, design, and management. Such a perspective dictates certain conceptual and procedural requirements concerning the methodology of policy and research, in that they are no longer considered as separate fields of inquiry, but are merged so as to satisfy common goals. A preliminary list of procedural guidelines for conducting inquiry in the spirit of the above discussion will be proposed next.

An integrative approach to policy and theory development in landscape aesthetics

Several characteristics of an ideal procedure for conducting inquiry have been mentioned in the preceding section on policy, planning, and design as well as in theory development. To recapitulate, maintaining contextual realism, simultaneous consideration of all relevant value dimensions (including aesthetics), and clarification of informational aspects concerning the impact of potential alternative actions, and allowing for change and development in the judgmental responses of the participants have been emphasized. To integrate the creative, judgmental, and purely informational aspects of the decision problem, a nonadversarial, logical, and integrative process is required. This process ought to encourage critical self-reflection based on perceived impacts of one's own decisions, and it should provide knowledge about others' values, judgments, and decisions, and provide opportunity for debate. The outcomes of such a process will be planning and design recommendations and an understanding or shared perception of the problem context that is instrumental in making those particular recommendations.

The representation of the problem context as perceived by the participants becomes the critical concept to be dealt with in designing procedures to develop policy as well as generate theory. This representation, which will be referred to as the "situational model," becomes the cognitive aid for decision and judgment formulation for problem resolution. Once the aid is constructed, it becomes the surrogate decision environment for the participants. Therefore, the particular characteristics of the model, as well as the various successive versions of it during the construction process, can provide valuable information from a theory-formulation point of view. The situational model, as such, is an important component of the proposed approach.

The other two components of such an approach are the human actors, or the participants who develop and use the model for policy formulation, and the interaction process through which the model is developed and used. To describe the approach in more detail requires that these components be further specified according to normative criteria to achieve a socially acceptable and rationally justifiable decision process. One must also specify cognitive criteria to acknowledge the theoretically relevant variables that influence aesthetic judgments and preferences while documenting the related evidence reliably. In the following sections, each of these components is individually discussed.

The human actors

It is required that the participants have a real, existing, or potential interest in the issue. This requirement is consistent with both the applied, or policy, and the theoretical perspectives. A participant may be characterized by the values she or he may display for various variables; these could include active involvement, knowledge of or access to relevant information, power or influence in decision making, and a share in the costs and benefits of alternative decisions. As such, the participants will fall into somewhat overlapping, broad categories of experts, public and private decision-making authorities, public and private organizations representing various interest groups, and other beneficiaries and victims of potential actions – that is, users.

The identification of the relevant interests will be influenced by the initial description of the problem context. The determination of the population for the selection of participants representing these interests will require different techniques, such as studies of aerial maps, content analysis of media and meetings, and sociometric surveys. The inclusion of new participants will result in the redefinition of the problem context and the possible addition of neglected interests. In most issues, however, the identification of participants will reach a closure relatively soon. Contrary to widespread belief among researchers who commonly depend on sophomore psychology students or "the general public" for their subject pool, satisfying this criterion of contextual realism by including real interests is not difficult.

The situational model

The elements of the situational model are the "attributes," or landscape properties, and the "outcomes," or consequences that are the end states to be reached through the implementation of the alternatives. Alternatives are feasible combinations of various values that the attributes may take.

In addition to attributes and outcomes, the model needs to contain (1) perceived relationships between attributes (e.g., when the level of an attribute enhances or limits the level of another attribute, or when attributes present themselves in very specific combinations), (2) relationships between attributes and outcomes (which may be objectively verifiable or may reflect an individual value or preference structure for an attribute with respect to achieving particular outcomes), (3) the degree of uncertainty associated with these stated relationships (reflected in the variance of individuals' perceptions of factual relationships, and the degree of confidence in the information provided), and (4) emphasis on the expected outcomes (i.e., the relative weights that individuals may assign to various consequences relating to the four major categories of pursuits that humans follow). These four elements in combination reflect the perceived problem context as structured and formulated as a decision problem.

The various decision models commonly utilized in the field of management science contain these elements either explicitly or implicitly. The typical landscape-assessment approach studies these elements (which are then termed *norms, attitudes,* and *beliefs*) either in isolation or without the contextual realism and dynamics of actual decision making. The difference of the approach presented here as compared with both the decision and the landscape-assessment approaches is not the acknowledgment of these elements as crucial variables in determining people's aesthetic judgments (although this also is an important aspect not shared by the majority of research done in this area). A more important difference is the way in which the problem structure evolves and its elements are defined. In other words, the process of constructing the situational model, an activity commonly referred to as "the art of modeling" and left to the discretion of the "expert," receives special attention.

Design and use of the situational model via participant interaction

The main function of the situational model within the approach described here is to represent the collective appreciation of the problem context by the participants. The interaction process may be decomposed into three stages: information acquisition/cue pick-up, evaluation/action, and feedback/learning (Einhorn and Hogarth, 1982). These stages apply both when translating the situation into a model (verbal, formal, visual, or a combination) and when making a judgment or decision.

The first stage relates to the determination of the relevant attributes and to the compilation of the information for the derivation of the rest of the model interactions. The second stage involves the manipulation of the model to reach a satisfactory judgment or decision concerning future action. The last stage refers to both short- and long-term impacts of the judgment or decision or the action in order to restructure the situational model or to modify the evaluation rules.

The objectives in carrying out these tasks adequately may be summarized as constructing a shared perception of the situation, encouraging search and interpretation of the relevant information concerning facts and values, and documenting interactions among the participants during their negotiations in the process. Hence, the outcome of the process is situation-specific information encompassing the essential variables for the generation of a theory of landscape aesthetics. The characteristics of a procedure to achieve these objectives, and how it differs from both the mainstream planning and the social-science approaches in aesthetic assessment, will now be described by referring to these three stages.

Information acquisition/cue pick-up. During this stage, the issue or problem context is structured into a situational model containing the elements described previously. Empirical studies have shown that judgment and choice are sensitive both to the formal structure and to the content of a "task" that corresponds to the situational model, as defined earlier (Slovic, Fischoff, and Lichtenstein, 1977). Therefore, this stage is most important from a theory-generation as well as a planning point of view.

In the social-science approach to aesthetic evaluation, there is no comparable effort to achieve a holistic representation of the issues. When the immediate goal is to provide preference data in a particular planning context, the researcher may present the landscape alternatives as whole entities or present the respondents with a decomposed list of landscape properties. In either case, the stimuli are determined by the researcher. The interaction of the respondents with the representation is limited to passive and single responses. When the objective is to determine the salient properties of the landscape to which people respond, the decision context is absent. Furthermore, the investigators typically assume that the description of the landscape must be either in physical dimensions or in subjective evaluations, and most have chosen to use subjective accounts. As a result, a focus on the preperceptual or objective environmental properties is lost (Heft, 1981).

In the planning approaches where formal decision aids are used, the nature of the situational model depends on the type of decision model and on the talent and ingenuity of the analyst in representing the problem context such that it is most meaningful to the client. It also has been suggested that the usefulness of decision models may be largely due to clarifying the problem for the respondent or client rather than claims for rationality (Humphreys and McFadden, 1980). When the "client" is no longer a single person but represents a number of groups with complex and conflicting roles and objectives, and when the problem does not readily lend itself to preconceived or standard models (typical of most real-life issues), then problem structuring becomes an even more important stage in decision making. Direct involvement of the participants at this stage is facilitated through the application of special cognitive aids (Checkland, 1981; Sancar, 1982; Warfield, 1975). Ways of achieving "client" involvement in developing the situational model have been suggested, with emphasis on discovering relationships and patterns among attributes and between attributes and consequences. Influence diagramming (Owen, 1978) and structural modeling based on relationship matrices (Saaty, 1980; Warfield, 1976; Watson, 1978) are examples of appropriate techniques that may be used to develop a shared representation of the problem context without having to impose a preconceived model. With these techniques, it is possible to consider large numbers of statements containing attributes and relationships simultaneously, without neglecting any because of human information-processing limitations. As such, judgments can be withheld, and premature bounding during the problem-definition stage can be avoided.

From the perspective of this paper, the initial specification of attributes by the participants is of more interest. It has been suggested that the process of representing an object or an alternative by a number of attributes or features depends on the context and prior processes of selective attention and cue achievement (Tversky, 1977b). As a result, the participants perceive the same environment, but different individuals perceive different aspects of it. Due to its focus on objective (and, therefore, potentially manipulable) characteristics of the context, the Gibsonian theory of ecological perception provides an appropriate conceptual framework for organizing the initial structuring of attributes.

Within this conceptual framework, the differences in people's perceptions are due to differences in information pick-up as a function of past experience and familiarity with the decision issue, the availability and degree of uncertainty of factual information, and functional meaning of environmental features, or "affordances," to the participants. The construction of the situational model is analogous to an "affordance analysis," which in Gibson's terms provides an account of the functionally meaningful features of the environment (Heft, 1981). The determination of attributes may then be seen as the initial activity in such an analysis. These features need not be limited to individual objects and events, but may also include standing patterns of behavior or an issue setting.

Studies of cue pick-up under laboratory conditions have used tracing methods, such as verbal protocols and eye movements (Hayes, 1982; Payne, 1976). In actual problem-solving situations, the generation of a list of attributes receives little attention, with a few exceptions where individuals are systematically probed by a facilitator (Warfield, 1976). Another promising approach to achieve this task is the phenomenological exploration that uses introspection and self-report techniques. An evaluation of these techniques is not given here, since the intent is to emphasize the conscious use of such techniques and explicit recording of this stage.

Evaluation/action. Two major activities take place in this stage. First, the inter-relationships among the model elements are determined. Then these are used for the evaluation of alternative courses of action via an agreed-on aggregation rule.

To complete the first activity, the interrelationships among attributes, those between attributes and consequences, and the importance given to the consequences are studied. Subjective and objective interrelationships among attributes are assessed either by having the actors predict the levels of one attribute by the levels of another attribute or by using objective data and statistically estimating the ecological intercorrelations. This information is then used to avoid double counting of attributes if an additive aggregation rule is to be used in evaluating alternative courses of action.

The attribute–consequence relationships refer to the direction and degree of the effect that each attribute has on the relevant objectives or consequences. In some cases, the functional relationship may be straightforward and objectively verifiable. Since such information is factual, experts and users with experience can provide the necessary functional forms. (However, as will be discussed shortly, even the factual information is debatable, and participants may place different levels of confidence in them. Consequently, their judgments will be affected by the nature of the information presented, as well as by the credibility of the source.) Otherwise, the attributes are related to the consequences via the use of value curves.

Value curve refers to an individual's value or preference structure for each attribute with respect to relevant objectives in evaluating the relative desirability or utility of alternative courses of action. The participants specify the functional form of the correspondence between the values of an attribute and its utility in or desirability for achieving the relevant objective. The importance of any information to a participant's evaluation may be specified by using the concept of "weight." The consequences or objectives also are ranked or rated in order of importance. The degree of uncertainty perceived within the problem context is reflected by the variance in participants' estimates of factual relationships and alternatively in the variance of their stated level of confidence in the information provided by others within or outside the participating group.

Once the situational model is clarified, it is used to choose an appropriate course of action. The decision may involve choice among existing alternatives or a modification or combination of these, or it may even result in the generation of additional alternatives. The evaluation of the alternatives is achieved via the use of an aggregation (information-integration) strategy, which may involve formal/rational models, as in multiattribute-multiobjective decision methods, or a reflective-dialectical procedure. Decision-theoretical approaches differ in the type of formal manipulations used, such as weighting methods, sequential-elimination methods, mathematical-programming methods, and spatial-proximity methods (MacCrimmon, 1973). An example of a reflective-dialectical procedure is the modified jurisdictional model (Mason and Mittroff, 1981), which is used to debate and evaluate opposing points of view.

It has been suggested that when presented with a decision task (in this case, a situational model), individuals use different implicit or explicit aggregation strategies. Their choice is influenced by the nature of the task interactions (Einhorn and Hogarth, 1982). The aggregation procedure may also differ in terms of the participants' involvement with one another. When the degree of uncertainty experienced during the development of the situational model is relatively low, the participants may debate the appropriateness of different rules for aggregation to arrive at a group decision. This implies that the participants will obtain a clear idea of the mechanism through which the model elements are evaluated. Otherwise, such operations are unknown to the clients or participants but are at the disposal of the analyst, who provides them with an optimal solution. Research evidence suggests that the participants are unlikely to accept the results of a black-box operation even though the inputs may be generated by them (Clegg, 1984).

When the degree of uncertainty experienced during the model construction is high, the individual may find it difficult or impossible to advocate a certain aggregation procedure. This may happen when individuals are unsure of how their beliefs, attitudes, and judgments are related. Under such circumstances, and before an aggregation rule can be debated as a group, the individual may investigate his or her own judgment policies by interacting with the model and by changing the weights and functional forms of relationships, as well as investigate the types of aggregation procedures. The result will be a more stable representation of the decision context and a more enlightened debate over various judgment aggregation policies (Hammond, Stewart, Brehment, and Steinman, 1975).

Feedback/learning. Most decision modeling, as well as social-science approaches using surveys of public preferences, appear to be based on the assumption that reflection precedes decision or action. In contrast, the procedure described here acknowledges that most real problems involving aesthetic or other values are resolved through trial-and-error processes. Such processes require a sequence of decisions where people have the opportunity to adjust their decision strategies according to the impacts. In the procedure described here, information feedback concerning the impacts of participant decisions is generated via interactions with the model and among participants.

To summarize, the outcome of the integrative approach outlined in this paper is a detailed record of the human actors, various interactions of the situational model, the proposed action alternative, and the individual and group interactions that take place during the development and use of the situational model.

Conclusions

In spite of growing concern regarding the aesthetics of the environment, attempts to incorporate this concern into decisions affecting the public have not benefited from recent developments in the area of decision sciences, particularly soft-systems methodology. Theoretical developments in the area of landscape aesthetics also appear to lack a well-defined trend toward convergence, in spite of a large volume of assessment studies based on expert and/or public preferences.

In this paper, an alternative procedure has been proposed that aimed at integrating the scientific and interpretive aspects of human experience. This procedure, which may be referred to as a reflective-dialectical strategy of inquiry and choice, has been described in detail to show the knowledge-accumulation function of the approach. Its relevance to theory development has been explained through the use of the "grounded theory" concept, in which the "theory generation" itself is emphasized rather than "theory testing" based on retrospective case studies.

The approach described here is significantly different from the psychophysical approach. The four landscape-perception paradigms can be placed on a continuum of passive to active in their treatment of human actors and dimensional to holistic in their treatment of landscape properties. The psychophysical approach falls closer to the passive/dimensional end of this spectrum, since human actors are treated as passive respondents and the situational model is static and limited in its scope (Zube et al., 1982). The integrative approach, however, falls on the active/holistic end of the spectrum. Now, the human actors are active participants in the process, and the series of situational models that emerge through the interaction process are holistic and dynamic representations of the landscape properties. Because of these properties, the integrative approach eventually will lead to a universalistic theory, whereas the psychophysical approach has failed to do so. Therefore, the present approach is *not* an alternative procedure to develop and use a psychophysical mode. There are major differences in philosophy and underlying assumptions. In the psychophysical approach, the statistical verification of a situational model posed by the researcher is the ultimate product. In the integrative approach, the record of the process of emergence, development, acceptance, and use of the situational models for decision making is the main outcome. It is this collection of records that ultimately will contribute to the inductive derivation of a universalistic theory.

The reflective-dialectical inquiry framework, as outlined in this paper, defines the "general rules" to be observed in landscape-aesthetics research and policy planning. Further refinement and application of the components and mechanics of this framework are described in the context of ongoing community-development studies, where explicit suggestions as to how the technique can be worked into landscape research, planning, and management programs are given (Sancar, 1987).

32   Aesthetic regulation and the courts

Kenneth T. Pearlman

Introduction

This paper is an attempt to clarify developments in the land-use area for scholars in the field of environmental psychology as well as for others interested in the relationship between research on visual quality and the impact of that research on the judicial system. On the whole, this paper does not present major new factual findings, but argues for better understanding of the ways in which courts make decisions in this important area. It examines the development of the concept of public purpose in zoning as it is used in issues related to aesthetics and finishes by considering how the results of research in aesthetics can be used by courts and local governments in making decisions.

The Supreme Court and the concept of public response

The concept of public purpose is important in zoning and land-use decision making because it determines the extent to which government can have an impact on the private decisions of landowners and land users. The issue of public purpose goes all the way back to the landmark case of *Euclid* v. *Ambler Realty Co.*, 272 U.S. 365 (1926), in which the Supreme Court of the United States upheld the power of local communities to zone land. From the time of that case to the present, there has been a clear, if not always steady, march to an expanded concept of the police power.

In *Euclid,* the Court upheld the use of zoning as being related to problems of public welfare on what, today, seems like a rather narrow basis. After reciting the typical arguments in support of zoning (e.g., better fire and safety protection, prevention of traffic accidents, decrease of noise, a more favorable environment in which to raise children), the Court went on to argue that certain uses placed in the wrong neighborhoods, although entirely unobjectionable in others, can be very much like nuisances. Although the Court expressly denied that the issue of what is a nuisance may be controlling on the question, it concluded that "like a pig in the parlor instead of the barnyard [i]f the validity of the legislative classifi-

476

cation for zoning purposes be fairly debatable, the legislative judgment must be allowed to control . . ." (272 U.S. at 388). This presumed validity of legislative decisions means, as the Court pointed out, that some perfectly acceptable land-use arrangements will be excluded, but "[t]he inclusion of a reasonable margin to insure effective enforcement will not put upon a law, otherwise valid, the stamp of invalidity" (272 U.S. at 388–389).

Thus for the Court in 1926, zoning laws were designed primarily to deal with land-use conflicts or potential conflicts. Nonetheless, although the Court did not dwell on aesthetics, the opinion refers to the destruction of open space and attractive surroundings when a potentially conflicting use – say, an apartment house – is located in a single-family neighborhood. This problem is far removed from the question of whether a municipality can regulate the exterior of structures for conformity to community standards on style and other aesthetic qualities, but it contains the germ of the idea that attractiveness and land-use regulation are related, and this notion was to become more powerful over time.

The next significant foray by the Court into the question of aesthetics and the use of land came in *Berman* v. *Parker,* 348 U.S. 26 (1954), which upheld the use of governmental power to purchase land in eminent domain from one landowner and to resell the property to another private developer in order to develop the land according to a plan for redevelopment. While this case was not a zoning case, and thus not a direct precedent for the zoning situation, it has become an important link in the expansion of the concept of public purpose in zoning.

In *Berman,* the plaintiff argued that his structure, which was taken as part of the purchases toward the intended redevelopment plan, could not be taken because it was not slum housing and, further, that property could not be taken merely to make the community more attractive. Justice Douglas, writing the opinion for the Court, began his discussion of public welfare by noting that the traditional concerns of the police power (even though this was an eminent-domain case),[1] which includes such items as public safety, health, morality, peace and quiet, and law and order, were merely illustrative. The concept of the public welfare, Justice Douglas argued, is broad and inclusive:

The values it represents are spiritual as well as physical, aesthetic as well as monetary. It is within the power of the legislature to determine that the community should be beautiful as well as healthy, spacious as well as clean, well-balanced as well as carefully patrolled. (348 U.S. at 33)

Having decided that the purpose is within the power of the legislature, the Court went on to hold that the means to achieve this power are also for the legislature.

Any doubts that the Court would limit its concerns to cases in eminent domain were laid to rest twenty years later in the case of *Village of Belle Terre* v. *Boraas,* 416 U.S. 1 (1974), in which the Court, in an opinion also written by Justice

Douglas, upheld the right of a community to determine its life style and the kind of community it wished to be. In *Belle Terre,* the village restricted land use to one-family dwellings, *family* meaning more than two persons related by blood, marriage, or adoption or not more than two unrelated. Thus an unmarried couple could live in the same house, but not three unrelated people. Douglas noted that the case dealt with issues of social and economic legislation that have traditionally been upheld where they are supported by a rational basis. Although not a case dealing with aesthetics, *Belle Terre* reverberates with echoes of *Berman* v. *Parker:*

A quiet place where yards are wide, people few, and motor vehicles restricted are legitimate guidelines in a land use project addressed to family needs. . . . The police power is not confined to elimination of filth, stench, and unhealthy places. It is ample to lay out zones where family values, youth values, and the blessings of quiet seclusion, and clean air make the area a sanctuary for people. (416 U.S. at 9)

Having gone this far, the Court should have surprised no one by its subsequent statements on the power of communities to zone.

Of these cases, the most important case treating aesthetics is the famous, and confusing, case of *Metromedia, Inc.* v. *City of San Diego,* 453 U.S. 490 (1981). The case was confusing for a number of reasons. First, there was no majority opinion. Justice White, joined by Justices Stewart, Marshall, and Powell, wrote a plurality opinion that overturned a sign ordinance on unusual grounds (that the ordinance gave greater protection to commercial than to noncommercial speech, which is discussed later). Justices Brennan and Blackmun formed the rest of the majority in an opinion resting on the ground that the sign ordinance in question created a total ban on outdoor advertising. In addition, there were dissenting opinions from Chief Justice Burger and Justices Rehnquist and Stevens.

A second cause of confusion was the basis for the plurality decision, which held that a sign ordinance that prohibited most outdoor-advertising signs but allowed a greater number of commercial messages than noncommercial messages (most noncommercial messages were allowed) was invalid because the previous cases decided by the Court gave greater weight to display of noncommercial speech than to commercial speech.

For the purposes of this article, the issue that concerned the plurality is not of paramount importance. What is important is the attention that the judges gave to the role of aesthetics in determining the extent to which the police power could be used to regulate at the local level.

One of the arguments raised (among others) was that the San Diego restrictions on commercial speech were in violation of the First Amendment because they restricted speech and failed to directly advance governmental interests in traffic safety and the appearance of the city. The plurality quickly concluded that eliminating billboards is related to traffic safety. With respect to aesthetics, the Court said that it was not speculative to recognize that billboards, by their very nature, can be aesthetically harmful. This is so even though aesthetic judgments may be subjective. For the majority, the only reason to give close scrutiny to aesthetic claims by a city is to ensure that such claims are not being used as a mask for other, impermissible reasons, such as an intent to restrict speech in an unlawful manner. If such a situation is absent, then the city retains the right to determine what will and what will not affect the city's aesthetic interests. Moreover, the city can make fairly closely refined judgments. Thus it can conclude, as it did in *Metromedia,* that off-premises commercial speech is more destructive of city interests than is on-premises commercial advertising. Nor does the city need to provide a significant amount of support for such conclusions, and, indeed, the record in the *Metromedia* case was quite scanty.

The plurality's views on aesthetics were joined by the dissenting justices, often in even stronger terms. Chief Justice Burger, who dissented because he felt that the city had a right to choose the method of regulating signs as it did, nonetheless also felt strongly about a city's right to regulate where aesthetics is involved. The purpose of San Diego's regulation, he argued, was, *inter alia,* to avoid the disfigurement of the surroundings that billboard signs create. Burger argued that a city can conclude that every large billboard adversely affects the environment,

for each destroys a unique perspective on the landscape and adds to the visual pollution of the city. Pollution is not limited to the air we breathe and the water we drink; it can equally offend the eye and the ear. (453 U.S. at 561)

Moreover, aesthetic concerns as a means to an end are, like any other means used by a city to achieve its land-use goals, for the legislature to determine.[2]

Justice Rehnquist, also in dissent, went even further than the Chief Justice, for he believed that "aesthetic justification alone is sufficient to sustain a total prohibition of billboards within a community" (453 U.S. at 570, citing *Berman* v. *Parker*). Moreover, he did not limit this power to communities that may be especially concerned with aesthetics, such as Williamsburg, Virginia, or to communities with older, historic areas. Nor did he feel that the city, even in a case where First Amendment allegations are present, should have to demonstrate the impact on aesthetics that major sign restrictions may have. This is not the role of the judiciary: "Nothing in my experience on the bench has led me to believe that a judge is in any better position than a city or county commission to make decisions in an area such as aesthetics" (453 U.S. at 570).

Justice Stevens joined the other dissenters in believing that a community may select what it feels are its needs in terms of regulation of visual quality. His long opinion is devoted more to First Amendment issues; but he did feel that there is a community interest in pleasant surroundings, and he noted that "the character of the environment affects property values and the quality of life . . ." (453 U.S. at 552).

The principal disagreement with the positions expressing a strong view about aesthetics even where First Amendment interests are involved came from Justice Brennan, joined by Justice Blackmun, who concurred in the decision on the grounds that the ordinance effectively constituted a ban on all outdoor advertising. Brennan felt that a city can effect such a ban only if it can show that a sufficiently substantial governmental interest is directly furthered by the ban and that a more narrowly drawn ordinance would be less effective.

This is, of course, a constitutional judgment, but its implications for control of visual quality are significant as Brennan drew them out. It means that there are certain burdens on cities to make a case. One requirement that Brennan would impose in a case of this sort is that the city show that it is "seriously and comprehensively addressing aesthetic concerns with respect to its environment" (453 U.S. at 531). Since the city failed to address other aspects of an unattractive environment, its commitment to aesthetics was put in doubt. In other words, while the city need not attack all problems at once, the failure to do so suggests at least that there is no substantial commitment.

Brennan also argued that the burden may be different depending on the city, citing Williamsburg, for example, which has interests in "aesthetics and historical authenticity" (453 U.S. at 534), and the federal government's interest in Yellowstone National Park, where the very existence of signs would be inconsistent with the natural environment. However, he specifically declined to express a view on the burden that large urban areas may have to meet and whether they can meet that burden.

Justice Brennan's argument raises some interesting questions. While they cannot be taken entirely out of their context of the First Amendment, there is a strong suggestion that Brennan did not care to base judgments on aesthetics or that, at least, he felt some concern. The linking of aesthetics with historical authenticity at Williamsburg brings this out clearly, for it links with aesthetics something that has traditionally been viewed as more precise in determination. It is easier to decide what may be historically important than what may be a valuable contribution to visual quality, or so the argument goes. Similarly, the Yellowstone example is a cautionary flag, for here is a situation where it is so patently clear that signs interfere with the environment that nobody can seriously dispute it.

Yet the argument ignores the need for communities that have serious problems with visual quality to work on improvement within a context that is bound to be uncertain and even controversial. Nonetheless, a community may well feel that it can deal with only a few problems at one time and that the program it has adopted represents the best, or at least a reasonable try, to do what is possible. Indeed, it may be more important for urban areas to work on this than for pristine areas that already have substantial support for what they are doing (I do not wish to imply the need for less-than-eternal vigilance to protect our heritage). Brennan's argument would certainly make this difficult, at least in cases where communication is involved. Nonetheless, even Brennan was sympathetic to concerns about visual quality, and it is clear from *Metromedia* that the Court as a whole recognizes the importance of these concerns.

Another important decision, which in fact preceded *Metromedia* by three years, was the Supreme Court's decision in *Penn Central Transportation Company* v. *City of New York,* 438 U.S. 104 (1978). This case involved the question of whether New York City's Landmarks Preservation Law could result in the taking of property – that is, an economic impact on one's property so great as to render the regulation unconstitutional. In this particular instance, the Landmarks Preservation Commission had designated Grand Central Terminal a landmark and had rejected plans to build a multistory office building on top of the station.

While much of the decision was devoted to the issue of whether the owners of the building retained sufficient economic value so that the regulation would not be considered a taking (they did, the Court held), the Court had some words to say about aesthetics. Citing *Belle Terre* and *Berman,* the Court noted that it had frequently upheld land-use controls that may enhance the quality of life by preserving the character and desirable aesthetic features of a city. The Court noted several features with respect to landmark laws that are important. The Court found that landmark designations are no more a matter of taste than are traditional zoning regulations. Indeed, they concluded that "there is no basis whatsoever for a conclusion that the courts will have any greater difficulty identifying arbitrary or discriminatory action in the context of land mark regulation . . ." (438 U.S. at 133).

The Court also emphasized the comprehensive nature of the preservation scheme, which included over 400 individual landmarks and 31 historic districts. Essentially, the Court felt that the ordinance embodies a comprehensive plan to preserve structures of historic or aesthetic interest and thus is not arbitrary. Whether this is closer to the comprehensive requirements desired by Justice Brennan in *Metromedia* than is the discussion of the plurality is open to question. While Brennan did call for a comprehensive attack on the issues related to aesthetics, it is clear that the New York ordinance does not require that New York handle all the problems related to aesthetics at one time. In this case, there was a landmarks ordinance but not a sign ordinance, whereas the reverse was true in *Metromedia*. The Court would thus seem to be arguing that "comprehensive" is part of a well-thought-out plan that deals with a problem within the power of the city to regulate.

One might inquire why Justice Brennan agreed with the majority in *Penn Central* but not in *Metromedia*. The answer would appear to be based not on aesthetics (although one could argue that the New York ordinance, with a commission and a concern for historical as well as aesthetic considerations, is different), but on the fact that Brennan felt that the sign ordinance in *Metromedia* represented a complete ban on one kind of speech and that under such circumstances needed to be justified to a greater extent than might otherwise be the case.

If Brennan's viewpoint were followed, it would have the curious impact of creating, or possibly creating, different sets of standards for certain types of aesthetic regulation. Large-scale regulation of signs would be more problematic, since such regulation would, under Brennan's view, be enforceable only in the context of a large-scale attack on undesirable visual characteristics of a city. This could be a serious problem because, for reasons of fairness in a political context, it is often difficult to regulate some signs and not others, and a city may well feel that the only way to deal with the problem is to treat it as a whole and not allow signs. At the same time, there are First Amendment problems (see also Pearlman, 1984) and they should not be minimized. A less-than-total approach may be advisable and still enable a city to handle questions of visual quality. In any event, Justice Brennan's agreement with the majority in *Penn Central* suggests, perhaps, that he too is quite willing to defer to local judgments in aesthetics unless there are First Amendment questions involved.

Further evidence that this is likely to be the case involves the recent case of *Members of City Council* v. *Vincent,* 104 S. Ct. 2118 (1984), in which the Supreme Court upheld a provision of the Los Angeles City Code prohibiting the posting of signs on public property. The impact of this case on the First Amendment questions is important but beyond the scope of this paper. However, the majority of the Court reiterated the strong support that aesthetic judgments have in deciding police-power cases: "The problem addressed by this ordinance – the visual assault on the citizens of Los Angeles presented by an accumulation of signs posted on public property – constitutes a significant substantive evil within the City's power to prohibit" (104 S. Ct. at 2130).

The Court, however, left the area a bit murky because as justification for this judgment, it also noted that the relevant interests are also psychological and economic, since the "character of the environment affects the quality of life and the value of property in both residential and commercial areas" (104 S. Ct. at 2130). This, of course, leaves open the issue of whether aesthetic justifications alone are sufficient to support a decision under the police power.

Justice Brennan, in dissent and joined this time not only by Justice Blackmun, but also by Justice Marshall, repeated his *Metromedia* arguments that special care needs to be taken in First Amendment areas. He raised concerns about the subjectivity of aesthetic judgments that make them difficult for a court to review and said that careful scrutiny should be used where total bans (and, in effect, First Amendment issues) are involved. Again presumably, when the First Amendment is not a serious problem, Brennan, too, would give significant weight to local judgments if the government can show that the regulations are unrelated to the restriction of speech. This can possibly be inferred from the fact that Marshall did not agree with the plurality in *Metromedia,* presumably because he did not feel that a total ban was present in the case (it should be remembered that Brennan's opinion in *Metromedia* was based on his belief that a total ban was involved).

It was mentioned at the beginning of the *Penn Central* discussion that the case was decided on the taking issue. In such a situation, the questions involve such issues as whether the owner's investment expectations have been sufficiently frustrated so as to result in an overregulation of property or whether the benefit to the public is outweighed by the burden on the property owner. The taking issue is a complex one and beyond the scope of this article. But the discussion of aesthetics in a taking context led the Court to draw some conclusions that are of importance to the use of the police power to zone or regulate for purposes of aesthetics.

First, the Court made it clear that aesthetic regulations can apply both to districts and to individual buildings. There is no need to develop historic or aesthetic districts that will be subject to special regulation. Landmark laws that apply to only selected parcels are acceptable, even though there may not be some geographical section of the city requiring special treatment. This allows aesthetics to be used as a basis for the police power, even though a basic concept of zoning – subjecting land areas to common controls on the basis of certain perceived characteristics – is not being followed when buildings are selected on a building-by-building basis. To be sure, the New York City law was a comprehensive piece of legislation and involved a commission and the city council in such decisions, which were further circumscribed by the specification of standards to be used. But, still, each building becomes, in effect, a case on its own facts.

Nor was the consequence of this approach – that certain landowners would be more clearly affected than others – a barrier. All landowners, even the specifically burdened, share in the benefits deriving from the ordinance, and it is for the government to determine whether such a benefit outweighs the burden. This aspect of the opinion is open to some question. It is true that these owners share in the benefit, but other owners of land are subject to only a lesser order of burdens – traditional zoning regulations. Owners of landmark properties are regulated in another way as well, and this fact should, at least arguably, add some force to their argument. It would have perhaps been better had the Court recognized this difference but simply concluded, as it did elsewhere in the opinion, that the economic loss was not of a sufficient magnitude to warrant a taking.

Nor is it a problem that the law does not guarantee application of aesthetic regulations to all similarly situated properties. In this case, the New York law applies to over 400 individual landmarks, as the Court noted. The problem here is whether a historic-landmarks law must cover a large number of buildings. In any event, 400 buildings in a city the size of New York is really not a very large number. Again, from a taking point of view, it is perhaps better to concentrate on the degree of loss, which may be greater in some cases of nonaesthetic regulation than in instances of aesthetic regulation. Unfortunately, the Supreme Court frequently raises issues as taking questions that appear to be really problems of reasonableness, and this has a long history in Supreme Court decisions. The issue ought to be whether it is reasonable to deal with only a limited number of buildings at a time. Such doctrinal problems aside, the case comes firmly down on the side of landmark regulation, even where such regulation is clearly less than total or truly comprehensive.

What can be said in summary of the Supreme Court decisions? First, it is clear that they are strongly supportive of community efforts to control the aesthetics of the environment. The Court really feels that it is for the community to determine what is to be regulated and what is not to be regulated. Problems of measurement – What is beautiful? What is attractive? – are not considered to be difficult problems as long as the effort to make the environment more attractive is a serious one.

For the most part, only the First Amendment looms as a deterrent to successful adoption of some forms of aesthetic regulation. But even here, sign regulation will be acceptable as long as there is no perceived total ban on signs and as long as communities take care not to favor commercial over noncommercial speech.

Challenges based on other grounds are not likely to succeed before the Court. Thus the Court clearly feels that zoning for aesthetics serves a valid public purpose. Further, the means chosen to achieve that end will, by and large, not be scrutinized by the Court, or at least not more carefully than any other means are generally scrutinized by courts.[3] Finally, the Court is unlikely to find a taking as a result of aesthetic regulation when some reasonable economic value remains in the property.

The Supreme Court decisions do not wholly dispose of all issues. First, as has been noted, the *Metromedia* case is open to severe confusion because of the divergent opinions and the lack of a majority opinion. Similarly, *Penn Central* raises some doctrinal issues in connection with taking jurisprudence that make it less than the ideal case to follow as gospel, although this may be more troubling to a lawyer than to a planner or an environmental psychologist.

Another issue that the Supreme Court appears to dispose of, but really does not, is the question of the extent to which aesthetics alone can be a basis for regulation. As we shall see in the discussion of state cases, below, many states require that zoning legislation be grounded on some basis other than aesthetics alone – for example, the promotion of property values. Unfortunately, the Supreme Court's guidance has been unclear here. Despite the language in *Berman* that communities can decide to be "beautiful as well as healthy, spacious as well as clean, well-balanced as well as carefully patrolled" (348 U.S. at 33), the case does not discuss to what extent these aesthetic considerations can be isolated. In addition, *Berman* was an eminent-domain case, and whether pronouncements about eminent domain can be transferred directly to regulation cases is to be approached with some caution.

Metromedia again suffers from the lack of a majority opinion. There is some language in the plurality opinion suggesting that billboards can be an aesthetic harm (453 U.S. at 511–512); but the ordinance contained also the goal of traffic safety, and so other ends were served. Moreover, *Penn Central* involves a conclusion by a New York City commission that landmark preservation provides a benefit both economically and to the quality of life. Finally, *Vincent* brings in issues of economics and psychology and quality of life in support of aesthetic judgments, leaving uncertain the extent to which aesthetics alone will suffice to uphold such judgments.

The question of whether aesthetics alone can support regulation is, however, much discussed among the state cases and in many ways has been the most important topic of debate in this area, and it is to the state cases that we now turn.

State cases

The state cases fall into several categories (for details, see Bufford, 1980). First, there are the states that have no reported cases deciding the issue of the validity of regulation on the basis of aesthetics alone. These states include Alabama, Alaska, Arizona, Georgia, Idaho, Nevada, Oklahoma, South Carolina, South Dakota, and Wyoming. There are currently eighteen jurisdictions that accept regulation based on aesthetics alone: Arkansas, California, Colorado, Delaware, District of Columbia, Florida, Hawaii, Massachusetts, Michigan, Mississippi, Montana, New Jersey, New Mexico, New York, Ohio, Oregon, Utah, and Wisconsin. Only nine states prohibit regulation based solely on aesthetics: Illinois, Maryland, Nebraska, North Carolina, Rhode Island, Tennessee, Texas, Vermont, and Virginia. Finally, there are fourteen states in which the issue is undecided: Connecticut, Indiana, Iowa, Kansas, Kentucky, Louisiana, Maine,

Minnesota, Missouri, New Hampshire, North Dakota, Pennsylvania, Washington, and West Virginia.[4]

The listing of states suggests an important factor or, rather, the lack of one. While one might classify the states without decisions as relatively rural in nature, the other categories are generally not so easily classifiable. Nor are they classifiable on perceived political and social differences among the states. Thus of the states that allow regulation based on aesthetics alone, the list includes such diverse jurisdictions as Utah and Mississippi, on the one hand, and New York and California, on the other. Similarly, the states that disapprove of aesthetic regulation without additional supporting reasons include such disparate states as Texas and Virginia, on the one hand, and Rhode Island and Vermont, on the other. This is not surprising if one remembers that state legal doctrines are indeed the product of specific courts under specific circumstances. This phenomenon is observed in other areas of land use, in many cases with even sharper contrast (Pearlman, 1984). What it means is that it will be very difficult to predict the acceptability of aesthetic regulation and explain it on sociological or political terms. Each jurisdiction must be treated *sui generis*. This becomes especially important when one asks whether courts are likely to be receptive to arguments based on social-science data and evidence with regard to support of aesthetics (see below, for a discussion of this topic).

A second point, on which there is some clear evidence, is that the trend is to increased acceptance of regulations based on aesthetics. Thus of the states authorizing such regulation, Arkansas expressed agreement in 1983, California in 1979, Colorado in 1978, Delaware in 1964, District of Columbia in 1954, Florida in 1960, Hawaii in 1967, Massachusetts in 1975, Michigan in 1975, Mississippi in 1974, Montana in 1977, New Jersey in several cases during the 1970s, New Mexico in 1982, New York in 1977, Ohio in 1968, Oregon in 1965, Utah in 1975, and Wisconsin in 1955. Most of the decisions came in the 1970s, and only two jurisdictions date back before the 1960s.

There has been a slowing down of late, but this is probably to be expected. The big rush to decision came during the 1970s, when the environmental movement was strong and aesthetics was beginning to receive recognition as a factor of major importance in overall environmental quality. For example, the early 1970s saw the enactment of the National Environmental Policy Act, which provides for a federal commitment to environmental quality, as well as important pollution-control and land-use legislation (e.g., Coastal Zone Management Act of 1972). As the environmental movement tends to reach a more stabilizing viewpoint, where gains are being consolidated but few advances made, it is understandable that decisions in the area of aesthetics have also been less frequent.

Nonetheless, as more jurisdictions receive cases in which the issue of aesthetic regulation is to be faced (and as the next section makes clear, it is not always easy to find such a case), there is reason to believe that the trend will continue. However, it may become more difficult to sell the notion to the remaining states unless something more than "This is the way the community decided" is to be found when recalcitrant courts ask how the validity of aesthetic judgments can be evaluated. Similarly, as local jurisdictions become more emboldened and pass legislation of increasingly comprehensive nature, the courts in jurisdictions that accept aesthetic regulation may still become more insistent on reasoned support for such decisions.

The notion that aesthetics alone may or may not be a foundation for police-power regulation sounds simple enough, but in fact it is fraught with difficulties. First of all, what does it mean? Typically, communities passing legislation based on aesthetics will base their legislation not only on aesthetics, but on other factors as well. Most typically. these factors will include the retention of proper-ty values and often such other factors as, in the case of a sign ordinance, traffic safety. It is often very difficult to know what was the major motivating factor in many of these ordinances, and, to be safe, communities will frequently throw in other factors, especially in jurisdictions where courts have not come down on the side of aesthetic regulation. At times, however, a decision cannot be avoided.

New Mexico and Arkansas have been the most recent states to uphold aesthet-ic zoning, and the New Mexico decision of *Temple Baptist Church, Inc.* v. *City of Albuquerque,* 98 N.M. 138, 646 P.2d 565 (1982 [all page references to the P.2d citation]), shows the pitfalls of reliance on other factors. In this case, the city of Albuquerque had a sign ordinance that sought to regulate the number, size, and height of commercial and noncommercial signs. The ordinance con-tained the stated intent that the ordinance was to control and abate unsightly use of land and buildings and to enhance the appearance of the landscape. In addi-tion, the ordinance had stated nonaesthetic reasons to help control congestion in the streets and to promote the health, safety, and convenience of the citizens. A manual of sign regulations subsequently issued by the city stated that the most important goal behind the ordinance was to promote the general welfare of the citizens. This is, no doubt, a rather general intent.

Whatever the principal intent of the ordinance, the trial court that heard the case found that the sign ordinance was not reasonably related to traffic safety and that the ordinance had to be judged on the basis of aesthetics alone. The trial judge concluded that such support could not be found in New Mexico law.

The Supreme Court of New Mexico reversed the decision. First, it reviewed the history of legal disputes over the issue of aesthetic regulation. While it recognized that some other jurisdictions had not been persuaded, it found that aesthetic considerations alone can justify the exercise of the police power. How-ever, the court emphasized that this does not give blanket approval to any set of aesthetic regulations. Thus such regulations still need to be construed for their reasonableness in relation to aesthetic purposes. This, of course, can be an issue of major importance. In the Albuquerque case, however, the court did not pursue the question to any great extent, simply concluding that it could not say that the number, height, and size regulations did not aid in creating or preserving an overall desirable ambience in Albuquerque.[5]

Crucial issues

Aesthetics and other concerns

The issues raised in the Albuquerque case — Is aesthetics alone a basis for the police power? How does one determine the reasonableness of any given regulation? — are in one form or another going to be present in any similar case. They raise several questions. How does one determine when a regulation is in fact based on aesthetics alone? If a regulation purports to be based on considerations other than aesthetics, can one say that these considerations are derivative in nature? For example, if property values are alleged to be part of the basis of the ordinance, are they not really the result of the ordinance's aesthetic-based content? Based on that, if the ordinance improves the visual quality of an area, but property values decline, is the ordinance more likely to be successfully attacked or has it nonetheless achieved its fundamental purposes? Another important question is how can one provide support for ordinances based on aesthetics alone. Courts do defer to community decisions in this area, but many are uncomfortable about not being able to find strong support for aesthetic judgments. Further, if stronger support can be found, it may make communities more willing to pass such ordinances and to do so openly, not bringing spurious considerations into play when they disguise the real reasons for passage of an ordinance. As in the Albuquerque case, although the ordinance seemed clearly designed for aesthetic reasons, it was supported by the city in part on other grounds. Another court may have been concerned about the whole ordinance and felt that the city had not carefully considered what it was doing. It may well be better to bring aesthetic reasons to the fore and not cloud the issue of the city's judgment by raising grounds not really at issue.

From the point of view of the community, it is very easy to "throw in" supporting reasons other than aesthetics. Indeed, many communities, attempting to pass a comprehensive ordinance for aesthetic regulation, will undoubtedly be advised by counsel to add additional reasons. This is not bad advice, for as we have seen, even where courts have upheld aesthetic regulation without such elements, it is often helpful to rely on them as well.

But this makes it difficult to understand and discern the true motivation of the community. In many cases, it is doubtful whether other considerations are actually taken into account. This may be especially true in administrative decisions. Professor Beverly Rowlett (1981) has noted that although legislation involving architectural-design-review legislation typically relies on considerations of property-value retention as well as of aesthetics, the extent to which the design review boards operating under that legislation actually rely on it is unclear. When one considers that all too frequently members on such boards make decisions on the basis of what "looks good" and perhaps not necessarily on the ordinance, it may be preferable to keep the ordinance as direct and forthright as possible. In any event, where the true considerations of the framers of the ordinance or of the administrative decision makers are hidden, it becomes difficult for courts to evaluate the impacts of local decision making.

Furthermore, it can be argued that nonaesthetic considerations are different from aesthetic considerations in this area because the aesthetic considerations are primary and the nonaesthetic aspects are derivative. Even if a community expresses the view of protection of property values, the ordinance really deals with aesthetics, just as a traffic-safety ordinance will protect property values, but one is more likely to want to examine the latter ordinance in the context of whether it protects traffic safety.

Even if courts can discern the true reasons and can find that economic considerations played a role, there is still the difficult problem of relating the protection of property values to aesthetics. It is difficult in the first instance to know whether there will be a positive economic impact from aesthetic legislation. Professor George Lefcoe (1974) has argued that it may be difficult, if not impossible, to determine the impact of aesthetics on property values. Unlike the case of ordinary property appraisal, where like properties are compared with one another, it is almost impossible for courts to compare the impacts of aesthetic regulation because one cannot compare identical or similar situations. Does an unaesthetic use adversely affect property values if located in one area compared with a similar area with no such nonaesthetic uses? This is a problem using the traditional analysis performed by courts, but, as discussed, the use of social-science techniques may assist courts in deciding such questions.

Another obvious problem is that one can postulate situations in which it is quite likely that certain nonaesthetic uses may well have a positive impact in economic terms. Thus if a community in an older historic area decides to allow a modern office building to locate in that area, even though it does not conform aesthetically, the result may be increased property values for everybody. Indeed, given the nature of certain businesses, many areas may suffer inevitable deterioration unless totally new types of structures can be permitted. This is not always the case, fortunately, but it does show the difficulty of finding a positive correlation between aesthetics and property values.

The role of social-science research

The economic impacts of aesthetic regulation may be no more objectively discernible than the aesthetic judgments on which they are based. But what of the aesthetic judgments themselves? To what extent can they be said to be objectively supported? Does it matter?

In many cases, as far as the courts are concerned, it does not matter. The courts typically use a presumption of validity toward a local ordinance. Thus if a community desires that all buildings in a certain area conform to colonial architecture, the court is unlikely to probe beyond that judgment. Yet commentators have argued that such deference in cases of aesthetic regulation is unwarranted (see Rowlett, 1981). Whether they are correct is not at issue here, but as the basis for aesthetic decisions begins to be in doubt in a number of instances, it behooves communities to do all they can to support such judgments. Nobody is likely to argue that a junkyard is attractive, but such items as unity of design and diversity of the visual environment may be debated to an extent that may cause courts to wonder. In addition, as the major decisions are made by the community in the legislative or administrative process, it would be helpful to know if local bodies can have more assurance that their decisions have some basis and thus are more likely to be upheld in court.

This raises the question of the extent to which social-science research can be useful as an underlying support for such decisions. Unfortunately, this is an issue that has not apparently been discussed by the courts, so there is little judicial guidance in the area of aesthetics. Social-science research has, of course, been used in other areas, most notably in *Brown* v. *Board of Education*, 347 U.S. 482 (1954), in which the Supreme Court overturned the concept of separate but equal schools for racial minorities in large part because studies by Gunnar Myrdahl and others had shown that blacks perceived themselves as being on the outside of the system and their learning suffered correspondingly (347 U.S. at 494–495). Undoubtedly, there have been other instances, but not in the area of aesthetics.

What types of issues might "experts" on aesthetics bring to the discussion? It is not my intention here to deal specifically with the validity of social-science research, and, indeed, that is not within my expertise. But it is important, I think, for social scientists to be able to provide courts and local legislative bodies with some justifications for their decisions.

One important way of doing this is to increase the use of empirical research on community preferences and attitudes. This is useful to determine how specific communities view certain issues. While legislative bodies are considered to have all the powers of the citizenry as a whole, well-designed studies would provide considerable support for their decisions, not as a vote, but as the informed choice of community values. Does the community really desire an area with only colonial homes, for example? Why? Can one get at the true concerns of the citizens?

Beyond specific controversies, social-science research may be useful in general issues. To what extent is there agreement on what kinds of land-use combination are perceived as aesthetic? To what extent is there an understanding of why this is so? Is there a basis for courts to approve aesthetic controls beyond the obvious keep-out-the-junkyard situations other than by examining the expressed desires of the legislatures and deferring to them? It need not be the case that the research conclusions point unequivocally in one direction. Indeed, in any event, that would be unlikely outside of perhaps a few narrow circumstances. But if local communities used expert testimony to show what was clear and what was not, where the evidence seemed to point and where it did not, then they would still have to make legislative judgments based on imperfect knowledge, but at least the courts would know that the local communities had given attention to what they were doing and why they had decided as they had.

There is, of course, always the danger that local communities will be misled by experts who may have ideological axes to grind. More likely, communities will not understand what the expert is saying or the limitations of the data and research. But even there, the effort made is more likely to convince a court that what was decided was done in a comprehensive manner. Moreover, where it is clear that the community failed to understand the expert testimony, this will make it easier for a court to come to its own reasoned decision on upholding or not upholding the community decision. The court will be better able to evaluate the interests that government had in passing the regulation. It will at worst have a better understanding of the issues involved. That should lead to better decision making, if done carefully.

"Carefully" is the key. To do this, lawyers and social scientists need to consider together the application of social-science data and knowledge in the area of aesthetics to local decision making and the courtroom situation. Courts are always weighing the public interest against the impacts on the individual in zoning cases. Anything that can be done to increase the understanding of the impact of legislation on these often-competing aspects of zoning cases can make the weighing process less of a guessing game than it currently is in the area of aesthetic judgments. The federal and state courts have opened up the field of aesthetic regulation in recent years. Now is the time to improve the quality of decisions in order to consolidate the gains.

Notes

1. There is some dispute in law over whether the power of eminent domain is different from the police power. This is of no critical importance here, since we are concerned with only public purpose, which appears as an issue in both situations.

2. Burger's position on the First Amendment issue is that as long as the regulation is neutral with respect to content and as long as other means of communication are open, then there is no First Amendment violation.

3. In most instances, courts defer to legislative judgments on land-use issues; see *Euclid* v. *Ambler Realty Co.*, cited in the text.

4. This list includes states where there is an indication that the state would approve or disapprove of regulation based *dicta* – that is, where the statements were not necessary to the decision in the case and therefore not binding.

5. The court also rejected challenges on grounds of taking, because the ordinance required signs to become conforming within five years, and on First Amendment grounds, concluding that the ordinance was a reasonable time, place, and manner restriction and thus permissible under the First Amendment.

References

Ackoff, R. (1975). Does quality of life have to be quantified? In C. W. Churchman (Ed.), *Systems and management annual*. New York: Wiley.

Adorno, T. W. (1983). *Aesthetic theory* (G. Lenhardt, Trans.). London: Routledge & Kegan Paul.

Alston, W. P. (1964). *Philosophy of language*. Englewood Cliffs, N.J.: Prentice-Hall.

Altman, I., and Chemers, M. (1980). *Culture and environment*. Monterey, Calif.: Brooks/Cole.

Altman, I., and Gauvain, M. (1981). A cross-cultural and dialectic analysis of homes. In L. S. Liben, A. H. Patterson, and N. Newcombe (Eds.), *Spatial representation and behavior across the life span* (pp. 283–320). New York: Academic Press.

Anderson, E. (1978). *Visual resource assessment: Local perceptions of familiar natural environments*. Unpublished doctoral dissertation, University of Michigan, Ann Arbor.

Anderson, T. W., Zube, E. H., and MacConnell, W. P. (1976). Predicting scenic resource values. In E. H. Zube (Ed.), *Studies in landscape perception* (Publication No. R-76-1, pp. 6–70). Amherst: University of Massachusetts, Institute for Man and the Environment.

Angier, R. P. (1903). The aesthetics of unequal division. *Psychological Review Monographs* 4, 541–561.

Appleton, J. (1975a). *The experience of landscape*. London and New York: Wiley.

Appleton, J. (1975b). Landscape evaluation: The theoretical vacuum. *Transactions of the Institute of British Geographers* 66, 120–123.

Appleton, J. (1978). *The poetry of habitat*. Hull, Eng.: University of Hull.

Appleton, J. (1980). *Landscape in the arts and sciences*. Inaugural lecture, 1979. Hull, Eng.: University of Hull.

Appleton, J. (1982). Pleasure and the perception of habitat: A conceptual framework. In B. Sadler and A. Carlson (Eds.), *Environmental aesthetics: Essays in interpretation*. Victoria: University of British Columbia Press.

Appleton, J. (1984) (reprinted this collection). Prospects and refuges revisited. *Landscape Journal* 8, 91–103.

Appleyard, D. (1969). Why buildings are known. *Environment and Behavior* 1, 131–156.

Appleyard, D. (1976). *Livable urban streets: Managing auto traffic in neighborhoods*. Washington, D.C.: Government Printing Office.

Appleyard, D., and Lintel, M. (1972). The environmental quality of city streets: The residents' viewpoint. *Journal of the American Institute of Planners* 38, 84–101.

Arnheim, R. (1966a). Gestalt psychology and artistic form. In L. Law Whyte (Ed.), *Aspects of form* (pp. 196–208). Bloomington and London: Indiana University Press.

Arnheim, R. (1966b). *Towards a psychology of art*. London: Faber and Faber.

Arnheim, R. (1977). *Dynamics of architectural form*. Berkeley: University of California Press.

Arrow, K. J. (1951). *Social choice and individual values*. New York: Wiley.

Baker, A., Davies, R. L., and Sivadon, P. (1959). *Psychiatric services and architecture*. Geneva: World Health Organization.

Balchen, B. (1974). Environment, ecology, energy: Added dimensions to practice. *American Institute of Architects Journal* 61, 15–17.

Barker, R. G. (1965). Explorations in ecological psychology. *American Psychologist* 20, 1–14.

Barrell, J. (1972). *The idea of landscape and the sense of place, 1730–1840: An approach to the poetry of John Clare.* New York: Cambridge University Press.

Barthes, R. (1967). *Elements of semiology* (A. Lavers and C. Smith, Trans.). London: Cape.

Bartlett, M. S. (1941). The statistical significance of canonical correlations. *Biometrika* 32, 29–38.

Bartlett, M. S. (1950). Tests of significance in factor analysis. *British Journal of Statistical Psychology* 3, 77–85.

Baum, H. S. (1980a). Analysts and planners must think organizationally. *Policy Analysis* 6, 480–494.

Baum, H. S. (1980b). Sensitizing planners to organization. In P. Clavel, J. Forester, and W. W. Goldsmith (Eds.), *Urban and regional planning in an age of austerity* (pp. 279–309). New York: Pergamon Press.

Beck, R. (1970). Spatial meaning and properties of the environment. In H. Proshansky, W. Ittelson, and L. Rivlin (Eds.), *Environmental psychology: Man and his physical setting* (pp. 131–141). New York: Holt, Rinehart and Winston.

Berger, A., and Good, L. (1963). Architectural psychology in a mental hospital. *Journal of the American Institute of Architects* 40, 76–80.

Berk, R. A. (1979). Generalizability of behavioral observations: A clarification of interobserver agreement and interobserver reliability. *American Journal of Mental Deficiency* 83, 460–472.

Berleant, A. (1970). *The aesthetic field.* Springfield, Ill.: Thomas.

Berleant, A. (1973). Aesthetic function. In D. Riepe (Ed.), *Phenomenology and natural existence* (pp. 183–193). Albany: State University of New York Press.

Berleant, A. (1978). Aesthetic paradigms for an urban ecology. *Diogenes* 103, 1–28.

Berlyne, D. E. (1960). *Conflict arousal and curiosity.* New York: McGraw-Hill.

Berlyne, D. E. (1967). Arousal and reinforcement. In D. Levine (Ed.), *Nebraska Symposium on Motivation.* Lincoln: University of Nebraska Press.

Berlyne, D. E. (1971). *Aesthetics and psychobiology.* New York: Appleton-Century-Crofts.

Berlyne, D. E. (1972). Ends and means of experimental aesthetics. *Canadian Journal of Psychology* 26, 303–325.

Berlyne, D. E. (1974a). The new experimental aesthetics. In D. E. Berlyne (Ed.), *Studies in the new experimental aesthetics* (pp. 1–25). Washington, D.C.: Hemisphere.

Berlyne, D. E. (1974b). Verbal and exploratory responses to visual patterns varying in uncertainty and redundancy. In D. E. Berlyne (Ed.), *Studies in the new experimental aesthetics* (pp. 121–158). Washington, D.C.: Hemisphere.

Berlyne, D. E. (1974c). Novelty, complexity and interestingness. In D. E. Berlyne (Ed.), *Studies in the new experimental aesthetics* (pp. 175–180). Washington, D.C.: Hemisphere.

Berlyne, D. E. (1974d). Concluding comments. In D. E. Berlyne (Ed.), *Studies in the new experimental aesthetics* (pp. 305–332). Washington, D.C.: Hemisphere.

Berlyne, D. E. (1975). Dimensions of perception of exotic and pre-Renaissance paintings. *Canadian Journal of Psychology* 29, 151–173.

Berlyne, D. E. (1976). The new experimental aesthetics and the problem of classifying works of art. *Scientific Aesthetics* 1, 85–106.

Berlyne, D. E., and Ogilvie, J. (1974). Dimensions of perception of paintings. In D. E. Berlyne (Ed.), *Studies in the new experimental aesthetics* (pp. 181–226). Washington, D.C.: Hemisphere.

Bernstein, R. J. (1978). *The restructuring of social and political theory.* Philadelphia: University of Pennsylvania Press.

Biddle, J. (Ed.). (1980). *Old and new architecture: Design relationship.* Washington, D.C.: Preservation Press.

Birkhoff, G. D. (1933). *Aesthetic measure.* Cambridge, Mass.: Harvard University Press.

Blackwood, L. G., and Carpenter, E. H. (1978). The importance of anti-urbanism in determining residential preferences and migration patterns. *Rural Sociology* 43, 31–43.

Blake, P. (1960). *Le Corbusier: Architecture and form.* Baltimore: Pelican Books.

Bloch, E. (1979). *Das Prinzip Hoffnung* (3 vols). Frankfurt: Suhrkamp-Verlag.

Bohlman, H., and Dundas, M. (1980). Local control of architecture: Is it legal? *Real Estate Law Journal* 9, 17–29.

Bollnow, O. F. (1961). Lived-space. *Philosophy Today,* 5, 31–39.

Bosanquet, B. (1892). *A history of aesthetics.* New York: Macmillan.

Boulding, K. (1956). *The image.* Ann Arbor: University of Michigan Press.

Bourdieu, P. (1974). *Zur Soziologie der symbolischen Formen.* Frankfurt: Suhrkamp-Verlag.

Bourne, L. E., Jr. (1966). *Human conceptual behavior.* Boston: Allyn & Bacon.

Bowsher, A. (1978). *Design review in historic districts.* Washington, D.C.: Preservation Press.

Brace, P. (1980, June). Urban aesthetics and the courts. *Environmental Comment,* 16–19.

Brock, H. W. (1977). Desiderata of normative public policy models. In *Proceedings of the Lawrence Symposium on Systems and Decision Sciences* (pp. 276–280). Berkeley, Calif.: Lawrence Hall of Science.

Brolin, B. (1980). *Architecture in context.* New York: Van Nostrand Reinhold.

Brolin, B. (1982). Personal correspondence to L. N. Groat.

Brotherton, I. (1979). Prospect-refuge theory: Is it hazardous? *Landscape Research* 4, 13–16.

Brown, D. S. (Undated). The meaningful city. In *Civic design miscellany* (pp. 53–55). Philadelphia: University of Pennsylvania Press.

Brownlow, T. (1983). *John Clare and the picturesque landscape.* New York: Oxford University Press.

Bruner, J. S., Goodnow, J. J., and Austin, G. A. (1956). *A study of thinking.* New York: Wiley.

Brunswik, E. (1956). *Perception and the representative design of psychological experiments.* Berkeley: University of California Press.

Brush, R. O. (1979). Attractiveness of woodlands: Perceptions of forest landowners in Massachusetts. *Forest Science* 25. 495–506.

Buechner, R. D. (Ed.). (1971). *National park, recreation and open space standards.* Washington, D.C.: National Recreation and Park Association.

Bufford, S. (1973). Beyond the eye of the beholder: Aesthetics and objectivity. *Michigan Law Review* 71, 1438–1463.

Bufford, S. (1980). Beyond the eye of the beholder: A new majority of jurisdictions authorize aesthetic regulations. *University of Missouri at Kansas City Law Review* 48, 126–166.

Bugelski, B. R. (1956). *Psychology of learning.* New York: Holt.

Buhyoff, G. J., Wellman, J. D., Harvey, H., and Fraser, R. A. (1978). Landscape architects' interpretations of people's landscape preference. *Journal of Environmental Management* 6, 285–292.

Burchard, J. E., and Bush-Brown, A. (1966). *The architecture of America: A social and cultural history.* Boston: Little, Brown.

Calhoun, J. B. (1971). Space and strategy of life. In E. H. Esser (Ed.), *Behavior and environment* (pp. 329–387). New York: Plenum Press.

Calvin, J. S., Dearinger, J. A., and Curtin, M. E. (1972). An attempt at assessing preferences for natural landscapes. *Environment and Behavior* 4, 447–470.

Canter, D. (1969). An intergroup comparison of connotative dimensions in architecture. *Environment and Behavior* 1, 37–48.

Canter, D. (1972). *People and buildings: A brief overview of research.* Paper presented at the British Psychological Society Occupational Psychology Annual Conference, Warwick.

Canter, D. (1977). *The psychology of place.* London: Architectural Press.

Canter, D., and Canter, S. (1971). Close together in Tokyo. *Design and Environment* 2, 60–62.

Canter, D., and Thorne, R. (1972). Attitudes to housing: A cross-cultural comparison. *Environment and Behavior* 4, 3–32.

Capek, M. (1961). *Philosophical impact of contemporary physics*. Princeton, N.J.: Van Nostrand.

Carlhian, J. P. (1980). Guides, guideposts and guidelines. In J. Biddle (Ed.), *Old and new architecture: Design relationship* (pp. 49–68). Washington, D.C.: Preservation Press.

Carlson, A. A. (1977). On the possibility of quantifying scenic beauty. *Landscape Planning* 4, 131–172.

Carlson, A. A. (1984). On the possibility of quantifying scenic beauty: A response to Ribe. *Landscape Planning* 11, 49–65.

Carp, F. M. (1970). Correlates of mobility among retired persons. In J. Archea and C. Eastman (Eds.), *EDRA Two: Proceedings of the Second Annual Environmental Design Research Association Conference* (pp. 171–182). Pittsburgh: Carnegie-Mellon University.

Carp, F. M., Zawadaski, R. T., and Shokrin, H. (1976). Dimensions of urban quality. *Environment and Behavior* 8, 239–265.

Carr, S. (1971). *City lights and signs: A pilot study* (Boston Redevelopment Authority). Cambridge, Mass.: MIT Press.

Carr, S., and Schissler, D. (1969). The city as a trip. *Environment and Behavior* 1, 7–35.

Carroll, J. D., and Chang, J. J. (1966). *A general index of nonlinear correlation and its application to the problem of relating physical and psychological dimensions*. Paper presented at the meeting of the Psychometric Society, New York.

Carroll, J. D., and Chang, J. J. (1970). Analysis of individual differences in multidimensional scaling via an *N*-way generalization of Eckart-Young decomposition. *Psychometrika* 35, 283–319.

Carroll, J. D., and Chang, J. J. (1977). *The MDS(X) series of multidimensional scaling programs: PREFMAP program* (Report No. 38). Edinburgh: University of Edinburgh Program Library Unit.

Cass, R. C., and Hershberger, R. G. (1972). *Further research toward a set of semantic scales to measure the meaning of designed environments*. Unpublished paper, Arizona State University, Tempe.

Cavaglieri, G. (1980). The harmony that can't be dictated. In J. Biddle (Ed.), *Old and new architecture: Design relationship* (pp. 37–48). Washington, D.C.: Preservation Press.

Chapin, F. S., and Hightower, H. C. (1965). Household activity patterns and land use. *Journal of the American Institute of Planners* 31, 222–231.

Charlesworth, W. R. (1976). Human intelligence as adaptation: An ethological approach. In L. B. Resnick (Ed.), *The nature of intelligence* (pp. 147–168). Hillsdale, N.J.: Erlbaum.

Checkland, P. (1981). *Systems thinking, systems practice*. Chichester, Eng.: Wiley.

Clamp, P., and Powell, M. (1982). Prospect-refuge theory under test. *Landscape Research* 7, 7–8.

Clegg, S. (1984). *Aesthetic evaluations in a public decision-making context: Design alternatives for Seminole Highway, Madison*. Unpublished master's thesis, University of Wisconsin, Madison.

Coastal Zone Management Act (CZMA). (1972). 1451 (CZMA 302) congressional hearings, and 1452 (CZMA 303) Congressional declaration of policy. Public Law 89-454 Title III, 302 and 303, as added Public Law 92–583.

Cohon, J. L. (1978). *Multiobjective programming and planning*. New York: Academic Press.

Collins, J. B. (1969). *Some verbal dimensions of architectural space perception*. Unpublished doctoral dissertation, University of Utah, Salt Lake City.

Collins, J. B. (1971, September). *Scales for evaluating the architectural environments*. Paper presented at the annual convention of the American Psychological Association, Washington, D.C.

Colquhoun, A. (1967). Typology and design method. *Arena: Journal of the Architectural Association* 83, 11–14.

Cook, R. J. (1983). *Community appearance: A model for the preparation of legal and effective standards*. Unpublished master's thesis, University of Wisconsin, Madison.

Cooper, C. (1972). Resident dissatisfaction in multifamily housing. In W. M. Smith (Ed.), *Behavior, design and policy aspects of human habitats* (pp. 119–146). Green Bay: University of Wisconsin Press.

Cooper, C. (1974). The house as symbol of self. In J. Lang, C. Burnette, W. Moleski, and D. Vachon (Eds.), *Designing for human behavior: Architecture and the behavioral sciences* (pp. 130–146). Stroudsburg, Pa.: Dowden, Hutchinson and Ross.

Corlese, C. F., and Jones, B. (1977). The sociological analysis of boom towns. *Western Sociological Review* 8 (1), 76–90.

Council on Environmental Quality. (1970). *Environmental quality: The First Annual Report of the Council on Environmental Quality.* Washington, D.C.: Government Printing Office.

Craik, K. H. (1966). *Environmental display adjective checklist.* Unpublished paper, University of California, Institute of Personality Assessment and Research, Berkeley.

Craik, K. H. (1968). The comprehension of the everyday physical environment. *Journal of the American Institute of Planners* 34, 29–37.

Craik, K. H. (1971). The assessment of places. In P. McReynolds (Ed.), *Advances in psychological assessment* Vol. 2. Palo Alto, Calif.: Science and Behavior Books.

Craik, K. H. (1981a). Comments on "The psychological representation of molar physical environments" by Ward and Russell. *Journal of Experimental Psychology: General* 110, 158–162.

Craik, K. H. (1981b). Environmental assessment and situational analysis. In D. Magnusson (Ed.), *Towards a psychology of situations: An interactional perspective* (pp. 37–48). Hillsdale, N.J.: Erlbaum.

Craik, K. H., Appleyard, D., and McKechnie, G. E. (1980). *Impressions of place: Effects of media and familiarity among environmental professionals* (Technical report). Berkeley: University of California, Institute of Personality Assessment and Research.

Cranz, G. (1982). *The politics of park design.* Cambridge, Mass.: MIT Press.

Crawford, D. W. (1976). Review of J. Appleton, *The Experience of Landscape. Journal of Aesthetics and Art Criticism* 34, 367–369.

Creelman, M. B. (1966). *The experimental investigation of meaning: A review of the literature.* New York: Springer.

Crumplar, T. (1974). Architectural controls: Aesthetic regulation of the urban environment. *Urban Lawyer* 6, 622–644.

Cullen, G. (1961). *The concise townscape.* London: Architectural Press.

Dale, P. S. (1972). *Language development: Structure and function* (2nd ed.). Hillsdale, Ill.: Dryden Press.

Daniel, T. C., and Boster, R. S. (1976). *Measuring landscape aesthetics: The Scenic Beauty Estimation method* (Research paper RM-167). Fort Collins, Colo.: U.S. Department of Agriculture, Rocky Mountain Forest and Range Experiment Station.

Daniel, T. C., and Vining, J. (1983). Methodological issues in the assessment of landscape quality. In I. Altman and J. F. Wohlwill (Eds.), *Human behavior and environment:* Vol. 6. *Behavior and the natural environment* (pp. 39–84). New York: Plenum Press.

DasGupta, P., Sen, A., and Margolin, S. (1972). *Guidelines for project evaluation.* New York: United Nations.

Davis, W. M. (1899). The geographic cycle. *Geographical Journal* 14, 481–504.

Dawkins, R. (1976). *The selfish gene.* New York: Oxford University Press.

Dawson, T. J. (1981, August). *Public participation in pesticide public policymaking in Wisconsin.* Paper presented at the National Plant Board Conference, Duluth, Minn.

Dearden, P. (1984). Factors influencing landscape preferences: An empirical investigation. *Landscape Planning* 11, 293–306.

Denman, A. (1981). *Design resourcebook for small communities.* Ellensburg, Wash.: Small Towns Institute.

Department of Natural Resources. (1981a). Correspondence/memorandum, from H. S. Druckenmiller to C. D. Besadny, February 2. Wisconsin Department of Natural Resources. File Ref. 1600.

Department of Natural Resources. (1981b). Correspondence/memorandum, from H. S. Druckenmiller to C. D. Besadny, October 6. Wisconsin Department of Natural Resources. File Ref. 1600.

Department of Natural Resources. (1982a). Item recommended for Natural Resources Board agenda, from H. S. Druckenmiller to C. D. Besadny, March 5. Wisconsin Department of Natural Resources.

Department of Natural Resources. (1982b). Correspondence/memorandum, from H. S. Druckenmiller to C. D. Besadny, April 16. Wisconsin Department of Natural Resources. File Ref. 1020.

Department of Natural Resources. (1983). Report of the Ad Hoc Committee on Scenic Beauty, from H. S. Druckenmiller to C. D. Besadny, January 4. Wisconsin Department of Natural Resources. File Ref. 1430.

Derouin, J. G. (1981). The Wisconsin model: A consensus approach to the resolution of environmental issues. *Wisconsin Academy Review* 28, 1.

Desbarats, J. M. (1976). Semantic structure and the perceived environment. *Geographical Analysis* 8, 453–467.

Devlin, A. (1976). The "Small Town" cognitive map: Adjusting to a new environment. In G. T. Moore and R. G. Golledge (Eds.), *Environmental knowing: Theories, research and methods* (pp. 58–66). Stroudsburg, Pa.: Dowden, Hutchinson and Ross.

Dewey, J. (1929). *Experience and nature.* London: Allen and Unwin.

Dewey, J. (1934). *Art as experience.* London and New York: Allen and Unwin.

Dickhut, K. E. (1983). *Acceptance of aesthetic regulations as a function of group type.* Paper presented at the Fourteenth Conference of the Environmental Design Research Association, Lincoln, Neb.

Downs, R. (1970). The cognitive structure of an urban shopping center. *Environment and Behavior* 2, 13–39.

Dunn, M. C. (1976). Landscape with photographs: Testing the preference approach to landscape evaluation. *Journal of Environmental Management* 4, 15–26.

Dunn, W. N., and Swierczek, F. W. (1977). Planned organizational change: Towards grounded theory. *Journal of Applied Behavioral Sciences* 13, 135–157.

Eckbo, G. (1964). *Urban landscape design.* New York: McGraw-Hill.

Edwards, A. T. (1946). *Good and bad manners in architecture.* London: Tiranti. (Original work published 1924)

Edwards, W. (1977). Use of multiattribute utility measurement for social decision making. In D. E. Bell, R. L. Keeney, and H. Raiffa (Eds.), *Conflicting objectives in decisions* (pp. 247–276). New York: Wiley.

Edwards, W. (1981). Reflections and criticisms of a highly political multiattribute utility analysis. In L. Cobb and R. Thrall (Eds.), *Mathematical frontiers of the social and policy sciences* (pp. 157–186). Boulder, Colo.: Westview Press.

Einhorn, H. J., and Hogarth, R. M. (1982). Behavioral decision theory: Process of judgment and choice. In G. R. Ungson and D. N. Brounstein (Eds.), *Decision making: An interdisciplinary inquiry* (pp. 15–41). Belmont, Calif.: Wadsworth.

Ertel, S. (1973). Exploratory choice and verbal judgment. In D. E. Berlyne and K. B. Madsen (Eds.), *Pleasure, reward and preference* (pp. 115–132). New York: Academic Press.

Europa yearbook: A world survey (Vol. 5). (1980). London: Europa Publications.

Evans, C. R., and Piggins, D. J. (1966). A comparison of the behavior of geometric shapes when viewed under conditions of steady fixation, and with apparatus for producing a stabilised retinal image. In D. Vernon (Ed.), *Experiments in visual perception* (pp. 293–298). London: Penguin Books.

Evans, D. R., and Day, H. I. (1971). The factorial structure of responses to perceptual complexity. *Psychonomic Science* 27, 357–359.

Evans, G. W., Jacobs, S. V., and Frazer, N. B. (1982). Adaptation to air pollution. *Journal of Environmental Psychology* 2, 99–108.

Evans, G. W., and Wood, D. (1981). Assessment of environmental aesthetics in scenic highway corridors. *Environment and Behavior* 12, 255–274.

Ewald, W. R., Jr., and Mandelker, D. R. (1977). *Street graphics: A concept and a system.* McLean, Va.: Landscape Architecture Foundation.

Eysenck, H. J. (1941). The empirical determination of an aesthetic formula. *Psychological Review* 48, 83–92.

Eysenck, H. J., and Tunstall, O. (1968). La personnalité et l'esthétique des formes simples. *Sciences de l'Art* 5, 3–9.

Farnsworth, P. R. (1954). A study of the Hevner adjective list. *Journal of Aesthetics* 13, 97–103.

Fechner, B. (1876). *Vorschule der Aesthetik.* Leipzig: Breitkopf und Hartel.

Feimer, N. (1984). Environmental perception: The effect of media evaluative context, and the observer sample. *Journal of Environmental Psychology* 4, 61–80.

Fenton, D. M. (1981). Visual sampling of environments: A methodological note. *Perceptual and Motor Skills* 53, 978.

Fenton, D. M. (1984). *Natural environmental perception and preference: An investigation of structure and meaning.* Unpublished doctoral dissertation, James Cook University of North Queensland, Townsville, Australia.

Fenton, D. M. (1985). Dimensions of meaning in the perception of natural settings and their relationship to aesthetic response. *Australian Journal of Psychology* 37, 325–339.

Festinger, L. (1957). *A theory of cognitive dissonance.* Evanston, Ill.: Row Peterson.

Fines, K. D. (1968). Landscape evaluation: A research project in East Sussex. *Regional Studies* 2, 41–55.

Fischoff, B., Lichtenstein, S., Slovic, P., Derby, S., and Keeney, R. (1981). *Acceptable risk.* Cambridge: Cambridge University Press.

Fisher, G. A., Heise, D. R., Bohrnstedt, G. W., and Lucke, J. F. (1985). Evidence for extending the circumplex model of personality trait language to self-reported moods. *Journal of Personality and Social Psychology* 49, 232–242.

Flaschbart, P. G., and Peterson, G. L. (1973). Dynamics of preference for visual attributes of housing environments. In W. F. E. Preiser (Ed.), *Environmental design research IV. Proceedings of the Fourth Annual Environmental Design Research Association Conference* (Vol. 1, pp. 98–106). Stroudsburg, Pa.: Dowden, Hutchinson and Ross.

Flynn, J. E. (1972–73). The psychology of light. A series of eight articles. *Electrical Consultant Magazine* 88 (12–89), 7.

Flynn, J. E. (1973). Concepts beyond the Illumination Engineers Society framework. *Lighting Design and Application Magazine* 3, 4–11.

Flynn, J. E., Spencer, T. J., Martyniuk, O., and Hendrick, C. (1973). Interim study of procedures for investigating the effect of light on impression and behavior. *Journal of the Illumination Engineers Society* 3, 87–94.

Flynn, J. E., Spencer, T. J., Martyniuk, O., and Hendrick, C. (1974). *The effect of lighting on human judgment and behavior* (Interim report, IERI Project 92-73).

Forester, J. (1980). Critical theory and planning practice. In P. Clavel, J. Forester, and W. W. Goldsmith (Eds.), *Urban and regional planning in an age of austerity* (pp. 326–342). New York: Pergamon Press.

Forester, J. (1982). Planning in the face of power. *Journal of the American Planning Association* 48, 67–80.

Frances, R. (1977). *Intérêt perceptif et préférence esthétique.* Paris: Editions du CNRS.

Fransella, F., and Bannister, D. (1977). *A manual for the repertory grid technique*. New York: Academic Press.

Frey, J. E. (1981). *Preference, satisfactions, and the physical environments of urban neighborhoods*. Unpublished doctoral dissertation, University of Michigan, Ann Arbor.

Friedman, J. (1973). *Retracking America: A theory of transactive planning*. Garden City, N.Y.: Doubleday.

Fromme, D. K., and O'Brien, C. S. (1982). A dimensional approach to the circular ordering of emotions. *Motivation and Emotion* 6, 337–363.

Gallagher, T. J. (1977). *Visual preference for alternative natural landscapes*. Unpublished doctoral dissertation, University of Michigan, Ann Arbor.

Gans, H. (1968). *People and plans: Essays on urban problems and solutions*. New York: Basic Books.

Gärling, T. (1976a). Multidimensional scaling and semantic differential scaling technique study of the perception of environmental settings. *Scandinavian Journal of Psychology* 17, 323–332.

Gärling, T. (1976b). The structural analysis of environmental perception and cognition: A multidimensional scaling approach. *Environment and Behavior* 8, 385–415.

Gass, S. I., and Sisson, R. L. (1975). *A guide to models in governmental planning and operations*. Potomac, Md.: Sauger Books.

Getz, D. A., Karow, A., and Kielbaso, J. J. (1982). Inner city preferences for trees and urban forestry programs. *Journal of Arboriculture* 8, 258–263.

Gibson, J. J. (1947). Motion picture testing and research. *Aviation Psychology Research Reports*, No. 7. Washington, D.C.: Government Printing Office.

Gibson, J. J. (1950). *The perception of the visual world*. Boston: Houghton Mifflin.

Gibson, J. J. (1960). The concept of stimulus in psychology. *American Psychologist* 15, 694–703.

Gibson, J. J. (1966). *The senses considered as a perceptual system*. Boston: Houghton Mifflin.

Gibson, J. J. (1977). The theory of affordances. In R. Shaw and J. Bransford (Eds.), *Perceiving, acting and knowing* (pp. 76–82). Hillsdale, N.J.: Erlbaum.

Gibson, J. J. (1979). *The ecological approach to visual perception*. Boston: Houghton Mifflin.

Gibson, J. J. (1982). Notes on affordances. In E. Reed and R. Jones (Eds.), *Reasons for realism: Selected essays of J. J. Gibson*. Hillsdale, N.J.: Erlbaum.

Gideon, S. (1941). *Space, time and architecture: The growth of a new tradition*. London: Oxford University Press.

Gilg, A. W. (1974). A critique of Linton's method of assessing scenery as a natural resource. *Scottish Geographical Magazine* 90, 125–129.

Glaser, B. G., and Strauss, A. L. (1967). *The discovery of grounded theory*. Chicago: Aldine.

Gobster, P. H. (1983). Judged appropriateness of residential structures in natural and developed shoreline settings. In D. Amadeo, J. Griffin, and J. Potter (Eds.), *EDRA 1983: Proceedings of the Fourteenth Environmental Design Research Association Conference* (pp. 105–119). Washington, D.C.: Environmental Design Research Association.

Goffman, E. (1959). *The presentation of self in everyday life*. Harmondsworth, Eng.: Pelican Books.

Goodman, R. (1971). *After the planners*. New York: Touchstone Press.

Gordon, D. A. (1952). Methodology in the study of art evaluation. *Journal of Aesthetics* 10, 338–352.

Gordon, S. I., and Nasar, J. L. (1984). *Development options for High Street across from the University* (Technical report to The Ohio State University Office of Research and Graduate Studies).

Greenbie, B. B. (1974). Social territory, community health, and urban planning. *Journal of the American Institute of Planners* 40, 74–82.

Greenbie, B. B. (1976). *Design for diversity: Planning for natural man in the neo-technic environment: An ethological approach*. Amsterdam: Elsevier Scientific.

Greenbie, B. B. (1981). *Space: Dimensions of the human landscape*. London: Yale University Press.

Gregory, R. L. (1966). *Eye and brain: The psychology of seeing*. New York: McGraw-Hill.

Gregory, R. L. (1969). On how little information controls so much behavior. In C. H. Waddington (Ed.), *Towards a theoretical biology, 2 sketches* (pp. 236–247). Edinburgh: Edinburgh University Press.

Grid Analysis Package (GAP) (2nd ed.). (1981). Manchester, Eng.: University of Manchester, Regional Computing Center.

Groat, L. (1982). Meaning in Post-Modern architecture. An examination using the multiple sorting task. *Journal of Environmental Psychology* 2, 3–22.

Groat, L. (1983) (reprinted this collection). *A study of the perception of contextual fit in architecture*. Paper presented at the Fourteenth Conference of the Environmental Design Research Association, Lincoln, Neb.

Groat, L. (1983a). The past and future of research on meaning in architecture. In D. Amadeo, J. Griffin, and J. Potter (Eds.), *EDRA 1983: Proceedings of the Fourteenth Environmental Design Research Association Conference* (pp. 29–35). Washington, D.C.: Environmental Design Research Association.

Groat, L. (1983b). Measuring the fit of new to old. *Architecture: The American Institute of Architects Journal* 72, 58–61.

Groat, L. (1984). Public opinion of contextual fit. *Architecture: The American Institute of Architects Journal* 73, 72–75.

Gropius, W. (1962). *Scope of total architecture*. New York: Collier.

Guldmann, J. M. (1979). Visual impact and location of activities: A combinatorial optimization methodology. *Socio-Economic Planning and Science* 13, 47–70.

Hall, E. T. (1966). *The hidden dimension*. Garden City, N.Y.: Doubleday.

Hall, R., Purcell, A. T., Thorne, R., and Metcalfe, J. (1976). Multidimensional scaling analysis of interior designed spaces. *Environment and Behavior* 8, 596–610.

Hammitt, W. E. (1978). *Visual and user preferences for a bog environment*. Unpublished doctoral dissertation, University of Michigan, Ann Arbor.

Hammond, K. R., Stewart, T. R., Brehment, B., and Steinman, D. O. (1975). Social judgment theory. In M. Kaplan and S. Schwartz (Eds.), *Human judgment and decision processes* (pp. 271–312). New York: Academic Press.

Handlin, D. P. (1972). The detached house in the age of the object and beyond. In W. J. Mitchell (Ed.), *Environmental design: Research and practice. Proceedings of the Third Environmental Design Research Association Conference* (Vol. 1, pp. 7-2-1–7-2-8). Los Angeles: UCLA.

Hardin, G. (1968). The tragedy of the commons. *Science* 162, 1243–1246.

Hare, F. G. (1975). *The identification of dimensions underlying verbal and exploratory responses to music through multidimensional scaling*. Unpublished doctoral dissertation, University of Toronto.

Harrington, D. L., and Preiser, W. F. E. (1980). Layperson assessments of visual and aesthetic quality. In R. Thorne and S. Arden (Eds.), *People and the man-made environment* (pp. 219–231). Sydney, Aus.: University of Sydney Press.

Harrison, J., and Sarre, P. (1975). Personal construct theory in the measurement of environmental images. *Environment and Behavior* 7, 3–58.

Hartshorne, R. (1939). *The nature of geography: A critical survey of current thought in the light of the past*. Lancaster, Pa.: Association of American Geographers.

Hayes, J. R. (1982). Issues in protocol analysis. In G. R. Ungson and D. N. Brounstein (Eds.), *Decision making: An interdisciplinary inquiry* (pp. 61–77). Belmont, Calif.: Wadsworth.

Hayward, D. G. (1974). Psychological factors in the use of light and lighting in buildings. In J. Lang, C. Burnette, W. Moleski, and D. Vachon (Eds.), *Designing for human behavior: Architecture and the behavioral sciences* (pp. 120–129). Stroudsburg, Pa.: Dowden, Hutchinson and Ross.

Haywood, R. C. (1966). Use of semantic differential dimensions in concept learning. *Psychonomic Science* 5, 305–306.

Hazard, J. (1962). Furniture arrangement and judicial roles. *ETC* 19, 181–88.

Heckhausen, H. (1964). Complexity in perception: Phenomenal criteria and information theoretic calculus: A note on D. E. Berlyne's "complexity effects." *Canadian Journal of Psychology* 18, 168–173.

Heft, H. (1981). An examination of constructivist and Gibsonian approaches to environmental psychology. *Population and Environment* 4, 227–245.

Heider, F. (1946). Attitudes and cognitive organization. *Journal of Psychology* 21, 107–112.

Helson, H. (1964). *Adaptation-level theory*. New York: Harper & Row.

Helwig, J., and Council, K. (1979). *S.A.S. user's guide*. Raleigh, N.C.: S.A.S. Institute.

Hendee, J. C., and Stankey, G. H. (1973). Biocentricity in wilderness management. *Bioscience* 23, 535–538.

Herbert, E. J. (1981). *Visual resource analysis: Prediction and preference in Oakland County, Michigan*. Unpublished master's thesis, University of Michigan, Ann Arbor.

Hershberger, R. G. (1969) (reprinted this collection). A study of meaning and architecture. In H. Sanoff and S. Cohn (Eds.), *EDRA 1: Proceedings of the First Annual Environmental Design Research Association Conference* (pp. 86–100). Raleigh: North Carolina State University.

Hershberger, R. G. (1971, April). *Predicting the meaning of designed environments*. Paper presented at the annual meeting of the Western Psychological Association, San Francisco, Calif.

Hershberger, R. G. (1972). Toward a set of semantic scales to measure the meaning of designed environments. In W. J. Mitchell (Ed.), *Environmental design: Research and practice. Proceedings of the Third Environmental Design Research Association Conference* (Vol. 1, pp. 6-4-1–6-4-10). Los Angeles: UCLA.

Hershberger, R. G. (1974). Predicting the meaning of architecture. In J. Lang, C. Burnette, W. Moleski, and D. Vachon (Eds.), *Designing for human behavior: Architecture and the behavioral sciences* (pp. 147–156). Stroudsburg, Pa.: Dowden, Hutchinson and Ross.

Hershberger, R. G., and Cass, R. (1974) (reprinted this collection). Predicting user responses to buildings. In G. Davis (Ed.), *Field applications* (pp. 117–134), Vol. 4 of H. Carson (Ed.), *Man–environment interactions: EDRA 5: Evaluations and application*. Washington, D.C.: Environmental Design Research Association.

Herzog, T. R., Kaplan, S., and Kaplan, R. (1976). The prediction of preference for familiar urban places. *Environment and Behavior* 8, 627–645.

Herzog, T. R., Kaplan, S., and Kaplan, R. (1982). The prediction of preference for unfamiliar urban places. *Population and Environment: Behavioral and Social Issues* 5, 43–59.

Hesselgren, S. (1969). *The language of architecture* (2nd ed.). Lund, Sweden: Studentlitteratur.

Hesselgren, S. (1975). *Man's perception of man-made environment*. Stroudsburg, Pa.: Dowden, Hutchinson and Ross.

Hester, R. T., Jr. (1984). *Planning neighborhood space with people*. New York: Van Nostrand Reinhold.

Hevner, K. (1935). Experimental studies of the affective value of colors and lines. *Journal of Applied Psychology* 19, 385–398.

Hevner, K. (1936). Experimental studies of the elements of expression in music. *American Journal of Psychology* 48, 246–268.

Hevner, K. (1937). The affective value of pitch and tempo in music. *American Journal of Psychology* 49, 621–630.

Heyligers, P. C. (1981). Prospect-refuge symbolism in dune landscapes. *Landscape Research* 6, 7–11.

Hilgard, E. R., and Bower, G. H. (1966). *Theories of learning* (3rd ed.). New York: Meredith.

Hine, T. (1978, June 26). Stalking the style of the Synagogue. *Philadelphia Inquirer*, sec. L, pp. 1–2.

Hinkle, D. N. (1965). *The change of personal constructs from the viewpoint of a theory of implications*. Unpublished doctoral dissertation, Ohio State University, Columbus.

Hinshaw, M., and Allott, M. (1972). Environmental preferences of future housing consumers. *Journal of the American Institute of Planners* 38, 102–107.

Hochberg, J. (1966). Representative sampling and the purposes of research: Pictures of the world and the world of pictures. In K. Hammond (Ed.), *The psychology of Egon Brunswik* (pp. 361–381). New York: Holt, Rinehart and Winston.

Hoelterhoff, M. (1985, April 23). Pei's pyramid: Revolution in Napoleon's court. *Wall Street Journal*, p. 28.

Honikman, B. (1972). An investigation of the relationship between construing the environment and its physical form. In W. J. Mitchell (Ed.), *Environmental design: Research and practice. Proceedings of the Third Environmental Design Research Association Conference* (Vol. 1, pp. 6-5-1–6-5-11). Los Angeles: UCLA.

Honikman, B. (1976). Personal construct theory and environmental meaning: Applications to urban design. In G. T. Moore and R. G. Golledge (Eds.), *Environmental knowing: Theories, research and methods* (pp. 88–98). Stroudsburg, Pa.: Dowden, Hutchinson and Ross.

Horayangkura, Y. (1978). Semantic dimensional structures: A methodological approach. *Environment and Behavior* 10, 555–583.

Howard, R. A. (1975). Social decision analysis. *Proceedings of the Institute of Electrical and Electronic Engineers* 63, 359–371.

Howard, R. B., Mlynarski, F. G., and Sauer, G. C., Jr. (1972). A comparative analysis of affective responses to real and represented environments. In W. J. Mitchell (Ed.), *Environmental design: Research and practice. Proceedings of the Third Environmental Design Research Association Conference* (pp. 6-6-1–6-6-8). Los Angeles: UCLA.

Hubbard, H. V., and Kimball, T. (1917). *An introduction to the study of landscape design.* New York: Macmillan.

Hudson, R. (1974). Images of the retailing environment: An example of the use of the repertory grid methodology. *Environment and Behavior* 6, 470–493.

Humphreys, P., and McFadden, W. (1980). Experience with MAUD: Aiding decisions structuring through reordering versus automating the decision rule. *Acta Psychologica* 45, 51–69.

Husserl, E. (1960). *Cartesian meditations.* The Hague: Nijhof. (Original work published 1931)

Israeli, N. (1928). Affective reactions to painting reproductions. *Journal of Applied Psychology* 12, 125–139.

Ittelson, W. H. (1976). Environmental perception and contemporary perceptual theory. In H. M. Proshansky, W. H. Ittelson, and L. G. Rivlin (Eds.), *Environmental psychology: People and their physical settings* (2nd ed.) (pp. 141–154). New York: Holt, Rinehart and Winston.

Ittelson, W. H. (1978). Environmental perception and urban experience. *Environment and Behavior* 10, 193–213.

Izumi, K. (1969). *Some psycho-social aspects of environmental design.* Mimeo.

James, W. (1962). *Psychology: The briefer course.* New York: Collier. (Original work published 1892)

James, W. (1904). A world of pure experience. *Journal of Philosophy, Psychology and Scientific Methods* 1, 533–543.

Jammer, M. (1957). *Concepts of force: A study in the foundations of dynamics.* Cambridge, Mass.: Harvard University Press.

Jencks, C., and Baird, G. (1969). *Meaning in architecture.* New York: Braziller.

Johnson, S. (1984, March 20). Architect urges blending of new with old. *News and Courier* (Charleston, S.C.), p. 6A.

Jordahl, H. C., Jr. (1983, July). Interview in Madison, Wis. [Mr. Jordahl is a former chairman of the Natural Resources Board.]

Kainz, F. (1962). *Aesthetics, the science* (H. M. Schueller, Trans.). Detroit: Wayne State University Press.

Kandinsky, W. (1964). Reminiscences. In R. L. Herbert (Ed.), *Modern artists on art.* Englewood Cliffs, N.J.: Prentice-Hall. (Original work published 1913)

Kant, I. (1951). *Critique of judgment.* New York: Hafner.

Kaplan, R. (1972). The dimensions of the visual environment: Methodological considerations. In W. J. Mitchell (Ed.), *Environmental design: Research and practice. Proceedings of the Third Environmental Design Research Association Conference* (pp. 6-7-1–6-7-5). Los Angeles: UCLA.

Kaplan, R. (1973). Predictors of environmental preference: Designers and "clients." In W. F. E. Preiser (Ed.), *Environmental Design Research IV. Proceedings of the Fourth Annual Environmental Design Research Association Conference* (Vol. 1, pp. 265–274). Stroudsburg, Pa.: Dowden, Hutchinson and Ross.

Kaplan, R. (1975). Some methods and strategies in the prediction of preference. In E. H. Zube, R. O. Brush, and G. G. Fabos (Eds.), *Landscape assessment: Values, perceptions and resources* (pp. 118–129). Stroudsburg, Pa.: Dowden, Hutchinson and Ross.

Kaplan, R. (1977a). Down by the riverside: Informational factors in waterscape preference. *Proceedings: River Recreation Management and Research Symposium* (USDA General Technical Report NC-28, pp. 285–289). Chicago: North Central Forest Experiment Station.

Kaplan, R. (1977b). Preference and everyday nature: Method and application. In D. Stokels (Ed.), *Perspectives on environment and behavior* (pp. 235–250). New York: Plenum Press.

Kaplan, R. (1978a). Participation in environmental design: Some considerations and a case study. In S. Kaplan and R. Kaplan (Eds.). *Humanscape: Environments for people* (pp. 427–438). Belmont, Calif.: Duxbury.

Kaplan, R. (1978b). The green experience. In S. Kaplan and R. Kaplan (Eds.), *Humanscape: Environments for people* (pp. 186–193). Belmont, Calif.: Duxbury.

Kaplan, R. (1979a). Visual resources and the public: An empirical approach. In G. H. Elsner and R. C. Smarden (Eds.), *Proceedings of our National Landscape: A conference on applied techniques for analysis and management of the visual resource* (General Technical Report PSW-35, pp. 209–216). Berkeley: U.S. Department of Agriculture Pacific Southwest Forest and Range Experiment Station.

Kaplan, R. (1979b). A methodology for simultaneously obtaining and sharing information. In T. C. Daniel and E. H. Zube (Eds.), *Assessing amenity resource values* (General Technical Report RM-68, pp. 58–66). Fort Collins, Colo.: U.S. Department of Agriculture, Rocky Mountain Forest and Range Experiment Station.

Kaplan, R. (1983). The role of nature in the urban context. In I. Altman and J. F. Wohlwill (Eds.), *Human behavior and environment: Vol. 6. Behavior and the natural environment* (pp. 127–161). New York: Plenum Press.

Kaplan, R. (1985). The analysis of perception via preference: A strategy for studying how the environment is experienced. *Landscape Planning 12,* 161–176.

Kaplan, R., and Herbert, E. H. (1987). Cultural and subcultural comparisons in preference for natural settings. *Landscape and Urban Planning 14.*

Kaplan, S. (1970). The role of location processing in the perception of the environment. In J. Archea and C. Eastman (Eds.), *EDRA Two: Proceedings of the Second Annual Environmental Design Research Association Conference* (pp. 131–134). Pittsburgh: Carnegie-Mellon University.

Kaplan, S. (1973). Cognitive maps in perception and thought. In R. M. Downs and D. Stea (Eds.), *Image and environment* (pp. 63–78). Chicago: Aldine.

Kaplan, S. (1975). An informal model for the prediction of preference. In E. H. Zube, R. O. Brush, and J. C. Fabos (Eds.), *Landscape assessment: Values, perceptions and resources* (pp. 92–101). Stroudsburg, Pa.: Dowden, Hutchinson and Ross.

Kaplan, S. (1977). Tranquility and challenge in the natural environment. In *Children, nature and the urban environment* (USDA General Technical Report NE-30, pp. 181–185). Upper Darby, Pa.: Northeast Forest Experiment Station.

Kaplan, S. (1978). Attention and fascination: The search for cognitive clarity. In S. Kaplan and R. Kaplan (Eds.), *Humanscape: Environments for people* (pp. 84–90). Belmont, Calif.: Duxbury.

Kaplan, S. (1979a) (reprinted this collection). Perception and landscape: Conceptions and misconceptions. In G. H. Elsner and R. C. Smarden (Eds.), *Proceedings of our National Landscape: A conference on applied techniques for analysis and management of the visual resource* (General Technical Report PSW-35, pp. 241–248). Berkeley: U.S. Department of Agriculture, Pacific Southwest Forest and Range Experiment Station.

Kaplan, S. (1979b). Concerning the power of content-identifying methodologies. In T. C. Daniel and E. H. Zube (Eds.), *Assessing amenity resource values* (General Technical Report RM-68, pp. 4–13). Fort Collins, Colo.: U.S. Department of Agriculture, Rocky Mountain Forest and Range Experiment Station.

Kaplan, S., Bufford, S., Lewis, L., Kaplan, R., and Horsbrugh, P. (1978). *Everyday natural environments as a design resource.* Paper presented at the Eighth Environmental Design Research Association Conference, Champaign-Urbana, Ill.

Kaplan, S., and Kaplan, R. (Eds.). (1978). *Humanscape: environments for people.* Belmont, Calif.: Duxbury.

Kaplan, S., and Kaplan, R. (1982). *Cognition and environment: Coping in an uncertain world.* New York: Praeger.

Kaplan, S., Kaplan, R., and Wendt, J. S. (1972). Rated preference and complexity for natural and urban visual material. *Perception and Psychophysics* 12, 354–356.

Kaplan, S., and Talbot, J. F. (1983). Psychological benefits of wilderness experience. In I. Altman and J. F. Wohlwill (Eds.), *Human behavior and environment:* Vol. 6. *Behavior and the natural environment* (pp. 163–203). New York: Plenum Press.

Kaplan, S., and Wendt, J. S. (1972). Preference and the visual environment: Complexity and some alternatives. In W. J. Mitchell (Ed.), *Environmental design: Research and practice. Proceedings of the Third Environmental Design Research Association Conference* (pp. 6-8-1–6-8-5). Los Angeles: UCLA.

Kasmar, J. V. (1970) (reprinted this collection). The development of a usable lexicon of environmental descriptors. *Environment and Behavior* 2, 153–169.

Kasmar, J. V., Griffin, W. V., and Mauritzen, J. H. (1968). Effects of environmental surroundings on outpatients' mood and perception of psychiatrists. *Journal of Consulting and Clinical Psychology* 32, 223–226.

Kasprisin, R., and Stouder, F. (1983a). Small town studio: Creston, Washington. *Western Planner* 4, 1 and 9.

Kasprisin, R., and Stouder, F. (1983b). *Small town design studio.* Seattle: Kasprisin–Hutnik Partnership.

Keeney, R., and Raiffa, H. (1976). *Decisions with multiple objectives: Preferences and value trade-offs.* New York: Wiley.

Kelly, G. A. (1955). *The psychology of personal constructs.* New York: Norton.

Kepes, G. (1944). *Language of vision.* Chicago: Thiebold.

Kepes, G. (Ed.). (1966). *Sign, symbol, image.* New York: Braziller.

Klausner, S. Z. (1971). *On man in his environment.* San Francisco: Jossey-Bass.

Kling, V. A. (1959). Space: A fundamental concept in design. In C. E. Goshen (Ed.), *Psychiatric architecture* (pp. 21–22). Washington, D.C.: American Psychiatric Association.

Klosterman, R. E. (1976). Towards a normative theory of planning. Doctoral dissertation series, Cornell University, Ithaca, N.Y.

Koffka, K. (1935). *Principles of Gestalt psychology.* New York: Harcourt Brace.

Krasner, L. (Ed.). (1980). *Environmental design and human behavior: A psychology of the individual in society.* New York: Pergamon Press.

Kreimer, A. (1977). Environmental preferences: A critical analysis of some research methodologies. *Journal of Leisure Research* 9, 88–97.

Krumholz, N., Cogger, J. M., and Linner, J. H. (1975). The Cleveland policy planning report. *Journal of the American Institute of Planners* 4, 298–304.

Kruskal, J. B. (1964). Multidimensional scaling by optimizing goodness of fit to a nonmetric hypothesis. *Psychometrika* 29, 1–27.

Küller, R. (1972). *A semantic model for describing perceived environment* (Document No. 12). Stockholm: National Swedish Institute for Building Research.

Küller, R. (1975). Beyond semantic measurement. In B. Honikman (Ed.), *Responding to social change* (pp. 181–197). Stroudsburg, Pa.: Dowden, Hutchinson and Ross.

Lang, J., Burnette, C., Moleski, W., and Vachon, D. (Eds.). (1974). *Designing for human behavior: Architecture and the behavioral sciences.* Stroudsburg, Pa.: Dowden, Hutchinson and Ross.

Langer, S. (1953). *Feeling and form.* New York: Scribner.

Langer, S. (1963). *Philosophy in a new key.* Cambridge, Mass.: Harvard University Press.

Laut, P. et al. (Eds.). (1977). *Environment of south Australia.* Canberra: Commonwealth Scientific and Industrial Research Organization, Division of Land Use Research.

Law, F., and Zube, E. H. (1982). *The effects of foreground detail and framing elements on response.* Paper presented at the Thirteenth International Conference of the Environmental Design Research Association (Workshop on Environmental Aesthetics), College Park, Md.

Law Whyte, L. (1972). On the frontiers of science: This hierarchical universe. *Architectural Digest* 10, 611.

Lefcoe, G. (1974). *Land development law: Cases and materials* (2nd ed.). Discussed in Rowlett 1981, *q.v.*

Leff, H. L., Gordon, L. R., and Ferguson, J. G. (1974). Cognitive set and environmental awareness. *Environment and Behavior* 6, 395–447.

Leff, H. S., and Deutsch, P. S. (1973). Construing the physical environment of professionals and lay persons. In W. F. E. Preiser (Ed.), *Environmental Design Research IV. Proceedings of the Fourth Annual Design Research Association Conference* (Vol. 1, pp. 284–297). Stroudsburg, Pa.: Dowden, Hutchinson and Ross.

Levin, J. E. (1976). *Riverscape preference: Onsite and photographic reactions.* Unpublished master's thesis, University of Michigan, Ann Arbor.

Lévy-Leboyer, C. (1982). *Psychology and the environment* (D. Canter and I. Griffiths, Trans.). Beverly Hills, Calif.: Sage. (Original work published 1979)

Lewin, K. (1936). *Principles of topological psychology.* New York: McGraw-Hill.

Lindblom, C., and Cohen, D. (1979). *Useable knowledge: Social science and social problem solving.* New Haven, Conn.: Yale University Press.

Lingoes, J. C. (1972). A general survey of the Guttman-Lingoes nonmetric program series. In R. N. Shepard, A. K. Romney, and S. K. Nerlove (Eds.), *Multidimensional scaling* (Vol. 1, pp. 52–68). New York: Seminar Press.

Lingoes, J. C. (1973). *The Guttman-Lingoes nonmetric program series.* Ann Arbor, Mich.: Mathesis Press.

Linstone, H., and Turoff, M. (Eds.). (1975). *The Delphi method: Techniques and applications.* Reading, Mass.: Addison-Wesley.

Linton, D. L. (1968). The assessment of scenery as a natural resource. *Scottish Geographical Magazine* 84, 218–238.

Lipps, T. (1900). Aesthetic einfühlung. *Zeitschrift für Psychologie und Physiologie der Sinnesorgane* 22.

Locasso, R. M. (1971). *The description of an architectural space from* in-situ *ratings vs. memory.* Unpublished master's thesis, Western Washington State University, Bellingham.

Locasso, R. M. (1976). *The influence of a beautiful vs. an ugly interior environment on selected behavioral measures.* Unpublished doctoral dissertation, Pennsylvania State University, State College.

Lorenz, K. Z. (1972). The role of Gestalt perception in animals and human behavior. In L. Law Whyte (Ed.), *Aspects of form*. Bloomington and London: Indiana University Press.

Lowenthal, D. (1968). The American scene. *Geographical Review* 58, 61–68.

Lowenthal, D., and Riel, M. (1972). The nature of perceived and imagined environments. *Environment and Behavior* 4, 189–207.

Lozano, E. (1974) (reprinted this collection). Visual needs in the environment. *Town Planning Review* 43, 351–374.

Lu, W. (1980). Preservation criteria: Defining and protecting design relationships. In J. Biddle (Ed.), *Old and new architecture: Design relationship* (pp. 186–202). Washington, D.C.: Preservation Press.

Luce, R., and Raiffa, H. (1957). *Games and decisions*. New York: Wiley.

Lundholm, H. (1921). The affective tone of lines: Experimental researches. *Psychiatric Review* 28, 43–60.

Lynch, K. (1960). *The image of the city*. Boston: MIT Press.

MacCallum, R. C. (1978). Recovery of structure in incomplete data by ALSCAL. *Psychometrika* 44, 69–74.

McCannell, D. (1976). *The tourist: A new theory of the leisure class*. New York: Schocken Books.

MacCrimmon, K. R. (1973). An overview of multiple objective decision making. In J. L. Cochrane and M. Zeleny (Eds.), *Multiple criteria decision making* (pp. 18–44). Columbia: University of South Carolina Press.

McHarg, I. (1962). The ecology of the city. *American Institute of Architects Journal* 39, 101–103.

Magnusson, D. (Ed.). (1981). *Towards a psychology of situations: An interactional perspective*. Hillsdale, N.J.: Erlbaum.

Magnusson, D., and Endler, N. S. (Eds.). (1977). *Personality at the crossroads*. New York: Erlbaum.

Mandler, G. (1967). Organization and memory. In K. W. Spence and J. T. Spence (Eds.), *The psychology of learning and motivation* (pp. 327–372). New York: Academic Press.

Marans, R. W. (1976). Perceived quality of residential environments: Some methodological issues. In K. H. Craik and E. H. Zube (Eds.), *Perceiving environmental quality: Research and applications* (pp. 123–147). New York: Plenum Press.

Marans, R. W., and Rogers, W. (1973). Evaluating resident satisfaction in established and new communities. In R. W. Burchall (Ed.), *Frontiers of planned unit development: Synthesis of expert opinion* (pp. 197–227). New Brunswick, N.J.: Rutgers University Center for Urban Policy Research.

Marcuse, H. (1964). *One-dimensional man*. Boston: Beacon Press.

Martyniuk, O., Flynn, J. E., Spencer, T. J., and Hendrick, C. (1973). *The effect of environmental lighting on impression and behavior*. Paper presented at the conference of Illumination Engineers. (Study Group A, Symposium), Lund, Sweden.

Maruyama, M. (1979). Mindscapes: Meta-principles in environmental design. *World Future Society Bulletin*, Fall.

Maslow, A. (1954). *Motivation and personality*. New York. Harper & Row.

Maslow, A. H., and Mintz, N. L. (1956). Effects of esthetic surroundings: I. Initial short-term effects of three esthetic conditions upon perceiving "energy" and "well-being" in faces. *Journal of Psychology* 41, 247–254.

Mason, R. O., and Mittroff, I. I. (1981). *Challenging strategic planning assumptions*. New York: Wiley.

Masterson, I. (1981). Starting small: One or two people can make the design difference. *Small Town* 12, 51–52.

Matthews, W. H. (1976). *Resource materials for environmental management and education*. Cambridge, Mass.: MIT Press.

Mehrabian, A., and Russell, J. A. (1974). *An approach to environmental psychology.* Cambridge, Mass.: MIT Press.

Meining, D. (1968). *The great Columbia Plain: A historical geography, 1805–1910.* Seattle: University of Washington Press.

Melnick, R. (1977). The regional character of towns: Notes from the Flint Hills of Kansas. *Small Town* 8, 4–10.

Meltsner, A. J. (1979). Political feasibility and policy analysis. *Public Administration Review* 34, 859–867.

Merleau-Ponty, M. (1964). *The primacy of perception.* Evanston, Ill.: Northwestern University Press.

Michelson, W. R. (1966). An empirical analysis of urban environmental preferences. *Journal of the American Institute of Planners* 32, 355–360.

Michelson, W. R. (1970). *Man and his urban environment.* Reading, Mass.: Addison-Wesley.

Michelson, W. R. (1976). *Man and his urban environment: A sociological approach* (2nd ed.). Reading, Mass.: Addison-Wesley.

Midgley, M. (1978). *Beast and man: The roots of human nature.* Ithaca, N.Y.: Cornell University Press.

Miller, J. K. (1975). The sampling distribution of and a test for the significance of the bi-multivariate redundancy statistic: A Monte Carlo study. *Multivariate Behavioral Research* 10, 233–244.

Milord, J. T. (1978). Aesthetic aspects of faces: A (somewhat) phenomenological analysis using multidimensional scaling methods. *Journal of Personality and Social Psychology* 37, 205–216.

Mintz, N. L. (1956). Effects of esthetic surroundings: II. Prolonged and repeated experience in a "beautiful" and an "ugly" room. *Journal of Psychology* 41, 459–466.

Montgomery, R. (1966). Comment on "Fear and the house-as-haven in the lower class." *Journal of the American Institute of Planners* 32, 31–36.

Moore, G. T. (1977). *Holism, environmentalism and ecological validity* (Working paper WP 77-1). University of Wisconsin–Milwaukee, Center for Architecture and Urban Planning.

Moore, G. T., and Golledge, R. G. (1976). Environmental knowing: Concepts and theories. In G. T. Moore and R. G. Golledge (Eds.), *Environmental knowing: Theories, research and methods* (pp. 3–30). Stroudsburg, Pa.: Dowden, Hutchinson and Ross.

Moos, R., Harris, R., and Schonborn, K. (1969). Psychiatric patients and staff reactions to their physical environment. *Journal of Clinical Psychology* 25, 322–324.

Morris, C. (1938). *Foundation of a theory of signs.* Chicago: University of Chicago Press.

Munsinger, H. L., and Kessen, W. (1964). Uncertainty, structure and preference. *Psychological Monographs* 78, 9.

Nasar, J. (1979). The evaluative image of a city. In A. Seidel and S. Danford (Eds.), *Environmental design: Research, theory and application. Proceedings of the Tenth Environmental Design Research Association Conference* (pp. 38–45). Washington, D.C.: Environmental Design Research Association.

Nasar, J. L. (1980). The influence of familiarity on responses to visual qualities of neighborhoods. *Perceptual and Motor Skills* 51, 635–642.

Nasar, J. L. (1981). Visual preferences of elderly public housing residents: Residential street-scenes. *Journal of Environmental Psychology* 1, 303–313.

Nasar, J. L. (1983). Adult viewers' preferences in residential scenes: A study of the relationship of environmental attributes to preference. *Environment and Behavior* 15, 589–614.

Nasar, J. L. (1984) (reprinted this collection). Visual preferences in urban street scenes: A cross-cultural comparison between Japan and the United States. *Journal of Cross-Cultural Psychology* 15, 79–93.

National Environmental Policy Act (NEPA). (1969). Public Law 91-190, Eighty-third Stat. 852–856.

Neisser, U. (1976). *Cognition and reality: Principles and implications of cognitive psychology.* San Francisco: Freeman.

Nellis, L. (1980). Planning with rural values. *Small Town* 10, 20–24.

Neutra, R. (1954). *Survival through design.* New York: Oxford University Press.

Neutra, R. (1956). *Life and human habitat.* Stuttgart: Alexander Koch.

Neutra, R. (1958). *Realismo biológico: Un nuevo renacimiento humanístico en arquitectura.* Buenos Aires: Nueva Vision.

Newman, O. (1972). *Defensible space: Crime prevention through urban design.* New York: Macmillan.

Nijkamp, P. (1979). *Multidimensional spatial data and decision analysis.* Chichester, Eng.: Wiley.

Norberg-Schulz, C. (1965). *Intentions in architecture.* Cambridge, Mass.: MIT Press.

Norberg-Schulz, C. (1971). *Existence, space and architecture.* New York: Praeger.

Nysdet, L. (1981). A model for studying the interaction between the objective situation and the person's construction of the situation. In D. Magnusson (Ed.), *Towards a psychology of situations: An interactional perspective* (pp. 375–392). Hillsdale, N.J.: Erlbaum.

Office of Financial Management. (1980). *The pocket data book.* Olympia: Office of Financial Management, State of Washington.

O'Hare, D. (1976). Individual differences in perceived similarity and preference for visual art: A multidimensional scaling analysis. *Perception and Psychophysics* 20, 445–452.

O'Hare, D. (1981). Cognition, categorization and aesthetic responses. In H. Bonarius and R. Holland (Eds.), *Personal construct psychology: Recent advances in theory and practice* (pp. 147–155). London: Macmillan.

O'Hare, D., and Gordon, I. E. (1977). Dimensions of the perception of art: Verbal scales and similarity judgments. *Scandinavian Journal of Psychology* 18, 16–70.

Olds, J., and Milner, P. M. (1954). Positive reinforcement produced by electrical stimulation of septal area and other regions of rat brain. *Journal of Comparative and Physiological Psychology* 47, 419–427.

Oostendorp, A., and Berlyne, D. E. (1978a) (reprinted this collection). Dimensions in the perception of architecture: Identification and interpretation of dimensions of similarity. *Scandinavian Journal of Psychology* 19, 73–82.

Oostendorp, A., and Berlyne, D. E. (1978b). Dimensions in the perception of architecture: A measure of exploratory behavior. *Scandinavian Journal of Psychology* 19, 83–89.

Oostendorp, A., and Berlyne, D. E. (1978c). Dimensions in the perception of architecture: Multidimensional preference scaling. *Scandinavian Journal of Psychology* 19, 145–150.

Orians, G. (1980). Habitat selection: General theory and applications to human behavior. In J. S. Lockhard (Ed.), *Evolution of human social behavior.* New York: Elsevier North Holland.

Osgood, C. E. (1971). Explorations in semantic space: A personal diary. *Journal of Social Issues* 27, 5–64.

Osgood, C. E., Suci, G., and Tannenbaum, P. (1957). *The measurement of meaning.* Urbana: University of Illinois Press.

Osgood, C. E., and Tannenbaum, P. H. (1955). The principles of congruity in the prediction of attitude change. *Psychological Review* 62, 42–55.

Osmond, H. (1959a). The history and social development of the mental hospital and the relationship between architecture and psychiatry. In G. E. Goshen (Ed.), *Psychiatric architecture* (pp. 7–9). Washington, D.C.: American Psychiatric Association.

Osmond, H. (1959b). The relationship between architecture and psychiatry. In G. E. Goshen (Ed.), *Psychiatric Architecture* (pp. 9–20). Washington, D.C.: American Psychiatric Association.

Owen, D. (1978). The use of influence diagrams in structuring complex decision problems. In *Proceedings of the Second Lawrence Symposium on Systems and Decision Sciences* (pp. 278–284). Berkeley, Calif.: Lawrence Hall of Science.

Palmer, J. F. (1978). An investigation of the conceptual classification of landscapes and its application to landscape planning issues. In S. Weidermann and J. R. Anderson (Eds.), *Priorities for environmental design research: Part 1* (pp. 92–103). Washington, D.C.: Environmental Design Research Association.

Palmer, J. F. (1980). *Cross-cultural perceptions of international landscape scenes* (Grant No. 323-03-210-80). Research Foundation of the State University of New York.

Pardee, J. D. (1983). *Predicting the scenic beauty of a river environment using low level aerial photography.* Paper presented at the Fourteenth Conference of the Environmental Design Research Association, Lincoln, Neb.

Parducci, A. (1968). The relativity of absolute judgments. *Scientific American* 219, 84–90.

Paris, D. C., and Reynolds, J. F. (1983). *The logic of policy inquiry.* New York: Longman.

Paulson, R. (1975). *Emblem and expression: Meaning in English art of the eighteenth century.* London: Thames and Hudson.

Paulson, R. (1976–77). Towards the Constable Bicentenary: Thoughts on landscape theory. *Eighteenth-Century Studies* 10, 245–261.

Paulson, R. (1982). *Literary landscapes: Turner and Constable.* New Haven, Conn., and London: Yale University Press.

Payne, J. W. (1976). Task complexity and contingent processing in decision making: An information search and protocol analysis. *Organizational Behavior and Human Performance* 16, 366–387.

Pearlman, K. T. (1984). Zoning and the First Amendment. *Urban Lawyer* 16, 217–257.

Pearlman, K. T. (1985). Zoning religious uses. *Land Use Law and Zoning Digest* 37, 3–7.

Penning-Rowsell, E. C. (1973). *Alternative approaches to landscape appraisal and evaluation.* Enfield, Eng.: Middlesex Polytechnic.

Pervin, D. E. (1978). Definitions, measurements and classifications of stimuli, situations and environments. *Human Ecology* 6, 71–105.

Peterson, G. L. (1967). A model of preference: Qualitative analysis of the perception of visual appearance of residential neighborhoods. *Journal of Regional Science* 7, 19–31.

Peterson, G. L., and Neumann, E. S. (1968). Modelling and predicting human responses to the visual recreation environment. *Journal of Leisure Research,* 219–237.

Pevsner, N. (1936). *Pioneers of the modern movement.* London: Faber & Faber.

Pitt, D. G. (1976). Physical dimensions of scenic quality in streams. In E. H. Zube (Ed.), *Studies in landscape perception* (Publication No. R-76-1). Amherst: University of Massachusetts, Institute for Man and the Environment.

Plutchik, R. (1980). *Emotion: A psychoevolutionary synthesis.* New York: Harper & Row.

Poffenberger, A. T., and Barrows, B. E. (1924). The feeling value of lines. *Journal of Applied Psychology* 8, 187–205.

Pollman, A. W., and Wishnick, H. L. (1980). What determines where small town residents shop? *Small Town* 11, 18.

Porteous, D. (1982). Approaches to environmental aesthetics. *Journal of Environmental Psychology* 2, 53–60.

Porter, T., and Mikellides, B. (1976). *Color for architecture.* New York: Van Nostrand Reinhold.

Posner, M. L. (1973). *Cognition: An introduction.* Glenview, Ill.: Scott, Foresman.

Pratt, C. C. (1961). Aesthetics. *Annual Review of Psychology* 12, 71–92.

Preiser, W. F. E., and Hall, R. (1980). *The Sydney aesthetics study.* Unpublished manuscript, University of Sydney.

Priestley, T. (1983). The field of visual analysis and resource management: A bibliographic analysis and perspective. *Landscape Journal* 2, 52–59.

Proshansky, H. (1978). The city and self-identity. *Environment and Behavior* 10, 147–169.

Purcell, A. T. (1984a). The organization of the experience of the built environment. *Environment and Planning B* 11, 173–192.

Purcell, A. T. (1984b). Multivariate models and the attributes of the experience of the built environment. *Environment and Planning B* 11, 193–212.

Purcell, A. T. (1984c). Aesthetics, measurement and control. *Architecture Australia,* 19–38.

Purcell, A. T., and Lamb, R. J. (1984). Landscape perception: An examination and empirical investigation of two central issues in the area. *Journal of Environmental Management* 19, 31–63.

Rabinowitz, C. B., and Coughlin, R. E. (1971). *Some experiments in quantitative measurement of landscape quality* (Regional Science Research Institute Discussion Paper No. 43). Philadelphia: Regional Science Research Institute.

Rainwater, L. (1966). Fear and the house-as-haven in the lower class. *Journal of the American Institute of Planners* 32, 23–31.

Rainwater, L. (1970). *Behind ghetto walls.* Chicago: Aldine.

Ranney, D. B., and Nasoff, J. K. (1972). *Water quality management: An analysis of institutional patterns* (Water Resources Center). Madison: University of Wisconsin Press.

Rapoport, A. (1971). Some observations regarding man–environment studies. *Art* 2, 1.

Rapoport, A. (1977). *Human aspects of urban form: Towards a man–environment approach to urban form and design.* New York: Pergamon Press.

Rapoport, A., and Hawkes, R. (1970). The perception of urban complexity. *Journal of the American Institute of Planners* 36, 106–111.

Rapoport, A., and Kantor, R. E. (1967). Complexity and ambiguity in environmental design. *Journal of the American Institute of Planners* 33, 210–222.

Rasmussen, S. E. (1959). *Experiencing architecture.* Cambridge, Mass.: MIT Press.

Ray, K., et al. (1980). *Contextual architecture.* New York: McGraw-Hill.

Redding, M. J. (1973). *Aesthetics in environmental planning.* Washington, D.C.: Government Printing Office.

Redl, F., and Wineman, D. (1952). *Controls from within.* Glencoe, Ill.: Free Press.

Rein, M. (1976). *Social science and public policy.* New York: Penguin Books.

Ribe, R. G. (1982). On the possibility of quantifying scenic beauty: A response. *Landscape Planning* 9, 61–75.

Ritter, J. (1974). Landschaft – Zur Funktion des Aesthetischen in der modernen Gesellschaft. In J. Ritter (Ed.), *Subjektivität.* Frankfurt: Suhrkamp-Verlag.

Ritzville Planning Commission. (1978). *Community planning survey.* Ritzville, Wash.: Ritzville Planning Commission.

Rosenberg, M. J., and Abelson, R. P. (1960). An analysis of cognitive balancing. In C. I. Hovland and I. L. Janis (Eds.), *Attitude, organization and change* (pp. 112–283). New Haven, Conn.: Yale University Press.

Ross, R. T. (1938). Studies in the psychology of the theatre. *Psychological Record* 2, 127–90.

Rowlett, B. (1981). Aesthetic regulation under the police power: The new general welfare and the presumption of constitutionality. *Vanderbilt Law Review* 34, 603–651.

Russell, J. A. (1980). A circumplex model of affect. *Journal of Personality and Social Psychology* 39, 1161–1178.

Russell, J. A. (1983). Pancultural aspects of human conceptual organization of emotion. *Journal of Personality and Social Psychology* 45, 1281–1288.

Russell, J. A., and Bullock, M. (1986). On the dimensions preschoolers use to interpret facial expressions of emotion. *Developmental Psychology* 22, 97–102.

Russell, J. A., and Lanius, U. (1984). Adaptation level and the affective appraisal of environments. *Journal of Environmental Psychology* 4, 119–135.

Russell, J. A., and Pratt, G. (1980). A description of the affective quality attributed to environments. *Journal of Personality and Social Psychology* 38, 311–322.

Russell, J. A., and Snodgrass, J. (In press). Emotion and the environment. In D. Stokols and I. Altman (Eds.), *Handbook of environmental psychology*. New York: Wiley.

Russell, J. A., and Ward, L. M. (1981). The psychological representation of molar physical environments. *Journal of Experimental Psychology: General* 110, 121–152.

Russell, J. A., and Ward, L. M. (1982). Environmental psychology. *Annual Review of Psychology* 33, 651–688.

Russell, J. A., Ward, L. M., and Pratt, G. (1981). Affective quality attributed to environments: A factor analytic study. *Environment and Behavior* 13, 259–288.

Ryan, T. A. (1960). Significance tests for multiple comparisons of proportions, variances and other statistics. *Psychological Bulletin* 67, 318–328.

Saaty, T. L. (1980). *The analytic hierarchy process*. New York: McGraw-Hill.

Sadler, B., and Carlson, A. (Eds.). (1982). *Environmental aesthetics: Essays in interpretation*. Victoria: University of British Columbia Press.

Sage, A. (1977). *Methodology for large scale systems*. New York: McGraw-Hill.

St. Louis is revising housing complex. (1972, March 19). *New York Times*, p. 32A.

Samuelson, D. J., and Lindauer, M. S. (1976). Perception, evaluation and performance in a neat and messy room by high and low sensation seekers. *Environment and Behavior* 8, 291–306.

Sancar, F. H. (1977). *A model of planning methodology for community participation and urban development*. Unpublished doctoral dissertation, Pennsylvania State University, State College.

Sancar, F. H. (1980). A dynamic simulation formulation based on user preferences. *ORSA/TIMS Bulletin* 10, 35.

Sancar, F. H. (1982). Formal aides to model formulation using worth assessment and Delphi. *Dynamica* 8 (pt. 2), 90–95.

Sancar, F. H. (1987). Implementation and evaluation of a modeling approach to community development planning: Meeting the challenge to paradigm breakdown. In R. L. Eberlein (Ed.), *Proceedings of the 1987 International System Dynamics Conference* (pp. 446–461). Shanghai, China: System Dynamics Society.

Sanoff, H. (1981). Involving small town citizens in design decisions: A North Carolina approach. *Small Town* 12, 30–33.

Santayana, G. (1896). *The sense of beauty*. Reprint, New York: Dover.

Schiller, F. (1967). *Naive and sentimental poetry, and On the sublime* (J. A. Elias, Trans.). New York: Ungar.

Schlosberg, H. (1954). Three dimensions of emotion. *Psychological Review* 6, 81.

Schrag, C. O. (1969). *Experience and being*. Evanston, Ill.: Northwestern University Press.

Schroeder, H. W. (1984). Environmental perception rating scales: A case for simple methods of analysis. *Environment and Behavior* 13, 573–599.

Schumacher, E. F. (1973). *Small is beautiful*. New York: Harper & Row.

Scott, G. (1935). *Architecture and humanism: A study of the history of taste* (2nd ed.). London: Constable.

Scully, V. (1962). *The earth, the temple and the gods*. London: Yale University Press.

Scully, V. (1969). *American architecture and urbanism:* New York: Praeger.

Seaton, R., and Collins, J. (1971). *Architectural simulations as stimuli*. Paper presented at the annual meeting of the Western Psychological Association, San Francisco, Calif.

Seaton, R., and Collins, J. (1972). Validity and reliability of ratings of simulated buildings. In W. J. Mitchell (Ed.), *Environmental design: Research and practice. Proceedings of the Third Environmental Design Research Association Conference* (pp. 6-10-1–6-10-12). Los Angeles: UCLA.

Sennett, R. (1970). *The uses of disorder: Personal identity and city life*. New York: Random House (Vintage Books).

Shafer, E. L., Jr. (1969). Perception of natural environments. *Environment and Behavior* 1, 71–82.

Shafer, E. L., Jr., Hamilton, J. F., and Schmidt, E. (1969). Natural landscape preference: A predictive model. *Journal of Leisure Research* 1, 1–19.

Shafer, E. L., Jr., and Mietz, J. (1969). Aesthetic and emotional experience rate high with Northeast wilderness hikers. *Environment and Behavior* 1, 187–197.

Shafer, E. L., Jr., and Richards, T. A. (1974). *A comparison of viewer reactions to outdoor scenes and photographs of those scenes* (Forest Service Research Paper NE-302). Upper Darby, Pa.: U.S. Department of Agriculture Northeast Forest Experiment Station.

Shafer, E. L., Jr., and Thompson, R. C. (1968). Models that describe use of Adirondack campgrounds. *Forest Science* 14, 383–391.

Shepard, R. N., Romney, K., and Nerlove, S. B. (Eds.). (1972). *Multidimensional scaling*. New York: Seminar Press.

Sieberling, N., and Masterson, I. (1981). Mini-parks spring up in an Iowa downtown. *Small Town* 12, 34–39.

Siegel, S. (1956). *Nonparametric statistics for the behavioral sciences*. New York: McGraw-Hill.

Simon, H. A. (1945). *Administrative behavior: A study of decisionmaking process in administrative organizations*. New York: Free Press.

Sitte, C. (1965). *City planning according to artistic principles* (G. R. Collins and C. C. Collins, Trans.). London: Phaidon Press. (Original work published 1889 as *Der Stadtbau nach seinen künstlerischen Grundsätzen*)

Slovic, P., Fischoff, B., and Lichtenstein, S. (1976). Cognitive processes and societal risk-taking. In J. S. Carroll and J. W. Payne (Eds.), *Cognition and social behavior*. Hillsdale, N.J.: Erlbaum.

Slovic, P., Fischoff, B., and Lichtenstein, S. (1977). Behavioral decision theory. *Annual Review of Psychology* 28, 1–39.

Smardon, R., Palmer, J., and Felleman, J. (1986). *Foundations for visual project analysis*. New York: Wiley.

Smith, C. W. (1959). Architectural research and the construction of mental hospitals. In G. E. Goshen (Ed.), *Psychiatric architecture* (pp. 10–15). Washington, D.C.: American Psychiatric Association.

Smith, P. (1977). *The syntax of cities*. London: Hutchinson.

Smith, R. G. (1959). Development of a semantic differential for use with speech-related concepts. *Speech Monograph* 26, 263–272.

Smith, R. G. (1961). A semantic differential for theater concepts. *Speech Monograph* 28, 1–8.

Snider, J. G., and Osgood, C. E. (Eds.). (1969). *Semantic differential technique*. Chicago: Aldine.

Sofranko, A. J., and Williams, J. D. (1980). *Rebirth of rural America*. Ames: Iowa State University, The North Central Regional Center for Rural Development.

Solar, G. (1979). *Das Panorama und seine Vorentwicklung bis zu Hans Conrad Escher von der Linth*. Zurich: Orell Fussli Verlag.

Sommer, R. (1969). *Personal space: The behavioral basis of design*. Englewood Cliffs, N.J.: Prentice-Hall.

Sonnenfeld, J. (1966). Variable values in the space and landscape: An inquiry into the nature of environmental necessity. *Journal of Social Issues* 4, 71–82.

Sonnenfeld, J. (1967). Environmental perception and adaptation level in the Arctic. In D. Lowenthal (Ed.), *Environmental perception and behavior* (pp. 42–59). Chicago: University of Chicago, Department of Geography.

Sonnenfeld, J. (1969). Equivalence and distortion of the perceptual environment. *Environment and Behavior* 1, 83–100.

Sorte, G. J. (1971). *Perception av landskap*. Licentiate's dissertation, Landbruksbokhandeln /Universitetsforlaget, As, Norway.

Spreiregen, P. D. (1965). *Urban design: The architecture of towns and cities*. New York: McGraw-Hill.

Springbett, B. M. (1960). The semantic differential and meaning in non-objective art. *Perceptional Motivation Skills* 10, 231–40.

Steel, F. (1981). *The sense of place*. Boston: CBI Publications.

Steinitz, C. (1968). Meaning and the congruence of urban form and activity. *Journal of the American Institute of Planners* 34, 235–247.

Stewart, D., and Love, A. (1968). A general canonical correlation index. *Psychological Bulletin* 70, 160–163.

Stolnitz, J. (1960). *Aesthetics and philosophy of art criticism*. Boston: Houghton Mifflin.

Stolnitz, J. (1961). On the origin of "aesthetic disinterestedness." *Journal of Aesthetics and Art Criticism* 20, 131–143.

Studer, R. (1982). *Normative guidance for the planning, design and management of environment-behavior systems*. Unpublished doctoral dissertation, University of Pittsburgh.

Subkoviak, M. J. (1975). The use of multidimensional scaling in education research. *Review of Educational Research* 45, 387–423.

Talbot, J. F., and Kaplan, R. (1984). Needs and fears: The response to trees and nature in the inner city. *Journal of Arboriculture* 10, 222–228.

Talbot, J. F., and Kaplan, R. (In press). Judging the sizes of urban open areas. Is bigger always better? *Landscape Journal*.

Tannenbaum, P. H., and McLeod, J. M. (1967). On the measurement of specialization. *Public Opinion Quarterly* 31, 27–37.

Taylor, L. (1968). *Urban–rural problems*. Belmont, Calif.: Dickenson.

Thayer, R. L., Hodgson, R. W., Gustke, L. B., Atwood, B. G., and Holmes, J. (1976). Validation of a natural landscape preference model as a predictor of perceived landscape beauty in photographs. *Journal of Leisure Research* 8, 292–299.

Thayer, S. (1980). The effect of facial expression sequence on judgments of emotion. *Journal of Social Psychology* 111, 305–306.

Thomas, F. (1972). *Law in action: Legal frontiers for natural resources planning* (Land Economic Monograph Series, No. 4). Madison: University of Wisconsin.

Tlusty, W., et al. (1983). *Minority report by members of the Ad Hoc Committee on Scenic Beauty*. Submitted to Natural Resources Board, January 11.

Toll, S. I. (1969). *Zoned America*. New York: Grossman.

Toner, W. (1979). Getting to know the people and the place. In J. Getzel and C. Thurow (Eds.), *Rural and small-town planning* (pp. 1–28). Chicago: Planners Press.

Trevarthen, S. (1978). Models of perceiving and modes of acting. In H. Pick and E. Saltzman (Eds.), *Modes of perceiving and processing information* (pp. 99–136). Hillsdale, N.J.: Erlbaum.

Trubek, L. G., and Trubek, D. M. (1979, October). *Civic justice through civil justice: A new approach to public interest advocacy in the United States*. Paper presented at the Conference on Access to Justice: Prospects for Future Action, Florence, Italy.

Tuan, Y.-F. (1973). Visual blight: Exercises in interpretation. In P. F. Lewis, D. Lowenthal, and Y.-F. Tuan (Eds.), *Visual blight in America* (Commission on College Geography Resource Paper No. 23, pp. 23–27). Washington, D.C.: Association of American Geographers.

Tuan, Y.-F. (1974). *Topophilia: A study of environmental perception, attitudes and values*. Englewood Cliffs, N.J.: Prentice-Hall.

Tunnard, C., and Pushkarev, B. (1981). *Man-made America: Chaos or control*. New York: Harmony Books.

Tversky, A. (1977a). On the elicitation of preferences: Descriptive and prescriptive considerations. In D. E. Bell, R. L. Keeney, and H. Raiffa (Eds.), *Conflicting objectives in decisions* (pp. 209–222). New York: Wiley.

Tversky, A. (1977b). Features of similarity. *Psychological Review* 84, 327–352.

Tversky, A., and Kahneman, D. (1981). The framing of decisions and the psychology of choice. *Science* 211, 453–458.

Tyng, A. (1975). *Simultaneous randomness and order: The Fibonacci-divine proportion as a universal forming principle.* Unpublished doctoral dissertation, University of Pennsylvania, Philadelphia.

Ulrich, R. S. (1973). *Scenery and the shopping trip: The roadside environment as a factor in route choice.* Unpublished doctoral dissertation, University of Michigan, Ann Arbor.

Ulrich, R. S. (1977). Visual landscape preference: A model and application. *Man–Environment Systems* 7, 279–299.

Ulrich, R. S. (1983). Aesthetic and affective response to natural environment. In I. Altman and J. F. Wohlwill (Eds.), *Human behavior and environment:* Vol. 6. *Behavior and the natural environment* (pp. 88–125). New York: Plenum Press.

Ulrich, R. S. (1984). View through a window influences recovery from surgery. *Science* 224, 420–421.

Unemployment rates in Washington. (1982). *State of the Region* 2, 13.

U.S., Bureau of the Census. (1980). *Census of population and housing: Preliminary population counts.* Washington, D.C.: Government Printing Office.

Van de Geer, J. P., Levelt, J. M., and Plomp, R. (1962). The connotation of musical consonance. *Acta Psychologica* 20, 308–319.

Veldman, J. (1967). *FORTRAN programming for the behavioral sciences.* New York: Holt, Rinehart and Winston.

Venturi, R. (1966). *Complexity and contradiction in architecture.* New York: Museum of Modern Art.

Venturi, R., Brown, D. S., and Izenour, S. (1972). *Learning from Las Vegas.* Cambridge, Mass.: MIT Press.

Vickers, Sir G. (1965). *The art of judgment.* New York: Basic Books.

Vielhauer, J. (1965). The development of a semantic scale for the description of the physical environment. *Dissertation Abstracts*, No. 66-759.

von Franz, M-L. (1968). The process of individuation. In C. Jung (Ed.). *Man and his symbols* (pp. 157–254). New York: Dell.

Walker, E. L. (1980). *Psychological complexity and preference: A hedgehog theory of behavior.* Monterey, Calif.: Brooks/Cole.

Wallace, A., and Carson, M. (1973). Sharing and diversity in emotional terminology. *Ethos* 1, 1–29.

Wallace, B. C. (1974). Landscape evaluation and the Essex coast. *Regional Studies* 8, 299–305.

Wapner, S., Cohen, S. E., and Kaplan, B. (1976). *Experiencing the environment.* New York: Plenum Press.

Wapner, S., Kaplan, B., and Cohen, S. E. (1973). An organismic-developmental perspective for understanding transactions of men and environments. *Environment and Behavior* 5, 255–289.

Ward, L. M. (1977). Multidimensional scaling of the molar physical environment. *Multivariate Behavioral Research* 12, 23–42.

Ward, L. M., and Russell, J. A. (1981a). Cognitive set and the perception of place. *Environment and Behavior* 13, 610–632.

Ward, L. M., and Russell, J. A. (1981b). The psychological representation of molar physical environments. *Journal of Experimental Psychology: General* 110, 121–152.

Warfield, J. N. (1975). *TOTOS: Improving problem-solving. Approaches to problem solving, Number 3.* Columbus, Ohio: Battelle Institute and the Academy of Contemporary Problems.

Warfield, J. N. (1976). *Societal systems: Planning, policy and complexity.* New York: Wiley.

Warr, P. B., and Knapper, W. (1968). *The perception of people and events.* New York: Wiley.

Watson, D., and Tellegen, A. (1985). Towards a consensual structure of mood. *Psychological Bulletin* 98, 219–235.

Watson, R. H. (1978). A useful tool for technological assessment? *Technological Forecasting and Social Change* 11, 165–185.

Webber, M. (1967). The urban place and the non-place urban realm. In M. Webber (Ed.), *Exploration into urban structure* (pp. 79–153). Philadelphia: University of Pennsylvania Press.

Webster's New World Dictionary (College ed.). (1959). Cleveland and New York: World.

Wener, R., Fraser, F. W., and Farbstein, J. (1983). Three generations of evaluation and design of correctional facilities. *Environment and Behavior* 17, 71–96.

Werner, H. (1963). On physiognomic perception. In G. Kepes (Ed.), *The new landscape in art and science* (pp. 280–283). Chicago: Cambridge Press.

Wheeler, L. (1984). The visual impact of 90 scenes: An intercultural comparison. *Perception of the Environment* 5, 7–8.

White, R. W. (1959). Motivation reconsidered: The concept of competence. *Psychological Review* 66, 297–333.

Whitehead, A. N. (1925). *Principles of natural knowledge.* New York: Cambridge University Press.

Whyte, W. H. (1980). *The social life of small urban spaces.* Washington, D.C.: Conservation Foundation.

Wiener, N. (1967). *The human use of human beings: Cybernetics and society.* New York: Avon Books.

Wilbur Planning Commission. (1979). *Community planning survey.* Wilbur, Wash.: Wilbur Planning Commission.

Wildrick, J. (1978). Introducing urban design to small communities. In A. Ferebee (Ed.), *Proceedings: First National Conference on Urban Design* (pp. 317–322). Washington, D.C.: RC Publications.

Wilkinson, P. F. (1983). *Urban open space planning.* Ontario: York University, Faculty of Environmental Studies.

Willard, L. D. (1980). On preserving nature's aesthetic features. *Environmental Ethics* 2, 293–310.

Wilson, E. O. (1975). *Sociobiology: The new synthesis.* Cambridge, Mass.: Harvard University Press.

Winer, B. J. (1971). *Statistical principles in experimental design* (2nd ed.). New York: McGraw-Hill.

Winkel, G., Malek, R., and Theil, P. (1970). A study of human responses to selected roadside environments. In H. Sanoff and S. Cohn (Eds.), *EDRA 1: Proceedings of the First Annual Environmental Design Research Association Conference* (pp. 224–240). Raleigh: North Carolina State University.

Wohlwill, J. F. (1968). Amount of stimulus exploration and preference as differential function of stimulus complexity. *Perception and Psychophysics* 4, 307–312.

Wohlwill, J. F. (1973). The environment is not in the head! In W. F. E. Preiser (Ed.), *Environmental Design Research IV. Proceedings of the Fourth Annual Environmental Design Research Association Conference* (Vol. 2, pp. 166–181). Stroudsburg, Pa.: Dowden, Hutchinson and Ross.

Wohlwill, J. F. (1974). *The place of aesthetics in the study of the environment.* Paper presented at the International Congress of Applied Psychology, (Symposium on Experimental Aesthetics and Psychology of Environment), Montreal.

Wohlwill, J. F. (1976). Environmental aesthetics: The environment as a source of affect. In I. Altman and J. F. Wohlwill (Eds.), *Human behavior and environment* (Vol. 1, pp. 37–86). New York: Plenum Press.

Wohlwill, J. F. (1979). What belongs where: Research on fittingness of man-made structures into natural settings. In T. C. Daniel and E. H. Zube (Eds.), *Assessing amenity resource values* (General Technical Report RM-68, pp. 48–53). Fort Collins, Colo.: U.S. Department of Agriculture, Rocky Mountain Forest and Range Experiment Station.

Wohlwill, J. F. (1982). Visual impact of development in coastal zone areas. *Coastal Zone Management Journal* 9, 225–248.

Wohlwill, J. F., and Harris, G. (1980). Responses to congruity or contrast for man-made features in natural-recreation settings. *Leisure Sciences* 3, 349–365.

Wohlwill, J. F., and Kohn, I. (1973). The environment as experienced by the migrant: An adaptation-level view. *Representative Research in Social Psychology* 4, 35–164.

Wohlwill, J. F., and Kohn, I. (1976). Dimensionalizing the environmental manifold. In S. Wapner, B. Cohen, and B. Kaplan (Eds.), *Experiencing the environment* (pp. 19–53). New York: Plenum Press.

Wolfe, M. R. (1975). Survey of major form-shaping strategies. M. R. Wolfe et al. (Eds.), *Urban design primer: Hawaii* (pp. 7–15). Honolulu: State of Hawaii Department of Planning and Economic Development.

Wolfe, T. (1981). From Bauhaus to our house. *Harper's* 26 (1573), 33–54; (1574), 31–59.

Wölfflin, H. (1955). *Principles of art history: The problem of the development of style in later art* (M. D. Hotlinge, Trans.). New York: Dover. (Original work published 1886)

Woodcock, D. M. (1982). *A functionalist approach to environmental preference.* Unpublished doctoral dissertation, University of Michigan, Ann Arbor.

Woods, W. (1971). *Simulation: A comparison of color film, black and white film, and video tape to reality in the simulation of architectural models.* University of British Columbia, Vancouver.

Wright, B., and Rainwater, L. (1962). The meanings of color. *Journal of General Psychology* 67, 89–99.

Yancey, W. L. (1971). Architecture, interaction, and social control: The case of a large scale public housing project. *Environment and Behavior* 3, 3–21.

Yeats, W. B. (1961). *Essays and introductions.* New York: Macmillan.

Young, F. W. (1968). *A FORTRAN IV program for nonmetric multidimensional scaling.* Chapel Hill: University of North Carolina, Psychometric Laboratory.

Young, F. W., and Lewyckyj, R. (1981). *ALSCAL-4 users guide* (2nd ed.). Chapel Hill: University of North Carolina, Psychometric Laboratory.

Yudell, R. J. (1977). Body movement. In K. C. Bloomer and C. W. Moore (Eds.), *Body, memory and architecture.* New Haven, Conn., and London: Yale University Press.

Zajonc, R. B. (1980). Feeling and thinking: Preferences need no inferences. *American Psychologist* 35, 151–175.

Zube, E. H. (1973). Rating everyday rural landscapes of the northeastern United States. *Landscape Architecture* 63, 370–375.

Zube, E. H. (1976). Perception of landscape and land use. In I. Altman and J. F. Wohlwill (Eds.), *Human behavior and environment* (Vol. 1, pp. 87–122). New York: Plenum Press.

Zube, E. H. (1984). Themes in landscape assessment theory. *Landscape Journal* 3, 104–110.

Zube, E. H., and Mills, L. V., Jr. (1976). Cross-cultural explorations in landscape perception. In E. H. Zube (Ed.), *Studies in landscape perception* (Publication No. R-76-1). Amherst: University of Massachusetts, Institute for Man and the Environment.

Zube, E. H., and Pitt, D. G. (1981). Cross-cultural perceptions of scenic and heritage landscapes. *Landscape Planning* 8, 69–87.

Zube, E. H., Pitt, D. G., and Anderson, T. W. (1974). *Perception and measurement of scenic resources in the southern Connecticut River Valley* (Publication No. R-74-1). Amherst: University of Massachusetts, Institute for Man and the Environment.

Zube, E. H., Pitt, D. G., and Anderson, T. W. (1975). Perception and prediction of scenic resource values of the northeast. In E. H. Zube, R. O. Brush, and J. G. Fabos (Eds.), *Landscape assessment: Values, perceptions and resources* (pp. 151–167). Stroudsburg, Pa.: Dowden, Hutchinson and Ross.

Zube, E. H., Pitt, D. G., and Evans, G. W. (1983). A lifespan developmental study of landscape assessment. *Journal of Environmental Psychology* 3, 115–128.

Zube, E. H., Sell, J. L., and Taylor, G. (1982). Landscape perception: Research, application and theory. *Landscape Planning* 9, 1–33.

Index of authors

519

Subject index

Subject index

529

public policy, xxii, xxiv, 228, 393, 434, 456–
58, 467–69, 474
aesthetics, xxi, 430, 435, 460
empirical research relevance, 100, 320, 393,
397, 422, 438, 439–40, 444, 445, 449,
459, 461, 462, 464, 490–91
participation, 435
relation to theory and research, 5, 393
special vs. public interest, 441–42

quality of features, *see* upkeep

refuge, 50, 103, 325, 347, 348, 350, 361–63;
see also prospect–refuge theory
relaxation
influences on, 312–15
salience in affect, 58, 102, 107, 219, 221,
226
research/practice gap, xxiv, 393, 395, 396,
397, 422
residential settings, 90, 101, 102, 200, 257,
275, 292, 293, 416, 454
single-family forms, 16–17
values, 21–22
rural settings, 101, 394
growth, 364, 365, 366
symbolic meaning, 81, 366

scene information, 48–49
semiology, 13–14
signs and billboards, 300, 405, 425–27, 478–
81, 482, 483, 484
simulation media, 43, 49, 173, 200–5, 235,
264, 276, 282, 304–5, 308, 320
size
estimates, 295
preference for, 290, 296, 297–98
symbol of status, 9, 17, 428
social environment, 66
social importance, expression of, 9–10, 15
sociocultural differences, 14, 276
factors, 16, 53, 142, 143, 366–67, 378,
388, 399, 400–1
space; *see also* openness; prospect
differences in perception, 189, 190
perception, 49, 85, 133, 157
phenomenological, 85, 87
salience in perception, 54–55, 115, 179,
180, 182, 345
spatial configuration, and meaning, 17
specific exploration, *see* exploratory behavior
stylistic ratings, 221–22
subjective measures, 117
suburban development, 416
surprise, 50, 62, 402, 403; *see also* collative
variables
survival, *see* evolution

symbol, definition, 14–15
symbolic meaning, 101
acquisition of, 23–25
sociocultural differences, 14
variables, 3, 16–19, 288
symbolism, 4, 38, 288, 289
lack of positive theory, 12, 14
research questions, 26
symbols
group differences, 16
human/nonhuman, 69
landscape, 67
nature, 72, 75–77, 78, 81
needs hierarchy, 15–16
rural, 81
sexual, 67–69
social, 66

texture, 51, 52, 344, 345
theory
comments on, 12, 14, 28, 29
importance, 44
relation to methods, 113

uncertainty and interest, 60; *see also* collative
variables
uniqueness, *see* novelty
upkeep, 9, 82
affective quality, 262, 270, 273, 286, 289
social significance, 9–10
urban setting, 6, 235, 263
case studies in urban design, 407–16
determinants of form, 452–53
need for nature, 72–73, 81

values, 20
variety, *see* complexity
vegetated scenes, 102
vegetated vs. man-made
affective quality, 364, 366, 372, 376, 377
salience in perception, 102–3, 338–39, 343
vegetation, 4, 8, 54, 74, 83, 286, 288, 340,
347, 366, 375
affective quality, 103, 110, 258, 262, 269,
273, 275, 292–96, 298, 323, 325, 340,
349
attitudes toward, 69–72
controls, 428
desired qualities, 79
large trees, 347, 350, 351, 356
urban, 72–73, 343
vehicles, response to, 262, 269, 271, 273, 287
verbal report, 121, 197, 212
visual overload, 300
visual quality, *see* environmental aesthetics

water, preference for, 114